20% Chance of Rain

Contents

Preface

Life is full of surprises. Some are good and some are not. As we travel through the time of our lives, we naturally develop tactics to deal with its joys and sorrows. From looking at the news media, a person could conclude that science, technology, and information are our personal saviors, prolonging our lives, protecting us from the behaviors of others, and even shielding us from the consequences of our personal actions. There's no doubt that the powers of the internet and the new information age have broadened our horizons. But technology has done little to change or alter what comes into view. The question is: How can each of us best avoid some of life's pitfalls and prepare to deal with those that penetrate our defenses?

There is no "one size fits all" tactic or recipe that works. Just as each of us is unique, so are our ethics, beliefs, aversions, and tolerances. The ways in which people deal with uncertainty seem simple on the surface, yet are extremely complex.

My motivation for this work was to write a book about risk that could be used as a supplement to a wide variety of risk-related courses and be interesting enough to be read as a standalone book. Risk is usually discussed in relation to a specific subject, such as environmental risk, transportation risk, health risk, food risk, and weather risk. The concept is blended with other subjects to inform or teach risk applications in specific areas. But risk is a subject all by itself. It permeates every part of our lives as a universal variable of human existence, transcending any specific application. This book provides a framework for you to apply to risk in both your professional and personal lives. Each chapter describes applied methods on how to analyze, assess, and manage risk. Together they give you tools to help you pack your personal and professional parachutes for some situations in which you need strategies to deal with unexpected events.

The information for this book came from various international organizations, books, journals, government reports, and newspaper articles. Much of the content is related to risk exposures discussed from the United States' perspective simply because these data sources are readily available. For readers in other parts of the world, the same methodologies are applicable to you by accessing your country's data sources, which should be similar to the ones used here. When possible, however, international data and examples are directly used to help readers around the world apply the principles discussed in the text. The international examples also help to show the great diversity in risk exposures and that there is a great deal of work that needs to be done—not all of it, by the way, in the developing world countries. After reading this book, you'll probably be able to identify additional news items that would either add to or fit into the chapters.

The title of this book came from a conversation with a gift store clerk at Sapphire Beach on St. Thomas. While vacationing there with my family, I noticed a weather report showing that a cold front was expected to pass over the region the next day. Since we'd planned to spend that day in a boat, I asked the clerk exactly what I should expect. She smiled and said, "Oh, there's a 20% chance of rain tomorrow." Since two young children were involved, I pursued the issue a little further. "And what does that usually mean?" I asked. She laughed and told me. "Don't worry. That's what they say every day. Maybe it will rain, and it maybe won't. In St. Thomas, there's always a 20% chance of rain." I was left speechless—for this simple response captured the essential element of the manuscript I was writing. Science really can make life appear more predictable, but it's never going to replace our common sense. I hope you agree.

RICHARD B. JONES

Acknowledgments

First and foremost, I would like to thank Janice Jones, who helped make this book a reality. Her technical insight, encouragement, patience, support, professional editorial expertise, and overall attention to detail were invaluable throughout the development and writing of this book.

My special thanks to the international law firm of Greenberg Traurig, LLP, for providing valuable referencing and data resources. In particular, I am indebted to two Houston office attorneys: Dwayne Mason, for many interesting risk and law discussions, and Alex Nowamooz, who was instrumental in helping me find and condense several of the chapter case studies.

I also would like to acknowledge the support and encouragement of many colleagues and friends. To attempt to name these wonderful people means I accept the risk of inadvertently omitting someone. This is a risk exposure I choose to eliminate by saying to everyone (and you know who you are!) THANK YOU.

R.B.J.

Chapter 1

Risk: Life's Question Mark

We also believe in taking risks, because that's how you move things along . . .

—Melinda Gates

Risk. The buzzword of our time. Myriad advances in medical science and modern technology might make you think our world is safer, more ordered, and more predictable than ever before in history, and we would be hearing less about risk. It's true that people are living longer, more productive lives. Being in one's seventies or even eighties is no longer looked upon as being in a time of looming death. Members of this demographic are going into space, flying at Mach 2 in jet fighters, and starring in new sitcoms. Yet along with all of the good brought about by technology has come the awareness that our universe remains a very unpredictable place.

We can and do change the future, but there's always a price, with both societal and individual costs in the equation. Technology has given us the ability to measure the intimate building blocks of life, to routinely visit the sanctity of space above our planet, and to control our lives in countless ways. But somehow, on the great scale of existence, along with all of these wonderful things has come the ability to see the dark as well as the light. The double-edged sword of technology that enables us comes with its price, albeit directly or indirectly.

Our ability to document and measure the frequency and severity of human tragedies and the bad things that happen generally exceeds our ability to know what to do about them. Changing regulations and laws will continue to be a mechanism by which we can prevent accidents involving the public. Still, somewhere in the process a decision always must be made on how much safety, security, or predictability is enough. There is always a cost involved. In a world where our abilities to see, communicate, and measure situations usually exceed the resources required to control the possible outcomes, we need a rationale to balance what we can do against what we can afford. This is the process of risk management.

What is risk? In simple, concise terms, *risk is the future's uncertainty*. It is a characteristic of life that everyone has in common. One might argue that other forms

20% Chance of Rain: Exploring the Concept of Risk, First Edition. Richard B. Jones.
© 2012 John Wiley & Sons, Inc. Published 2012 by John Wiley & Sons, Inc.

of life perceive the future and therefore they too must experience risk. While it's true, for example, that as the weather begins to change, some animals start storing food and others migrate, this is more instinct than decision-making. The concept of risk appears to be unique to humans. What separates us from the rest of the animal kingdom is precisely the characteristic that enables us to recognize the concept of risk: It is our ability to exercise rational thought. You might argue about the degree of rational thought possessed by some people, but as a group, this is what separates us from all other life forms. Our unique mental capabilities enable us to apply information from the past, react to the present, and plan for the future.

Yet the more we learn about our world, the more we learn about the plethora of ways we can be harmed by it.

Some of the things we fear and risk are of our own making, such as chemicals, cars, and planes. Other risks are from natural causes. The picture is blurred even further when we factor in the reporting of such events. Between the volume of information available and the style and motivations of today's media, we hear, see, and read only a small subset of what actually happens each day. There is no absolute scale to measure tragedy to determine what news gets reported and what news remains quiet. Reporters today can stream content of their choice to our TVs, computers, and web-enabled cellphones from around the world. What the media chooses to show us does have an effect on us. Research has shown that the more we're exposed to sensational and shocking content relative to our experience, the greater our perception that the world is a hazardous place.

Here's an example. In the early 1990s there was general perception that violent crime was a widespread national problem in the United States even though the reported crime statistics actually indicated a decreasing trend. When this issue was studied by social scientists, they found a correlation between people who believed crime was on the rise and the amount of violent and dramatic programming they viewed on TV. The 5-year study showed that over a large segment of the population, with varying crime rates, watching television news was correlated with increased fear of and concern about crime [1]. Also fueling the misperception was the amount of dramatic and violent prime-time TV programming watched per week. Even though the shows were fictional, their realism in part triggered the same, if not a stronger, reaction than did the actual news.

The media's influence on our perception and judgment is pervasive and subtle. Another study indicated that TV viewers watching medical dramas and news which had medical content responded with a loss of wellbeing and increased fears for personal health and for the health of those around them [2].

There is no doubt that communication media of all forms influences our perceptions of reality. But is the world more dangerous today than it was in the past? You might think so based on what we hear, see, and read. But that's not the case. Actually, we are safer today than at any other time in recorded history. In 1850 the life expectancy in the United States was 38.3 years, in 2010 it was 78.2 years: about a 40-year increase in 160 years. The risks presented by disease, transportation, and even crime have shown decreasing trends over the same time interval. Then how can we think that today's life has more hazards than ever before? Primarily it's because the news media has learned that fear sells more than safety, or to put it another way, harm sells more than good. Executives at the broadcast companies just didn't dream up this idea. These companies stay sensitive to public opinion through consulting firms that conduct surveys and perform market research to determine what viewers want to see, hear, and read. After all, stations with the highest ratings can demand the highest advertising prices and revenue generation is the ultimate motivator.

It's easy to point fingers at the journalistic and media press and blame them for our apparent misperceptions. But our world contains a diversity of cultures, technological sophistication, and infrastructure-related services that deliver different standards of care to their constituents. Consequently what's sensational or shocking for someone living in a country or region with high service levels may be interpreted as "nomal" or "routine" for someone viewing the events in a part of the world with lower standards of care. The tremendous diversity in safety, health, and crime risk levels can be seen in life expectancy differences shown in Figure 1.1 [3]. Monaco is apparently the safest, or lowest risk country, with a life expectancy of about 89.8 years. The country in the list with the highest risk is Angola, having a life expectancy of 38.5 years—just slightly higher than the United States in 1850.

So how are we supposed to know what to believe, when to be skeptical, when to discard information, and then what to do? Of course there is no "one size fits all" answer. Everyone's manner in dealing with life's uncertainties is different. Each of us makes choices in daily life according to countless different factors. Yet regardless

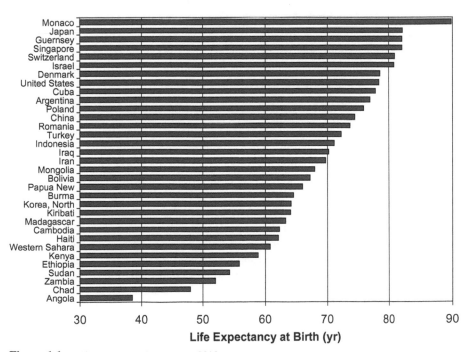

Figure 1.1   Life expectancy by country, 2010.

of the details, we all have in common the fact that there are limits: finite resources we can apply to make our futures more secure. Perhaps quality is free, but risk reduction isn't and sooner or later, like it or not, in one form or another, risk remains.

Reducing future's uncertainty is not something that we can do easily even as a global community. The tradeoff between money and benefits are very real issues. The 6.9 billion people alive today and the approximate 75 million additions every year each want a better place to live and grow. In the United States alone, there is a birth about every 8 seconds and a death every 11 seconds. Reducing uncertainty in our future is not getting any easier, and the challenges facing a growing population on a fixed amount of real estate show we don't have a choice any more. Balancing risk reduction and cost is something we all do, even corporations and governments. No one has what it takes to make the world totally secure and predictable.

Your life is a product of many factors. Some you control, some the government regulates, and some others don't easily fit into a category. The future will always be a question mark. You can't avoid uncertainty as long as you're alive, so it makes sense to figure out a strategy to deal with this variable common to everyone on the planet.

Another fact in the same category is the clear requirement that we make decisions—what to eat, what to wear, where to go, and what to do. And then there are the decisions that we make for others. So here lies the essence: If there is a 20% chance of rain today, do you carry an umbrella? How do you make decisions in the environment of uncertainty? From the context of technical problem calculations by scientists, mathematicians, and engineers, to the decisions made by you and me—we all make decisions about how we will manage the risks in our lives.

Decisionmaking involves analysis of information in some form and a choice selected from two or more alternatives. There are usually other factors to consider, including direct costs, opportunity costs, and related implications. There are also ethical issues to consider that reference the value system of the people involved. Ethical considerations are especially important when decisions are made containing inherent uncertainty in situations where finite resources exist. Risk management is one form of the decisionmaking process within the broader field of ethics. The outcomes vary depending on which philosophy and method you adopt in decisionmaking.

Since this entire book is about risk in decisionmaking, it makes sense to begin with a discussion about some of the ways decisions are made. The following principles can provide a frame of reference when you need to choose a course of action in the environment of finite resources and uncertainty. There are five ethical decision principles [4] discussed here. They are important in understanding risk management decisions in the context of ethics.

1. Utilitarian

The philosophy was developed primarily by Jeremy Bentham (1748–1832) and John Stuart Mill (1806–1873). The basic premise of this approach is that the action selected should provide the greatest good for the greatest number of people, or the least harm for the greatest number of people.

2. **Peoples' Rights**

This concept, developed by the 18th century thinker Immanuel Kant and his followers, says that a person's right to choose is an essential part of life. The inherent ability to choose freely is unique to humans. Kant believed that people have dignity based on their ability to choose freely and subsequently have the moral right to have these choices respected. It is a violation of human dignity to manipulate people as objects and in other ways in which they are not allowed freedom of choice.

There are other rights are also included in Peoples' Rights. For example:
- *The right not to be harmed*: We have the right not to be injured without knowingly doing something to deserve it.
- *The right to personal privacy*: We have the right to maintain unique ownership of information and of our "personal space" unless we choose to share it or our choice does not violate the rights of others.
- *The right to be treated fairly*: We have the right to receive goods and services as specified in a contract or agreement.
- *The right to the truth:* We have a right to be told the truth and to be informed about matters that significantly affect our choices.

In this principle, actions are unethical to the extent they violate the rights of others: the greater the violation of rights, the more serious the unethical action.

3. **Fairness or Justice**

The basis of this approach is Aristotle's view that "equals should be treated equally and unequals unequally." In other words, people should be treated fairly but people can receive different treatment based on different qualifications. For example, two workers of equal skill and experience should receive the same salary, and workers with less skill and experience should receive lower wages. Fair treatment in employment hiring for example means that "equals" should be interpreted relative to the duties and skills required to perform the work. There are situations where people are treated as "unequals" for good reasons. For example, blind people should not be allowed to drive buses.

Another application of this philosophy is in defining "distributional justice." Two people can be guilty of the same crime but can receive different punishments. Suppose, person #1 is a repeat offender and person #2 is not. Person #1 was the leader in the crime and person #2 played a smaller role. These situations suggest that unequal punishments may be appropriate.

The key word in Aristotle's statement is "equals." The approach is not a justification for favoritism or discrimination. To examine the degree in which this philosophy is being applied to a particular situation, test the notion that the groups are indeed equal in the relevant characteristics. This will tell you if the principle is being correctly applied.

4. Common Good

This principle has its origins in the early Greek philosophers, and presents a socialcentric ethical view. What is good for the community trumps the good of individuals. The community is composed of individuals who believe that their own wellbeing is closely connected to the wellbeing of the entire community. All members share this common belief. In short, the principle states: "What is ethical is what advances the common good."

The community could be a nation, a town, or a company. Situations where this approach is applied are military service, affordable healthcare, effective law enforcement, and low environmental emissions. This principle challenges us to think of ourselves as individuals who choose to work together for the purpose of achieving common goals that we could not accomplish as individuals.

5. Virtue

All ethics relate behavior to a set of standards, but this approach recognizes that even though humans are imperfect, we should strive to achieve certain ideals. It represents a moral compass to help improve behavior in a way that will achieve the fullest development of humanity. Virtues are attitudes and behaviors like honesty, courage, compassion, and integrity. In dealing with an ethical problem using the virtue principle, a relevant question is: What will promote my personal and also community character development?

These five principles are not mutually exclusive. They are references by which to compare your decision alternatives and to measure the nature of your actions. Basically, the ethical tenets can be tested by asking five questions:

1. Which alternative will do the most good for the most people?
2. Which alternative will respect the rights of the most people?
3. Which alternative has the least discrimination or favoritism and treats people equally?
4. Which alternative is the best to advance the common good?
5. Which alternative will promote and develop moral virtues?

Now let's consider some examples and see how to apply these ethical principles to test the efficacy of certain decisions.

Suppose you are a medical professional. You could be a licensed physician or just someone who has medical art skills that are not common to the population. You are walking down the street and the person in front of you suddenly collapses to the ground. By instinct, you rush over to the individual now lying unconscious on the ground and observe the symptoms of a heart attack. Someone in the gathering group calls for emergency services and the police. The person making the call tells everyone that an ambulance will be here within 5 minutes. You notice the person's breathing is subsiding and then stops. The faint siren of an ambulance can be heard in the distance. What should you do? There are several options but let's evaluate the two basic alternatives:

Option 1: Walk away to avoid any involvement in the situation eliminating the potential for liability from the person or person's family for damages allegedly incurred from your assistance.

Option 2: Start CPR immediately, giving orders to other persons to help in the procedure.

Now let's apply the five principles by asking the five questions.

1. *Which alternative will do the most good for the most people?* Option 1.
   Emergency services have liability protection and will arrive shortly. You, on the other hand, could suffer extremely high court costs and subsequent financial penalties from civil litigation. These costs do more harm by adding the high costs of malpractice insurance and by choosing not to lend aid, you help to keep these insurance costs down for the medical professional community.

2. *Which alternative will respect the rights of the most people?* Option 2.
   Each individual has the right not to be injured and the right to be treated fairly. You need to take action because you possess the skills that can aid the victim with care to help him mitigate harm.

3. *Which alternative has the least discrimination or favoritism and treats people equally?* Option 2.
   By not exercising your skills you are discriminating against the victim and favoring yourself. As part of your normal work duties, you would provide these services without question to the best of your ability. By choosing not to provide the same level of care, you are discriminating for your personal gain. This is unfair.

4. *Which alternative is the best to advance the common good?* Option 2.
   Which option would be good for the community? The fact that a passerby could (and would) save the life of a stranger is certainly the type of behavior that promotes strong community identification. Even if the victim died, the fact that a stranger courageously tried to help is a powerful message of the common goal of community safety and caring.

5. *Which alternative will promote and develop moral virtues?* Option 2.
   Answering this question is the easiest one of the five in this example. Providing needed assistance to a stranger, whether it's for a medical condition, a flat tire, or some change for a parking meter is emblematic of the virtue communicated in the Golden Rule: "Do unto others as you would have them do unto you."

Okay. Now that you have some experience with these ethical principles, let's apply them to a much larger and difficult scenario.

You are the risk manager of a small town of 10,000 people that is located just below a large dam. The winter brought above-average snowfall to the northern mountains and now that spring has arrived, the combination of melting snow and heavy rain is causing excessive stress on the dam structure. Late at night in the middle of a heavy rainstorm, you receive a call from the dam manager that the dam is going to fail within the hour. As a prudent risk manager you have emergency evacuation plans already in place and you proceed to quickly activate the emergency evacuation teams. Just before you make the first call to the teams, the dam manager calls you to let you know that one of the two roads out of town is blocked by a large mudslide. You re-evaluate you evacuation strategy and determine that you are completely sure you can safely evacuate about half of the town's population. This is Option 1. Another strategy, Option 2, indicates there is the possibility of saving everyone but also a possibility that everyone would perish. The odds are about 50:50 for saving or killing everyone. Time is growing short. There is no time to do any additional data collection and analysis. You need to make a decision, now! The longer you wait the more likely the dam will break and everyone will perish. Which option are you going to choose? Let's go through the questions and examine the ethics of the two options.

1. *Which alternative will do the most good for the most people?* Option 1.
   At least with this choice you are sure half of the town's population will survive.

2. *Which alternative will respect the rights of the most people?* Option 1.
   This moral action treats everyone the same. The fact that a single person died in the flood from the failed dam is random. You did not preselect him or her for death. You treated everyone equally.

3. *Which alternative has the least discrimination or favoritism and treats people?* Option 1.
   Apply Aristotle's statement, "Equals should be treated as equals and unequals unequally," to this situation. All 10,000 people equally share in the hazard. There is no special group that is exposed to a lower- or higher-hazard environment. By selecting Option 1, you have treated everyone the same and have saved 5,000 people. Of course, Option 1 also ensures 5,000 deaths.

4. *Which alternative is the best to advance the common good?* Option 2.
   As members of the same community, you believe that it is better to try to save everyone than it is to only save half of the town's people. You want to give everyone a chance to live.

5. *Which alternative will promote and develop moral virtues?* Option 2.
   All life is sacred and you believe it is immoral to commit half of the town to certain death. It is this ethic that you employ when you select Option 2, in which you have a chance of saving everyone.

I suspect that you probably disagree with some of my choices in these examples. Each person can look at a situation differently relative to his or her personal values so that there is no "right" or "wrong" response to the aforementioned situations. People respond to events based on their values and this is why risk management decisions for the same set of circumstances can be radically different. The five principles give you a structure by which to test the ethical quality of decisions in your value system. The decisions can be yours or others' who make decisions that influence your life. Ethical considerations are integral to decisionmaking, for no other purpose than to help you examine the moral quality of the decisions we make in our lives.

Notice that up to this point there is one blatant omission in this discussion: the law. None of the five principles refer to obeying the law as a tenet of ethical behavior. The ethical principles are much more insightful and broadly applicable than simple laws. And it's worth noting that ethical behavior may, depending on your value system, involve violating the law. Don't quote this book at your court trial. The legal system, established for the common good, does have a process by which to change laws. If you believe, for example, that a certain law is unethical, you can ask your congressional representatives to write a law to change it or you can get convicted of its violation and pursue your case through the legal system, perhaps all the way to the Supreme Court, to have it altered. The processes to change laws has also been established for the common good and it can work.

DISCUSSION QUESTIONS

1. After you read this book, you will possess knowledge not common to the general population. This knowledge gives you ethical duties and responsibilities. This is similar to the situation, for example, of a physician's responsibilities. In this case, medical emergencies on airplanes are a classic example where a physician's skills can influence a sick person's wellbeing. As an informed risk manager, you will observe certain behaviors, attitudes, and situations that can produce accidents, disease, and death. Ethically, do you have the moral responsibility to inform people of their increased risks?

2. Give an example of an ethical decision that requires you to violate criminal law. In your example, do you think the jury would find you innocent based on your values and ethics? Are there any well-known people who have done this?

3. Develop your own scenarios and decisions and defend your choices based on the five ethical principles discussed in the chapter.

4. From Figure 1.1, choose two countries and list 10 characteristics for each that are life-expectancy risk factors. Rank the factors from the highest to lowest risk and then estimate the percentage increase in life expectancy you might obtain if the top two factors were mitigated. How would you defend your estimates?

Case Study: Vaccines

The needs of the many outweigh the needs of the few.

> *Star Trek II: The Wrath of Khan, paraphrased from:*
> *John 11:49–50*
> *Aristotle, The Aim of Man*

Without a doubt, vaccines are one of the greatest achievements of the human race. The first vaccine was for smallpox, an infectious disease that has been tracked back to 10,000 B.C. in Northeastern Africa [5–6]. There is evidence that this killer was even well-known by the ancient Egyptian Dynasties (1570–1085 B.C.), and in China at the same time, through Indian Sanskrit texts [7]. The disease traveled to Europe and greatly influenced the development of western civilization through large-scale epidemics accounting for millions of deaths and countless millions more disfigured with lesion scars. The disease followed the spread of civilization to North America with similar devastating epidemics.

Then in 1774, Benjamin Jesty, a successful farmer in Downshay, England, noticed that milkmaids infected with cowpox, a less serious disease related to smallpox, were immune to subsequent outbreaks of smallpox that periodically swept through the area. He inoculated his wife and two young sons with pus from cowpox sores and observed their apparent immunization over time [8]. But Jesty's discovery was not communicated to the world.

Twenty-two years later in 1796, Edward Jenner, a country doctor from Gloucestershire, England, hypothesized the same connection between cowpox and small-pox immunity. Dr. Jenner performed several human inoculations using pus from cowpox sores and observed the same results as did Jesty. After a series of similar highly structured experiments, he published a book called *Inquiry into the Causes and Effects of the Variolae Vaccine*. His assertion "that the cowpox protects the human constitution from the infection of smallpox" laid the foundation for modern vaccinology [9]. After this information became communicated around the world, smallpox became a preventable disease. Jesty and Jenner probably did not know each other even though they where contemporaries but regardless, they are responsible for saving lives of countless millions of people in the future. The last case of smallpox in the United States was in 1949. The last naturally occurring case in the world was in Somalia in 1977 [10]. Smallpox has been eradicated from our planet, and this was just the beginning.

Today there are safe and effective vaccines routinely manufactured and delivered to doctors and healthcare centers, available for the following twelve infectious diseases:

Diphtheria, tetanus, pertussis (DTP)	Measles, mumps, rubella (MMR)
Haemophilus influenzae type b (Hib)	Meningococcal (MCV4, MPSV4)
Hepatitis A (HAV)	Polio (OPV or IPV)
Hepatitis B (HBV)	Pneumococcal conjugate (PCV)
Human papillomavirus (HPV)	Rotavirus (RV)
Influenza—each year at flu season	Varicella (VZV)

Additional vaccines, such as those for HIV, malaria, HPV, and others, are in various phases of testing.

From a societal perspective, immunizing the population, or the majority of the population, from these serious, debilitating, and potentially fatal diseases reduces pain, suffering, and healthcare costs for everyone. And the evidence of their effectiveness is irrefutable [11]. Here are the facts:

- Before 1985, Haemophilus Influenzae type b (Hib) caused serious infections in 20,000 children each year, including meningitis (12,000 cases) and pneumonia

(7,500 cases) [12]. Between 2002 and 2009, there were approximately 35 cases of Hib reported per year.

- In the 1964–1965 epidemic, there were 12.5 million cases of rubella (German measles). Of the 20,000 infants born with congenital rubella syndrome, 11,600 were deaf, 3,580 were blind, and 1,800 were mentally retarded as a result of the infection [13]. While localized outbreaks occur, especially in children too young to be vaccinated, today there are fewer than 25 cases reported each year [14].

- Before 1963, more than 3 million cases of measles and 500 deaths from measles were reported each year. More than 90% of children had had measles by age 15. In 2008, there were 16 cases [15].

- In 1952, polio paralyzed more than 21,000 people. There have been no reported cases in the United States since at least 2000.

- In the early 1940s, there was an average of 175,000 cases of pertussis (whooping cough) per year, resulting in the deaths of 8,000 children annually. In 2008, 13,278 cases were reported.

- In the 1920s, there were 100,000 to 200,000 cases of diphtheria each year and 13,000 people died from the disease. In the United States there was one reported case in 2000, and none since 2006.

Yet, these health benefits to society are not without human costs. There is another side to these medical success stories that is unsettling for some parents of young children and for some adults. Not everyone reacts favorably to vaccines.

Vaccines are biological agents designed to induce our immune system to produce disease antibodies. This is a delicate task of getting the body to produce a disease's antibodies, without actually giving the donor the full disease. This is done by injecting a weakened form of a particular germ, some kind of inactivated or killed germ, or a germ component. The body then produces antibodies that are designed to kill the germ in the future. Some vaccines require multiple inoculations and even some "booster" shots over time to maintain immunity. But once the body's immune system produces antibodies, it apparently remembers and responds in the future if the germ is detected again. In other words, we become immune to diseases without ever having them. This is what happens most of the time, but there are side effects ranging from minor soreness and rashes to permanent, long-term injuries and death.

How can this happen? Vaccine testing is an extremely detailed process, but not everyone can be tested. Due to an individual's specific genetic makeup and current health conditions, adverse reactions do happen. What I mean by this is there are cases where healthy people are vaccinated with the intent of reducing their disease risk, and instead they die or are permanently injured. There are also cases where people suffering from chronic, long-term health problems react unfavorably to vaccines and get sicker. Vaccine side effects are risks everyone takes when either they or their children receive an immunization.

Vaccines are especially important for schoolchildren where the close contact promotes disease transmission. Consequently, to reduce these risks all 50 U.S. states and Washington, D.C., have school-entry requirements for vaccines. Forty-eight allow exemptions for religious reasons (West Virginia and Mississippi are the only exceptions) and 21 states allow for personal-belief exemptions [16, 17]. Medical exemptions are allowed in all states. Homeschooled children are not subject to state vaccine requirements. As of 2010, the only exception was North Carolina, which does require vaccinations for homeschooled children [18].

The fraction of the population that suffers injury or death from vaccines is very, very small, yet if you are the victim or, worse yet, if it is one of your children, somehow the law of large numbers and the philosophy exhibited by the quotation cited at the beginning of this case study can be difficult to accept.

The U.S. government has taken action to give victims an opportunity to receive financial compensation for vaccine injury or death through the National Vaccine Injury Compensation Program. It is funded through an excise tax of $0.75 added to the cost of every administered vaccine dose [19]. The fund is designed as the legal mechanism for victims to receive compensation in order to protect vaccine manufacturers from the financial costs. If victims could sue manufacturers for damages, vaccine costs could be unaffordable for many people. Since it is in society's interest to have as many people as possible receive vaccines, the government is administering injury claims with its own lawyers and processes. Additional information is available at http://www.hrsa.gov/vaccinecompensation/omnibusproceeding.htm.

Between 1998 and January 2011, there were 13,693 cases filed with the Vaccine Injury Compensation Program with over $2.1 billion paid to 2,569 claimants. Over 40% of the 13,693 cases are related to claimants believing that vaccines caused their children's autism in spite of several scientific studies showing no causal relationship between autism and vaccines. The following legal proceedings describe one family's experience [20].

Jane was born on August 30, 1994. The pregnancy and first 15 months of life were normal. Following the standard schedule for infant vaccinations, Jane received the MMR (measles, mumps, rubella) vaccination at 15 months. The vaccines contained the mercury-based preservative called thimerosal. About two weeks later she saw her pediatrician for a fever and rash. The fever initially improved but then rose to a reported 105.7°F with additional symptoms of coughing, gagging, and vomiting. At the pediatrician's office she showed a fever of 100.3°F and had a "purulent postnasal drip." The diagnosis was "sinusitis v. flu" and antibiotics were prescribed. The symptoms subsided and the next visit to the pediatrician was at 18 months for a routine checkup. No significant health issues were observed but the pediatrician did note that Jane was "talking less since ill in January." Three months later the pediatrician noted "developmental delay suspected" and additional testing showed that Jane's brain development was abnormal. At 23 months of age, Jane was diagnosed with "severe autism" and "profound mental retardation." In addition to these neurological problems, she was also diagnosed with chronic constipation, diarrhea, gastro-esophageal reflux disease, erosive esophagitis, and fecal impaction. She has had seizures and displayed symptoms of arthritis and pancreatitis.

Her parents filed a claim with the Vaccine Injury Compensation Program. They claimed that the ethyl mercury in thimerosal used in the MMR vaccines damaged Jane's immune system. As a result, the vaccine-strain measles virus remained in her body, causing her to suffer inflammatory bowel disease and subsequent brain damage.

To obtain compensation under the program, claimants must show "by a preponderance of the evidence," that the vaccine caused the injury. A key piece of evidence would be revealed if the vaccine-strain measles virus could be detected in Jane's body. During a routine gastrointestinal procedure Jane underwent for her medical conditions, a biopsy was performed and the tissue sample was sent to a testing lab. The results came back positive: the vaccine-strain measles virus was detected. In addition to this "smoking gun" evidence, the parents also engaged six expert witnesses who testified in detail with their endorsement of the vaccine-strain measles causation theory.

The government's response was to examine the integrity of the testing lab's results. The lab, which is no longer in business today, was a for-profit, nonaccredited company established to support civil litigation against vaccine manufacturers in the United Kingdom. The government used several expert witness, some hired by vaccine manufacturers, to examine the testing lab's operational procedures. They concluded the lab's testing procedures were flawed and the test results were unreliable. In addition, the government witnesses testified to their belief that the vaccine cannot cause autism.

The judge, or "Special Master" in these types of cases, concluded that the evidence did not demonstrate that the MMR vaccine was related to the cause of Jane's medical conditions. His conclusion was primarily based on three detailed technical facts:

- The testing lab failed to publish the technical sequencing data to confirm the result validity.
- Other labs failed to replicate the results.
- The immunohistochemistry testing results were nonspecific to the measles virus genetic material.

(A complete description of these facts requires knowledge of the detailed microbiology involved in the testing protocol, so no more detail is given here.) The court denied the request for compensation.

The parents filed an appeal in which they supplied an additional witness who testified that the testing laboratory had a good reputation and that its work has been published in peer-reviewed medical journals. He also stated his opinion that the laboratory used proper procedures and took appropriate measures to avoid contamination. The government's experts, on the other hand, claimed they found a 20% error rate in the lab's test results, with duplicate samples sometimes even producing opposite results. They claimed the only explanation for the poor testing performance was contamination.

On appeal the court recognized the temporal relationship between the MMR vaccine, fever, and the later emergence of autism, but also said that this relationship is insufficient evidence to show causality and no new evidence was presented that contradicted the Special Master's initial ruling. The appeal was denied.

There are several scientific articles in published, peer-reviewed journals that essentially support the court's decision in this case. However, it is also acknowledged that science has not yet determined the cause (or causes) of childhood gastrointestinal disease and autism. This tragic and passionate controversy for many parents will continue until the causality is clearly understood.

Questions

1. If you were the lawyers for Jane, what would you have done differently?
2. Do you think the court ruling is fair? Why? Consider the public implications of your response.
3. Do any other countries have a vaccine compensation program similar to that of the United States?

ENDNOTES

1 Daniel Romer, Kathleen Hall Jamieson, and Sean Aday, "Television News and the Cultivation of Fear of Crime," *Journal of Communication*, March 2003, pp. 88–104.
2 Yinjiao Ye, "Beyond Materialism: The Role of Health-Related Beliefs in the Relationship Between Television Viewing and Life Satisfaction Among College Students," *Mass Communications and Society*, Vol. 13, Issue 4, September 2010, pp. 458–478.
3 Central Intelligence Agency, "Country Comparison: Life Expectancy." www.cia.gov (accessed January 2010).
4 Manuel Velasquez et al., "A Framework for Moral Decision Making," *Issues in Ethics*, Vol. 7, No. 1, Winter 1996.
5 S. Lakhani, "Early Clinical Pathologists: Edward Jenner (1749–1823)," *Journal of Clinical Pathology*, No. 45, 1992, pp. 756–758.
6 D. R. Hopkins, *Princes and Peasants: Smallpox in History*. Chicago: University of Chicago Press, 1983.
7 Stefan Riedel, "Edward Jenner and the History of Smallpox and Vaccination," *Proc (Bayl Univ Med Cent)*, Vol. 18, Issue 1, January 2005, pp. 21–25.
8 J. F. Hammarsten et al., "Who Discovered Smallpox Vaccination: Edward Jenner or Benjamin Jesty?" *Trans. Am. Clin. Climatol. Assoc.*, Vol. 90, 1979, pp. 44–55.
9 E. Jenner, *Inquiry into the Causes and Effects of the Variolae Vaccine*. London: Sampson Low, 1798, p. 45.
10 Centers for Disease Control and Prevention, Smallpox Disease Overview, February 6, 2007.
11 National Network for Immunization Information, http://www.immunizationinfo.org/parents/why-immunize (accessed January 29, 2010).
12 K. M. Bisgard, A. Kao, J. Leake, et al., "Haemophilus Influenzae Invasive Disease in the United States, 1994–1995: Near Disappearance of a Vaccine-Preventable Childhood Disease," *Emerging Infectious Diseases*, Vol. 4, Issue 2, 1998, pp. 229–237.
13 W. Atkinson, C. Wolfe, S. Humiston, R. Nelson, eds., *Epidemiology and Prevention of Vaccine-Preventable Diseases*. Atlanta: Centers for Disease Control and Prevention, 2000.
14 The Center for Vaccine Awareness and Research, Texas Children's Hospital, "Vaccines by Disease: Learning More about the Measles, Mumps, and Rubella Vaccine," 2010.
15 Centers for Disease Control and Prevention, "Selected Notifiable Disease Rates and Number of New Cases: United States, Selected Years 1950–2008," 2009, Table 27.
16 Johns Hopkins Bloomberg School of Public Health, Institute for Vaccine Safety, "Vaccine Exemptions," 2009. http://www.vaccinesafety.edu/cc-exem.htm. Accessed July 19, 2011.
17 The 21 states that allow exemptions for personal beliefs in schools and daycare centers are: Arkansas, Arizona, California, Colorado, Idaho, Louisiana, Maine, Michigan, Minnesota, New Mexico, North Dakota, Ohio, Oklahoma, Oregon, Pennsylvania, Texas, Utah, Vermont, Washington, and Wisconsin.
18 D. Khalili and A. L. Caplan, "Off the Grid: Vaccinations among Home-Schooled Children," *Journal of Law, Medicine, and Ethics*, Vol. 35, Issue 3, 2007, pp. 471–477.
19 Vaccine Injury Compensation Trust Fund, http://www.hrsa.gov/vaccinecompensation/VIC_Trust_Fund.htm (accessed February 1, 2010).
20 *Cedillo v. Secretary of Health Human Serv.*, 617 F.3d 1328 (Fed. Cir. 2010).

Chapter 2

Measurement: The Alchemist's Base

Although this may seem a paradox, all exact science is dominated by the idea of approximation. When a man tells you that he knows the exact truth about anything, you are safe in inferring that he is an inexact man. Every careful measurement in science is always given with the probable error ... every observer admits that he is likely wrong, and knows about how much wrong he is likely to be.

—Bertrand Russell

In medieval times, scientists mixed, stirred, and poured various ingredients together in their quests to make gold from base metals. Though these alchemists are gone, contemporary men and women have almost as difficult a task as they build concoctions of measurement facts in the search for their truths. Today, measurements and numbers constitute the potion. Just as two people can use the same ingredients yet bake different creations, people can and do use the same measurements to support different conclusions. The illusion we have today is that technology and technique hold the answers. Yet the more we mix, stir, and pour our measurements, the more we may mimic the ancient alchemist's vain quest. This chapter will begin to show you that in spite of our sophisticated technology, when it comes to measurements, we are not much more advanced than the alchemists.

Measurement is intoxicating. Regardless of where we live, we seem prone to the lure of its charm. Cars, golf carts, boats, and airplanes are equipped with satellite global positioning systems (GPS) showing drivers their movements, speed, and elevation over an onboard map display. Good fishing spots once only known to the seasoned veteran are now accessible to anyone who has a GPS locator. Some golf courses have specially equipped carts with GPS devices that measure the distance, in yards, from the cart to each hole. (The golfers are left on their own to measure the distance from the cart to the ball, unless they have a hand-held GPS device.) If you're not on a golf course and find yourself lost, don't worry; your GPS-equipped

---

20% Chance of Rain: Exploring the Concept of Risk, First Edition. Richard B. Jones.
© 2012 John Wiley & Sons, Inc. Published 2012 by John Wiley & Sons, Inc.

cell phone will help you navigate the roads and trails of your life, and may well show you the view along the way. While it's true that in many cases measurement technology has made our world more impersonal, even this is changing. It is now a common service for car manufacturers offer coupled GPS–cellphone-based systems to diagnose car problems remotely, help you find the nearest ATM machine, blink the lights or blow the horn of your car in a large parking lot if you forget where you parked, and automatically signal for help whenever an air bag deploys. All of this is possible because your car's position and status are being measured.

In many cases, measurements are proudly broadcast. Look around; you'll find measurements everywhere. Numbers on food packages measure the nutrients and contents, weather forecasting has its color radar pictures, there are gallons per mile statistics on new cars, energy usage statistics on appliances, and just look at all of the energy expended in public opinion polling. There is no doubt that measurement is often regarded as a numerical security blanket. For legal defense, for legal offense, for scientific advancement, for the competitive edge, and just to be nosy, we measure. It's that simple.

Let's look first at the one of the most detailed measurements in the world: the measurement of time. On the scale of computer operations, one second is a very, very long time. When it comes to computers, the faster you can divide time, the more instructions you can execute and hence the faster the whole computer works. To find better ways to divide (and measure) time, there is a small group of scientists around the world whose members measure time in incredibly minute fractions of a second.

In the 1950s, "atomic clocks" were invented using cesium atoms where the basic clock was a microwave transition between energy levels [1]. The radiation emitted a microwave of a specific frequency. The accuracy achieved was about 10^{-10} or 1 second in about 300 years. Not bad—but that was just the beginning. The quest for more accuracy continues today. In 2008, the U.S. National Institute of Standards and Technology unveiled NIST-F1 Cesium Fountain Atomic Clock with an accuracy of about 5×10^{-16}, or 1 second in about 60 million years and even higher accuracies measured in experimental prototypes [2].

The interesting part of this work is this question: when you are developing more accurate time measurements, how can you measure something that is potentially more precise than your current standards? The answer: Build two of them and compare their results. Gauging new levels of time accuracy is more than a simple technology challenge. As scientists split time into such small sections, gravitational and special relativity effects become important. And as relativity effects become more important, there may come a time where a complete rethinking of "timekeeping" becomes necessary.

Cybertime machines are a far cry from the first time measurement devices. The height of the sun in the sky, fractions of the moon, and beats of the human heart were among the first ways people measured relatively short times. Today, our everyday measures may not be as precise as cesium fountain clocks but that doesn't stop us from measuring seemingly everything in sight. Are we better people or, from a societal point of view, are we healthier, happier, more prosperous, and safer because we measure? Think about this question while you are reading this book.

There is no doubt that measurement is essential to understanding and dealing with risk. After all, unless you have "measured" in some way, you would have no understanding or awareness of your surroundings. On the surface, this discussion may sound similar to the old query: "If a tree falls in a forest and no one is around to hear it, does it make a sound?" For a risk-measurement version, we could ask, "If a person is not aware of a risk, does the risk exist?" The answer is yes. Suppose, for example, you were not aware (or in other words had no measure) of the life-saving effects of wearing seat belts. Whether you knew about seat belt safety or not, your risk is still greater than that of a person wearing the belt. In the practical perspective, what you don't know *can* hurt you.

But looking at the situation from another point of view, risk management is a rational decision. It requires information. Without measurement feedback to supply the information, there is no risk. This school of thought says that without measurements to supply information about uncertainty, there is no uncertainty, hence no risk. Here, what you don't know can't hurt you.

The flaw in this argument is that measurements do not change the actual risk; they only identify it and may give you some indication of its magnitude. What changes with measurement is our perception of the risk. When we "measure" automobile risk, the risk reduction associated with the behavior of wearing seat belts might help us make some of our risk management decisions concerning driving. If we did not know about seat belts, the risk associated with riding in a car wouldn't be less just because of our lack of knowledge. This is what measurements do for us. In a succinct statement: Measurements alter our perception of uncertainty.

Measurements, perception, and the human mind are inseparable. The seemingly simple process of measurement is a lot more complicated than it seems. We develop one set of measurements based on our initial beliefs; the results alter our perceptions, changing the measurements; and we go around the cycle again. The swamp really gets muddy when the measurements are biased by political agendas, personal prejudices, or the host of other factors that lurk in the dark recesses of our realities.

Not all is bleak, however. There are ways of navigating through the mud to sift out truth from fiction and the biased from the unbiased. There are things to look for: rules for how to differentiate the good from the bad, the precise from the perception of precision, and, of course, fact from fiction.

Measurement has always been and will always be a mix of art and science. Little has changed with the application of technology. Science and computers provide more choices, more types of measurement devices, more data from which to extract information, more capability for precision, more opportunity to succeed, and more opportunity to fail.

Let's look at a situation that shows exactly how confusing and even deceptive some apparent simple "measurements" can be. When you go to the supermarket to buy poultry and read the "FRESH" label, do you know what this means? Until recently it meant little. But for poultry, it now means the bird was never cooled below its freezing point of 26°F. That's simple enough, but the industry now has some new terminology that suggests one condition, but means another. For example, "never frozen" legally means the bird was never stored at 0°F or below. Thus store bought turkeys cooled to 1°F are not actually "frozen" according to the U.S. Department of Agriculture [3]. They say that birds stored between 26°F and 0°F may develop ice crystals but won't freeze all the way through. Unless butchered poultry has some secret internal heating mechanism, I don't see why birds at stored at 1°F would be "never frozen" yet poultry at 0°F qualify to be labeled "frozen." This example shows how even simple, yet subtle measurement terms can influence our lives by implying a certain characteristic to the consumer, yet can legally mean a very different thing. Let's go a little further into the depths of measurement.

MEASUREMENT SCALE

Measuring isn't always as easy as it sounds. Consider the "gas can paradox." Suppose you stop at a gas station to fill a 5-gallon gas can. You know it's a 5-gallon can because its volume is printed on its side. The can was empty when you started. You keep your attention on the pump meter and stop pumping when the pump indicates 5.00 gallons. You look at the can, and it's not full. You now transfer your attention to the can and stop pumping when the can is full. The pump meter now reads 5.30 gallons. Where's the error: In the pumping meter or in the size of the can? How could you resolve this problem outside of going to the Bureau of Standards? In essence you can't. Somewhere in your plan of action, you would have to rely on a standard.

Sometimes standards are not really standards. Here are some examples. If you are putting sugar in your coffee, a teaspoon is a teaspoon. However, if you are administering medication to your 3-year-old, would you be as cavalier in selecting a spoon out of the drawer? Are all so-called teaspoons the same? Check your silverware drawer. If yours is like mine, you'll find more than one size. And would you be more careful in filling the spoon with your 3-year-old's medicine than when filling the same spoon with sugar for your coffee?

Now suppose you go to Canada and ask for 5 gallons of gas. Since the Canadian gallon is different than the U.S. gallon, it's clear that the same names can be used for different standards.

Scale is one of the most important and least appreciated elements of measurement. Sometimes you can't do anything about it, but at least you should know a little about the concept. A measurement scale implies a standard frame of reference. A measurement without an accepted scale makes conditions ripe for misinterpretations.

There is at least one airline pilot and co-pilot who will always double-check the scale on their fuel load reports before flying. On July 13, 1983, they were flying a Boeing 767 from Ottawa to Edmonton when both engines shut down due to lack of fuel and glided (yes, jets can glide) to a safe landing at an abandoned airport [4]. The number on the fuel load report indicated the flight should have had sufficient fuel for the flight plus a required reserve. As it turned out, the fuel gauges where not functioning for the flight so the ground crew at Ottawa manually checked the fuel quantity with a drip stick. The ground crew then computed the amount of fuel using the standard factor of 1.77 lb/liter that was written on the fuel report form. This was the conversion factor for all other Air Canada aircraft at that time. Here's the problem: The new (in 1983) Boeing 767 is a metric-based aircraft and the correct conversion factor was 0.8 kg/liter. The flight crew thought they had over 20,000 kg of fuel when in fact they had just over 9,000 kg. At 35,000 ft (10.7 km) with both engines and cockpit navigation instruments shut down, the pilot glided 45 miles (72 km) to safe airport landing. As it turned out, the captain was an accomplished glider pilot, a skill not commonly applied in wide-body jet flying.

I'm sure Air Canada corrected their B-767 fuel forms and in a very short time everyone understood that a different measurement scale needed to be used with this aircraft. Here's another, and even a more dramatic example of "measurement scale risk" or "measurement scale uncertainty" in the unforgiving world (or universe) of aerospace operations.

On December 11, 1998, NASA launched the Mars Climate Observer [5] from the Kennedy Space Center. The 629 kg (1,387 lb) payload consisted of a 338 kg (745 lb) spacecraft and 291 kg (642 lb) of propulsion fuel. Its mission was to fly to Mars where the main engine would fire for 16 seconds, slowing down just enough to allow the spacecraft to enter an initial elliptical orbit with a periapsis (closest altitude) of 227 km above the surface. Over the next 44 days, "aerobraking" from atmospheric drag on the spacecraft would slow the spacecraft even more and place it in the correct altitude range for its scientific and Mars Lander support missions. That was the plan.

Over the nine months of traveling to Mars the spacecraft's position and rotation speed was corrected systematically via short impulse thruster burns. Data transmissions from the spacecraft would be fed into trajectory models here on earth and the results would tell the navigation team how to apply the onboard thrusters to make the small midcourse corrections. About a week before Mars orbit insertion, the navigation team determined that the orbit insertion periapsis would be in the 150–170 km range. Shortly before the main engine firing, a recalculation put the initial orbit periapsis at 100 km. While there was serious concern about the trajectory errors, since the lowest survivable altitude was 80 km and little time to determine the cause, the mission was not considered lost. However, 4 minutes and 6 seconds into Mars orbit insertion, the engine was ignited and all communications ended. The communication blackout was expected and it was predicted to last for 21 minutes. The blackout period started 41 seconds earlier than calculated and the spacecraft never responded again. The $324 million mission was lost.

The root cause? Spacecraft navigation information was received at one site and then transmitted to a special team that developed and ran the ground-based trajectory models. They would feed this data into their models and then give the spacecraft navigation team the required vector and thruster impulse information to make the desired course corrections. Just 6 days after the loss of the spacecraft, engineers noticed that the models underestimated the correct values by a factor of 4.45 and this discovery pointed to the problem. One pound of force equals 4.45 Newtons, or

1 lb-sec = 4.45 kg-sec. The spacecraft was designed and operated using the metric scale. The ground-based trajectory models were designed and operated using the English scale. Correcting the final known position, the spacecraft had actually entered an orbit periapsis of 57 km, which was unsurvivable.

Air Canada corrected their latent scale risk and NASA has made considerable program integration changes to ensure something as simple but deadly as data unit inconsistencies will not occur again. Considering the success of the International Space Station, the largest space station ever built, involving equipment, systems, and parts from 16 countries, it appears that measurement scale risk has been effectively managed. Yet with the world converting to metric measures, NASA has decided to stay on the imperial system (pounds and feet) for the next generation of shuttle launch and space vehicles. Their reasoning highlights a common problem in improving legacy systems. The retiring shuttle systems were designed with imperial units and the new shuttle replacement concepts use systems that are derived from the old shuttle. To convert to the metric scale would simply cost too much. So for economic reasons, NASA is going to remain on the imperial scale.

Programs that don't have these legacy issues will be developed by NASA using metric units. For example, all lunar operations programs will use metric units. So with only the United States, Liberia, and Burma remaining on the imperial measures, the moon will join the majority of earthbound countries in using metric measures. That "one small step for man" and all future steps on the moon will be measured in meters rather than in feet or yards.

Moving to the smaller side of measurement, chaos theory gave the subject a boost toward the esoteric and philosophical when it was first discovered. Chaos scientists realized that, depending on at what level you look at a system, you might see what you believe to be chaos [6]. However, by viewing the same system with a different scale, a definite order can be observed. Thus a seemingly random system really has order and can more easily be understood if we apply a different scale. In plain, down-to-earth English, scale is a generalization and elaboration of the old, well-known concept of seeing the forest or the trees. Or, in the case of chaos theory, seeing jagged lines or the shape of a snowflake—depending on your "scale" of vision.

Think about differences in scale when you look at ocean waves. The number of wavelengths you can observe will vary of course, depending on conditions. On the smallest scale, the wavelengths are less than a foot. On the larger scale, the wavelengths can be measured by movements of the tides.

Scale and measurement are intimately related. You can't have one without the other. Depending on the measurement scale you use, your results can be totally different. I think you would agree that you can't predict the weather very well just by looking out of your window; a satellite view gives a better perspective. Every measurement activity has its limitations because it is based upon a given scale. Often this scale can be traced back to a time measure. Just like playing a piece of music on the piano at different tempos, measurement results will change depending on the time scale and frame of reference that is adopted.

Something as simple as the accuracy of the measurement scale shouldn't be questioned, right? Yet as time changes, our frames of reference sometimes change. Of course, a meter is still a meter. Generally weights and lengths are invariant over time, but some other important scales have changed.

The Scholastic Aptitude Test (SAT), first administered in 1926, has changed its scoring methods. The Educational Testing Service (ETS), the test producer, used an interesting word to characterize the change in scoring. In 1995 they "recentered" both the math and verbal scores, citing facts such as that the abilities of the "average American" are poorer than they were in 1941 and scores have been steadily declining every year. In case you're wondering why 1941 was selected, it's because this was the last year the scores were "recentered." Somehow I don't think lowering the scale encourages test scores to go up! The lowering of the scores effectively awards the poorer performing students at the expense of the superior students. Thus a combined score (verbal + math) of 1200 in 1994 isn't quite the same as it is today [7]. The effects of the scale changes are complex and one can understand the technical justifications from reading the literature; regardless of the causes, however, the scale has changed.

The SATs are only one of several gauntlets traversed by students traveling the road to higher education. Another activity designed to qualify high school students for college are the Advanced Placement exams. These courses are taught like regular high school courses for a semester and a standardized test is given at the end of the course. The grading system is 1–5, where 1 and 2 are interpreted as failure and 3 or above may qualify the student to receive college credits. There are a lot of benefits of passing these exams and students have been taking these courses in record numbers. For example, in 2009, 2.9 million students took the exams [8]. This was one record, but there was a second record also set this year. Nationally, more than two out of every five students who took the exams (41.5%) failed, the highest fraction ever. To give you an idea of how many previously have scored poorly, in 2000, 35.7% failed and in 2005, 37.9% failed [9]. Forming a linear trend line from these numbers forecasts the 2009 failure percentage at 39.7%. The message here is the exam failure is increasing faster than at a linear pace. This is some background information for some recent developments in Advanced Placement exam grading.

In the past, students were penalized for guessing. The final score was computed as the number of correct answers minus a percentage of the number of incorrect answers. But starting in the May 2011 exams, scores will only include the number of correct answers [10]. In other words, the penalty for wrong answers (guessing) will be eliminated. This is how the Scholastic Aptitude Test (SAT) is already graded so this change does make the testing guidelines consistent. However, now that guessing has no penalty, the scores can't be any lower and in fact the total scores should go up. The net effect may not be any different since the 1–5 grading system can still fail the lower 40% of scores but regardless, the scale has changed. The College Board management stated that this change is one part of a major overhaul of their testing instruments and more improvements can be expected in the future.

Two scales that everyone has in common are time and money. Thanks to the magic of our financial markets and a host of other factors, the scale of monetary value is a function of time. By far the best example of changing scale occurred in the early part of 1997 and relates to how the government measures the changing worth of money.

A study unveiled at the end of 1996 from the now infamous Boskin Commission [11] (formally called the Advisory Commission to Study the Consumer Price Index), showed that the Consumer Price Index (CPI), one of the U.S. government's most important inflationary scales, overstated the cost of living by 1.1 percentage points. This finding was estimated to save about $133.8 billion dollars over the next five years. Since about 30% of the federal budget, including many entitlement programs, is indexed to the CPI, the *Wall Street Journal* believes the federal government would spend a trillion dollars less than expected over the next decade. You and I save money by tightening our belts and spending less. The federal government saves a trillion dollars by changing a formula. The actual effects of recomputing the CPI evaluated about ten years later suggested [12] that the changes should have been 1.2 to 1.3% in 1996. In 2009, the error was closer to 0.8% but since the CPI is a result of a complex statistical formula involving "stratified random sampling," there is an inherent uncertainty in any CPI evaluation that doesn't seem to be mentioned when CPI results are released [13]. Social Security, veteran benefits, and a myriad of other governmental programs are related to this index. The point? Just because a governmental agency has computed a new index value does not mean that it is an exact value. The correct strategy would be to release a range rather than a single number. When you hear and read about changes in economic indicators, think about the possible error or uncertainty in the results. In many cases, with a little scholarship you can also identify the ranges and decide for yourself what the changes really are—if they exist at all.

Measurement standards are not always esoteric subjects. They permeate almost every aspect of life in our global society. Consider the United States' "national pastime," major league baseball; more specifically, the size, weight, composition, and performance characteristics of a baseball itself.

The Rawlings Sporting Goods Co. is the official maker of baseballs for both major and minor leagues in the United States. Quality and uniformity are paramount. You may not think that baseball quality control is important, yet to a sport that captures the hearts of millions of people and produces billions of dollars in revenue, the business of baseball is more than a game. Just look at the home run hitting and base hit legacies of baseball's greatest players. If baseballs weren't held to tight manufacturing standards, these records would be meaningless.

To ensure that baseball measurements stay the same, Rawlings samples balls from each new production lot made at their plant in Costa Rica. In a room where the environment is maintained at 70°F and 50% humidity, each ball is fired at a speed of 58 mph at a wall of ash wood. This is the same wood as used for making Major League bats. The ball's speed on the rebound is measured and divided by the initial speed to compute ball's "coefficient of restitution" (COR). Major League rules state that the balls must rebound at 54.6% of their initial speed, plus or minus 3.2

percentage points. Any faster and the ball has too much spring, any less and the ball has too little.

According to Robert Adair, currently Professor Emeritus at the Yale University Department of Physics and the "official physicist" to the National League in the late 1980s, a batter connects with a home run power hit from a pitcher's fast ball of 90+ mph in about 1/1000th of a second. The collision between the ball and bat compresses the hard ball to about half of its original size.

In order to routinely manufacture these items, Rawlings uses techniques that are not widely advertised. Yet we know the basics. Each baseball requires about a quarter mile of wool and cotton thread tightly wound around a rubber-covered cork center. If the ball is wound too tightly, it has more bounce, but then exceeds its weight requirements since more material was used in the ball. This example illustrates some of the inherent checks and balances in the measurement process. The end result? Because measurement standards exist, the eligibility requirements to join baseball's record-breaking elite have changed little over the decades.

In recent years, the influence of performance-enhancing drugs has tainted some baseball record achievers. When questioned about the steroid influence on performance, Professor Adair doubted that the drugs have had a major effect. He stated that historically, weight lifting was not encouraged since the general (incorrect) belief was that the increase in muscle mass would slow reaction speeds. Today, weight training is common and baseball players are generally stronger and bigger than they were in the past. According to Professor Adair, 30 years ago the average baseball player's weight was about 170 lb. Today the average weight is over 200 lb [14]. So, at least for baseball performance, the measurement scale has not changed due to any changes in the implements of the game. The players have changed.

When performance records are broken, any hint of impropriety can taint the achievements regardless of the veracity of the claims. One thing is certain in baseball: the hitting records are not due to changes in the ball's performance measurements.

PRACTICAL LAWS OF MEASUREMENT

Aside from scale considerations there are a series of principles you can rely on to help you through the murky waters of measurement. I call them "laws" because as far as I am concerned they're true, even though no one has ever tried to prove them. Together they'll give you a tool kit to tackle measurement facts. Use them as a frame of reference and scale to help you decide for yourself how to interpret "the facts." We'll discuss them one by one, with some examples that show how the laws work.

Law #1: Anything can be measured

Did you know [15]. . .

Almost 50% more people fold their toilet paper than crumple it.

68% of Americans lick envelope flaps from left to right (the other 32% do it from right to left).

37% prefer to hear the good news before the bad news

28% squeeze the toothpaste from the bottom of the tube

38% sing in the shower or bath

17% of adult Americans are afraid of the dark

35% believe in reincarnation

17% can whistle by inserting fingers in their mouth

and finally:

20% crack their knuckles

Do you care? You can rely on the fact that somebody does. Just search the Internet for any subject and add the word "statistics" and you can see the diversity of measurements people develop and analyze. This is how new products are created. Market research is a highly competitive, multibillion-dollar industry that measures consumer preferences. We are exposed to its products continuously. For example, let's examine the fairly mundane subject of soap. What makes you buy one brand instead of another? Its scent? The amount of lather? Its shape? Color? Size? The design on the box? All of these characteristics are a function of consumer preferences that soap manufacturers measure on a continual basis.

Consumer preferences are but a small part of the measurement challenge. As we discuss risk management, you'll see how much of this subject is dependent on our collective judgments, opinions, and beliefs. The challenge we face is how to use the analytical and technical tools we've developed to measure intangible, transient variables such as public risk perceptions.

Technology, coupled with the powers of human creativity, mathematics, statistics, and science, provides the potential to measure virtually anything and everything. Even the most seemingly subjective processes can be measured once the desired results are clearly defined. This means that real measurement decisions are now much more difficult than in the past, and brings us immediately to our second law.

Law #2: Just because you can measure something doesn't mean you should

When my son was young, he was delighted every time he got to use my measuring tape. He accepted the tool without question and proceeded to measure everything in sight. Of course, as a 4-year-old, he also used the measuring device for some rather creative purposes for which it wasn't designed. He had neither an application, nor understanding of any of his "measurements," but the device was fun to use. This may remind you of certain adults when they are given new toys …

Science, on the other hand, has applications and understanding as solid foundations for measuring value. Every year the best of these works get awards. The most well-known and probably most prestigious of these are the Nobel awards. Noble laureates are the select few whose work has made the most significant advances in applications of our current tools toward the understanding of our universe and the advancement of high ideals. In a way, the Nobel Prize awards represent the ultimate assessment of measurement of an individual or team's scientific or general work performance.

Yet in early October every year, just before the Nobel Prizes are announced, there is another set of awards also given to researchers whose work represents similar ideals, albeit on the opposite side of the spectrum. These are the Ig Nobel Prizes. In 1991 the U.S.-based organization Improbable Research began awarding research or achievements that "makes people laugh, then think." The organization has grown in size and international popularity since then and its publications range from a journal-level periodical to web blogs. The Ig Nobel awards are given out by real Nobel Laureates, and represent scientific or otherwise works that "cannot or should not be reproduced." As you might expect, this "award," intended for the most part as a good-natured spoof, is not always well received. Here is the list of the 2009 "winners."

Biology: Fumiaki Taguchi, Song Guofu, and Zhang Guanglei of Kitasato University Graduate School of Medical Sciences in Sagamihara, Japan, for demonstrating that kitchen refuse can be reduced more than 90% in mass by using bacteria extracted from the feces of giant pandas [16].

Chemistry: Javier Morales, Miguel Apátiga, and Victor M. Castaño of Universidad Nacional Autónoma de México, for creating diamonds from liquid—specifically from tequila [17].

Economics: The directors, executives, and auditors of four Icelandic banks—Kaupthing Bank, Landsbanki, Glitnir Bank, and Central Bank of Iceland—for demonstrating that tiny banks can be rapidly transformed into huge banks, and vice versa—and for demonstrating that similar things can be done to an entire national economy.

Literature: Ireland's police service (An Garda Siochana), for writing and presenting more than fifty traffic tickets to the most frequent driving offender in the country—Prawo Jazdy—whose name in Polish means "Driving License."

Mathematics: Gideon Gono, governor of Zimbabwe's Reserve Bank, for giving people a simple, everyday way to cope with a wide range of numbers—from very small to very big—by having his bank print bank notes with denominations ranging from one cent ($.01) to one hundred trillion dollars ($100,000,000,000,000) [18].

Medicine: Donald L. Unger, of Thousand Oaks, California, for investigating a possible cause of arthritis of the fingers, by diligently cracking the knuckles of his left hand—but never cracking the knuckles of his right hand—every day for more than sixty (60) years. He never developed arthritis and therefore concluded that cracking knuckles does not cause arthritis [19].

Peace: Stephan Bolliger, Steffen Ross, Lars Oesterhelweg, Michael Thali, and Beat Kneubuehl of the University of Bern, Switzerland, for determining—by experiment—whether it is better to be smashed over the head with a full bottle of beer or with an empty beer bottle [20].

Physics: Katherine K. Whitcome of the University of Cincinnati, Daniel E. Lieberman of Harvard University, and Liza J. Shapiro of the University of Texas, for analytically determining why pregnant women don't tip over [21].

Public Health: Elena N. Bodnar, Raphael C. Lee, and Sandra Marijan of Chicago, Illinois, for inventing a brassiere that, in an emergency, can be quickly converted into a pair of protective face masks, one for the brassiere wearer and one to be given to some needy bystander [22].

Veterinary Medicine: Catherine Douglas and Peter Rowlinson of Newcastle University, Newcastle-Upon-Tyne, U.K., for showing that cows that have names give more milk than cows that are nameless [23].

As you can see there is an air of satire in the awards but there also is a very serious aspect of recognizing science and achievements that are on the edge of apparent technical or practical relevance. As eloquently stated in a recent article about the awards, "nature herself doesn't understand the meaning of trivial" [24].

Trivial or not, in the next law, we recognize that everyone (and every measurement) has their limitations.

Law #3: Every measurement process contains error

It is important to remember that the act of obtaining a measurement is a process in itself. Even though measurements may be accurate, the overall results of the measurement process may be wrong. There are always some errors involved in the measurement process. Success in measuring requires an understanding of the errors and the magnitude of each. Occasionally, error values turn out to be larger than the value of the measurement. This does not necessarily mean the results are useless. It does strongly suggest, however, the value is a "ball park" result, and not a statement of precision.

There is one line of work in which there is plenty of error, we all have an interest in accuracy, and being "in the ball park" isn't good enough. NASA's Near Earth Orbit (NEO) Program identifies and tracks comets and asteroids that pass relatively close to earth. The program's interests are scientific since these celestial objects are undisturbed remnants of the creation of the solar system about 4.6 billion ago. But NEO, in cooperation with the global astronomical community, also serves as an early warning service for Earth residents for objects that may be on a collision course. Most of the objects that enter our atmosphere have been small rocks providing no more than a streak of light across the sky. However, there are indications that other, larger rocks, a few kilometers wide, have had a significant influence on our world ecosystem and any collision with objects of this size today would change everyone's day.

Most of the time, we have plenty of warning of impending doom since objects large enough to produce Armageddon can be seen and tracked for years before they would be a problem. While all of us work, sleep, and play, there are people and computers searching the skies looking for new threats and tracking old ones.

On the morning of October 6, 2008, at 6:39 GMT, Richard Kowalski, at the Catalina Sky Survey, discovered a small near-Earth asteroid using the Mt. Lemmon 1.5 meter aperture telescope near Tucson, Arizona [25]. Kowalski reported the finding to the Minor Planet Center (MPC) in Cambridge, Massachusetts, the NASA-funded center that has the responsibility for naming new celestial objects and for the efficient collection, computation, checking, and dissemination of astrometric observations and orbits for minor planets and comets. The object was named 2008 TC3, the information was sent around the world to other astronomers, and the data collection activities began. In total, some 570 astrometric (positional) measurements were made by 26 observatories around the world, both professional and amateur. When scientists at NASA's Jet Propulsion Laboratory in Pasadena, California, compiled all of the telemetry information from the reporting sites, they realized that they had too much data [26]. Data collected from different telescopes requires that the time clocks on each instrument precisely agree in order to compute 2008 TC3's position at any instant. The differences in clock synchronization translate into errors in 2008 TC3's location. And for an object traveling at 17,000 mph, a small error in time means a big difference in position. The initial calculations showed that 2008 TC3 was going to hit Earth, but the big questions were exactly where and when. To reduce the time synchronization errors scientists did not use all of the collected data but selected only a dozen data points from each observatory. This method ensured that no one observatory's data and consequently no one observatory's error would dominate the results. The final calculations showed that 2008 TC3 was going to going to impact in the northern Sudan around 02:46 UT on October 7, 2008, just a little over 20 hours after being discovered. The differences between the actual and predicted times and locations are shown in Table 2.1. The quality of these calculations shows that proper management of errors can actually enhance the statistical results.

Table 2.1 2008 TC3 Entry Time and Location Model Comparison with Actual Data

Event	Atmospheric entry	Airburst detonation
Observed Event Time (UT)	02:45:40	02:45:45
Predicted Event Time (UT)	02:45:38.45	02:45:45.13
Observed Long./Lat. (deg.)	31.4E 20.9N	32.2E 20.8N
Predicted Long./Lat. (deg.)	31.412E 20.933N	32.140E 20.793N

Yet not all data and errors can be so elegantly managed. Here's an example that will influence how much money you will have in your back pocket or your pocketbook. In 1998, the Congressional Budget Office (CBO) predicted a $1.6 trillion in budget surplus over the next 11 years. They are the first to admit the prediction was made with many assumptions about what the future holds. The biggest problem that would invalidate the surplus estimate was the number and severity of economic recessions during the next 10 years. At the time the CBO made this prediction, they had no recessions in their 10-year forecast. This is where the magnitude of the error becomes very personal, at least to your finances. Economic experts claim that even a mild recession in one year could increase the deficit by $100 billion in a single year. A forecasting error of just 0.1 percentage point slower growth each year over the next 10 years would trim a cool $184 billion off the CBO projected surpluses.

Now fast-forward 10 years. The United States experienced a mild recession in 2001–2002 and shared a global recession starting in 2007 or 2008 (depending on which economist you believe), federal spending has accelerated to record amounts, and budget deficits are at record levels. These assumptions were not in the 1998 forecast.

Yet putting aside our economic theory differences, there is one thing we all can believe in: the uncertainty (or error) of any government forecast. Figure 2.1 displays an excellent representation of what we know implicitly, but the U.S. Congressional Budget Office statisticians presented the "data" in a very precise graph of measurement uncertainty. Even though the graph was published in January 2004, the basic lessons are the same today. The y-axis is the U.S. budget deficit (–)

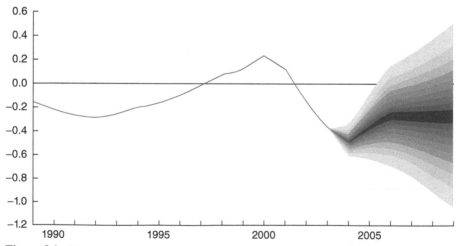

Figure 2.1   Uncertainty of the Congressional Budget Office's 2004 projections of the budget deficit or surplus under current policies.

or surplus (+) in trillions of dollars. The multishaded fan represents the range of possibilities. The baseline budget projections—the projections with the highest probabilities—fall in the middle of the fan. But nearby projections in the darkest part of the fan have nearly the same probability of occurring as do the baseline projections. Moreover, projections that are quite different from the baseline also have a significant probability of occurrence. On the basis of the historical record, any budget deficit or surplus for a particular year, in the absence of new legislation, could be expected to fall within the fan about 90% of the time and outside the fan about 10% of the time.

The lesson here is simple. Regardless of whether governmental projections are good or bad news, they are made under a set of assumptions about the future. If government economists could predict the future accurately, chances are they wouldn't be working for Uncle Sam or anyone else. Since even a small change can alter the numbers drastically, surplus results are only an outcome of a specific scenario and no one knows for sure if the measurement assumptions will turn out to be true. Another interesting property of Figure 2.1 is that the historical results are exact. We know exactly what happened, and consequently there is no uncertainty (or risk) in the past. As Figure 2.1 so eloquently displays, the only uncertainty (or risk) is in the future.

Judge for yourself the likelihood of the government's track record for economic forecast accuracy and you'll have an estimate of the likelihood of actually seeing this money. My opinion? This is one case where the size of the error makes politically meaningful surplus or deficit forecasts practically meaningless. And by the way, how many times have you read the final accuracy of budget forecasts after the time period has occurred?

Law #4: Every measurement carries the potential to change the system

Taking a measurement in a system without altering the system is like getting something for nothing—it just plain doesn't happen. In particle physics, this law is called the Heisenberg Uncertainty Principle [27]. In 1927, Dr. Werner Heisenberg figured out that if you try to measure the position of an electron exactly, the interaction disturbs the electron so much that you cannot measure anything about its momentum. Conversely, if you exactly measure an electron's momentum, then you cannot measure its position. Clearly, measurement affects the process here. Even when dealing with the fundamental building blocks of matter, measurement changes the system.

Here's a more common example. Consider the pressure valve in a bicycle tire. In order to move the needle to calibrate the distance corresponding to the internal tire pressure, kinetic energy must be released. This, in turn, reduces the pressure to be measured. Although the needle mechanism reduces the pressure a negligible amount, it is reduced nevertheless.

As you might expect, electrons and tires are not the only places where this law applies. It's virtually impossible to have any measuring process not affect what's being measured.

Here's another scenario. Nuclear power plants rely on large diesel-powered generators to supply emergency electrical power for reactor operation. The Nuclear Regulatory Commission requires that these generators be started once a month to ensure that they are in good working condition [28]. In reality, the testing itself degrades the useful life of many components. It is likely that testing at this high frequency actually causes the diesel engines to become more unreliable. Some people say that the diesels should be started and kept running. The controversy in this area goes on, but the paradox is real—excessive testing to give the user confidence that the equipment is reliable can actually make the equipment less reliable.

In daily life, this law also recognizes how people react to measurement. Consider the difference in telephone transaction styles between the systems and operators who provide telephone directory assistance and those who answer your calls to exclusive catalogue order services. It's easy to tell which of the two groups is measured on the number of calls they handle in a day. Each group is providing high-quality service by delivering exactly what the customer expects, yet behaviors differ based on the measurements selected to achieve the desired business results. Imagine the changes that would occur if the performance of Neiman Marcus operators was measured solely by the number of calls each handled per hour. Also, I have observed that in transacting other business over the telephone, when there is a recording saying, "Your call may be monitored for quality purposes," the people seem more pleasant and helpful. How about you?

Now let's consider an example of Law #4 from the roads. Some truck drivers are paid by the mile and others are on salary. Tight or impossible schedule promises often force freelance drivers to drive more aggressively. Here speeding, aggressive driving, and long hours at the wheel producing 100,000 miles per year are the behaviors required to survive in the business. Salaried drivers not subject to these tight time schedules produce a radically different driving behavior [29]. You can observe these different behaviors in trucks driving routinely on the interstates. See for yourself.

And nowhere has measurement changed behavior like it has in the world of television journalism. In the early days of TV, newscasting meant a reading of "just the facts" with accompanying on-site footage. The news was a necessary evil as it was often the loss leader for commercial broadcast companies. All of that changed with the success of "60 Minutes." Suddenly, news became profitable as advertising rates for this show jumped. The measure of news success moved from the mundane but accurate to the entertaining. And that's where we are. Today, newscasters must worry about ratings as much as other entertainers—sure makes it easier to understand some of the features, doesn't it?

In recent years, the world of measurement [30, 31] has been introduced to a new discipline, "Legal Citology." Its founder, Fred R. Shapiro, is an associate law librarian at Yale University. The legal profession uses past cases as building blocks in its work, much as physical scientists use laws and results from previously published experiments in their fields. Inherent in the application and understanding of peoples' work is the frequency in which people are cited in footnotes. Mr. Shapiro is counting the number of times authors are cited, and he publishes each author's list [32]. He began doing this in the mid-1980s and the "footnote fever" is running rampant. His count is a pure quantity rather than quality score. However, if you're a law professor, you'll want to score high on Shapiro's Citology scale. Some law schools are using the citation counts in hiring professors. The old academic phrase "publish or perish" has been changed by the Citology score. Now it's more like "publish something quotable or perish!"

Some people are studying the political aspects of the lists. Professors have noted that the citation counts of white men traditionally failed to acknowledge the work of minorities. Other "researchers" claim that minorities, feminists, and other groups cite each other excessively as a form of "payback."

Legal Citology measurement has changed the system in a manner I don't think Fred Shapiro and others ever imagined. One last comment: Shapiro's citation lists count authors in journals only from 1955 on, because of computer limitations. This is probably just as well, as professors who were working then are now retired. They probably wouldn't score high anyway. They didn't have PCs, word processors, email, or the Internet to assist in their quotable writing proliferation.

Law #5: The human is an integral part of the measurement process

There is absolutely no doubt that people influence the measurement process. That is what the Hawthorne Effect [33] and related phenomena are all about. The Hawthorne Effect is the label placed on a study by a Harvard business school professor during the late 1920s and early 1930s. Professor Elton Mayo studied how changes in lighting influenced worker productivity at the Hawthorne plant of Western Electric in Chicago, Illinois. He found that productivity increased not because of lighting changes, but because the workers became convinced of management's commitment to improve their working conditions and help them be more productive.

Sometimes, however, human interactions with measurements can have just the opposite result of those intended. In early 1994, Tomas J. Philipson, an economist at the University of Chicago and Richard A. Posner, a federal judge in Chicago and a leader in the fields of law and economics, published a book titled: *Private Choice and Public Health: The AIDS Epidemic in an Economic Perspective* [34]. In the text, the authors argue that widespread testing (measuring) for the AIDS virus may actually contribute to the spread of the disease. They contend "non-altruistic people who have tested positive have no incentive to choose safe sex. On the other hand, those people who have tested negative could use the results to obtain the less cumbersome and easier behavior of unprotected sex." The authors use an economics analogy, comparing HIV tests to credit checks routinely done on people who want monetary loans. Both are used to increase the likelihood that the person is a good "trading partner." Just as good credit checks lead to more loan activity, they argue that more HIV testing will lead to more unprotected sexual activity, thereby increasing HIV transmission. Mr. Philipson suggests that "If you didn't test people, they would be much more careful."

The authors also take care to state that they have no empirical proof that increased testing leads to, or positively correlates to, the spread of the disease. But they also argue that there isn't a good statistical case to suggest that widespread testing will reduce transmission either. In essence, the authors' basic premise regarding human response to testing (measurement) is that sexual activity is a rational activity. You can decide that one for yourself.

Summary Law #6: You are what you measure

Sometimes it is impossible to list everything you want to measure in detail. What you really want is to be successful. You have to define the primary measure of that success, and, in business, leave the details to the resiliency and creativity of your employees. Stock, stock options, bonuses, vacations, or a convenient parking spot are some typical motivational tools or "perks" applied to reward and encourage the behaviors required to achieve the desired measurement objectives. This is true for executive compensation as well. If a CEO is being measured on stock price or return on investment, profit, or shareholder equity, then his or her interests and management focus will vary. This is not surprising but it is interesting that from reading corporate annual reports you can often discern from the data which criteria are being used.

The business examples of this law, such as those mentioned in the last paragraph, are fairly straightforward. While there may be subtle details in how the law is applied, the general philosophy is the same. Yet not all applications share this simplicity.

For example, there are two words we use and hear frequently that imply a measurement result. Their use is so common, however, that we seldom question the process that was used, the interpretation of results, or the applicability of the conclusions to our own ways of doing things. These words are "best" and "worst."

Consider airline on-time performance measurements, published each month by the Department of Transportation to give the public and the airlines feedback on which airlines have and have not been dependable air carriers. Each flight's departure and arrival times are recorded, sent to a database, compiled by the government, and reported on a regular basis. In the highly competitive environment of airline service, the on-time statistics are powerful (free) marketing material for the "best" and bad news for the "worst." Yet there is more to this performance measurement than the public sees in the reports. To understand the issues we must go to back in history to the on-time results published for the first quarter of 1996 and shown in Table 2.2.

Following the release of these figures, Continental Airlines accused top-ranked Southwest Airlines of fudging its arrival data. These statistics are important because Continental gives bonuses to employees in months where they score in the top three. At Southwest, arrival times are used to determine pilot pay, the connected sequence

Table 2.2 Percentage of Flights Arriving within 14 Minutes of Schedule, First Quarter, 1996

Airline	January	February	March
Southwest	77.9	82.0	84.1
Alaska	72.0	72.4	76.9
Continental	69.4	77.9	77.0
Northwest	68.4	75.5	79.3
America West	68.1	69.1	71.9
American	61.2	72.8	77.8
United	60.1	70.5	79.4
USAir	59.2	69.9	72.3
Delta	49.3	70.8	69.2

Source: U.S. Department of Transportation.

of subsequent arrival and departure times at airports later in the flying day, and maintenance scheduling intervals. From this point of view, it is clear that major fudging by Southwest could not occur. However, the "on-time derby," as it has been called, has become so competitive that even small adjustments can change the airline ranking. In the last quarter of 1995, Continental and Southwest were separated by only three-tenths of a percentage point. With this small difference, some people felt that fudging could be done without messing up the airline's entire schedule. Continental has claimed "numerous inconsistencies" between Southwest's actual arrival times and the times reported to the Department of Transportation (DOT). This concern was not an isolated instance. It was prevalent throughout the industry for several years [35].

How could Southwest fudge the numbers? Let's look into how this data was collected. Most major air carriers had equipped their aircraft with an automated system that indicates arrival time by sending an electronic signal to the airline's computer system when the front door of the plane is opened. By definition, any flight that departs or arrives within 15 minutes of its scheduled time is declared "on time." The automation takes the human out of the measuring process. So, you might ask, what's the problem?

Southwest, Alaska, and America West Airlines were the only three major carriers that did not have the automated system. Pilots were responsible for reporting their arrival times—hence the opportunity for changes existed. Human actions were part of the measuring process, competing with automated systems that did not get a penalty or reward for the numbers they reported. Here, both motivation and possibility for error casts doubt on the measurement results. This is especially true when the results between the automated and human-driven systems are very close. Minute differences in watches or clocks could contribute to error even if the planes arrived at exactly the same time. Equipping Southwest's planes with the automated communications systems of the other airlines would have cost about $5.7 million. To eliminate doubt of Southwest Airline's performance, management looked at other lower-cost automated solutions they could implement to put them on an even playing field with their very unfriendly competitors.

This is why airline ground and plane crews work hard to get passengers seated and the entrance doors closed. When the door is closed or when the aircraft is moved from the gate, the departure is recorded. The additional time the aircraft may sit on the tarmac before taking off is not used to penalize the "on time" departure data.

Today, technology has leveled the playing field, at least for the data collection aspects of these statistics. Most airlines use what is called the Aircraft Communications Addressing and Reporting System (ACARS) to handle a wide range of functions—including departure and arrival data transmissions. Some utilize a laser guidance system call Docking Guidance System (DGS) that pilots use to park at passenger gates. The manual departure and arrival data transmission is only a backup mode today. The standardization of on-time data has virtually removed the issue that was prevalent in the 1990s among the major airlines.

Now, with the playing field leveled, how could an airline improve its on-time performance? It could maintain a highly reliable aircraft fleet, provide a robust training program to facilitate passenger loading and unloading, and simplify check-in procedures, or it could add a few minutes to its flight time for delays. After all, there is no penalty for being early!

Bill McGee, wrote an insightful article for *USA Today* in June 2009 [36] stating that unless the distance between seismic plates have shifted dramatically, airlines are now indicating it is taking longer to fly around the country. The CEO of American Airlines, Gerald Arpey, told the shareholders that American was making "strategic adjustments" in an effort to improve customer service. One of the ways American is doing this is to increase its block times.

In this article he reports that, for the route Miami (MIA) to Los Angeles (LAX), the posted block times were: in 1990, 5 hr 45 min; in 1995, 5 hr 25 min; in 2006, 5 hr 30 min and in 2009, 5 hr 40 min. For a shorter flight, from New York (LGA) to Washington, D.C. (DCA): in 1990, 1 hr 15 min; in 1995, 1 hr 00 min; in 2006, 1 hr 16 min; and in 2009, 1 hr 22 min.

It is safe to say that the 2009 aircraft are as fast as, if not faster than, the 1990 versions and that the increase in block time is probably not to due to climate change-induced headwinds. It appears the airlines have done some performance "recentering" of their own that has blurred the line between "best" and "worst."

It's easy to state the philosophy of "you are what you measure," but in situations where there are several attributes being measured, being the "best" in some of them does not necessarily make you the "best" overall. In these very real situations, a decathlon approach may be useful. For example, the "best airline" could be chosen from the summation of airline rankings over several categories, including percentage of on-time departures and arrivals, lost baggage, canceled flights, and customer complaints.

U.S. News and World Report uses the weighted attribute approach when it develops its list of different types of "Top 100" universities. In the ranking process, this publication employs a system of weighted factors that include student retention, faculty resources, student selectivity, financial resources, graduation rate, and alumni giving, along with a number of subfactors. Now if you don't care about alumni giving or some of the other weighting variables, then *U.S. News and World Report* rankings are, at best, a guideline in your assessment of school performance. Other published reports tell you the "best" places to live, retire, vacation, and work. They are all compiled from a detailed ranking process that, without an individual's knowing the details, makes the results useless in practice. The approach, however, is a valid measurement process for situations in which the entity being measured is composed of multiple variables or aspects. After all, you are a complex living organism with a unique personality and a plethora of unique characteristics. How could you expect to measure life's aspects with a simple tape measure?

Along these lines there is a related phenomenon that seems to be as old as history, where "you are what you measure" takes on genuine results. It is evident in seemingly every area of the human health spectrum, scientifically recognized as real, yet still evades detailed explanation and understanding. I know this introduction sounds ominous, but the process I am about to discuss is a lot more sophisticated and powerful than its simple name suggests. It is the placebo effect.

The word goes back to Hebrew and Chinese origins. Like those of most word etymologies, the string of historical connections of this word is long and diverse. The relatively modern phrase "placebo" has a Latin origin meaning "I shall please" and was used in the Roman Catholic Liturgy. A supplement to Vespers (the evening prayer) was read and prayed when a member of the religious community had died. The text began in Latin with the phrase *"Placebo Domino in regione vivorum,"* roughly translated as "I shall be pleasing to the Lord in the land of the living" [37].

Medicine has been cutting, poking, bleeding, sweating, and doing other "attention-getting" things to the human body for thousands of years. In many of these seemingly primitive treatment sessions, doctors knew that what they were doing was not directly related to curing the patient. The therapies were placebos. In spite of scientific evidence compiled through the second half of the 19th century, doctors continued to use them because they were pleasing to the patient. The advent of the 20th century saw dramatic improvement in eliminating these torture therapies. Nevertheless, doctors now recognize that placebos play a powerful role in contributing to patient cures. Yet despite almost a century of study, little is known about how placebos actually work.

There is even a great deal of discussion around the definition of a placebo. Here is one that was labeled an acceptable working definition:

A placebo is any therapeutic procedure (or component of any therapeutic procedure) which is given:

    (1) deliberately to have an effect, or

    (2) unknowingly and has an effect on a symptom, syndrome, disease, or patient but which is objectively without specific activity for the condition being treated.

> *The placebo is also used as an adequate control in research. The placebo effect is defined as the changes produced by placebos. [38]*

The working definition gives you the gist of what a placebo is all about. The general rule of thumb in drug-effect studies is that 20% of the placebo group participants improve from being treated with the "inactive material." However, there are studies where a remarkable 58% placebo improvement rate has been seen depending on the disease, the placebo used, and the suggestion of the authority [39].

Drug treatments are not the only realm in which the placebo effect is seen. Surgery patients who were enthusiastic about their procedure generally recover faster than do patients who are skeptics. Sure, there are other factors that contribute, but the perception of experience and the psychological factors inherent in dealing with rationally thinking human beings are integral, absolutely inseparable parts of the measurement process.

There are also related health effects for nondrug or -treatment situations. If you've been around any workplace recently where people are lifting things, you might see workers wearing support belts designed to reduce the chance of back injuries. In an occupational safety study, researchers found that just wearing the belt made people less prone to injury. This was especially true for the act of lifting, as workers could feel their muscles contracting around the belt area and this feedback would help them use proper techniques. They also found the interabdominal pressure obtained from holding one's breath had more effect on back support than wearing the belt [40].

Now the net effect is that people who wear the belts have a better chance of not having back injuries, but due to the placebo effect, it is difficult to determine if it is due to the belts or to changes in personal behavior. Since we are talking about psychological measurement, this is a perfect example of how Laws #4 and #5 can be combined. In researching this topic, I came across a simple, precise definition of a placebo [41], one with which I think you will agree: *Placebo: A lie that heals*.

The placebo effect demonstrates that the human power of perception can be greater than the powers of technology and science put together. Logical explanations of scientific cause and effect can be overwhelmed by the illogical, emotional, and other nonanalytical aspects of the human psyche. We can see the tremendous power of perception and how logical, scientifically acceptable measurement strategies can be ignored due to peoples' beliefs, values, and internal measurement systems. Summary Law #6 is more than corporate hype or a journalistic sound bite. You, the reader of these words, you are also what you measure.

DISCUSSION QUESTIONS

1. Develop a weighted attribute scoring system to determine the best to worst airline performance from the following data.

Carriers	Percent flights cancelled	Carriers	Average departure delay (minutes)	Carriers	Average arrival delay (minutes)	Carriers	Percent flights diverted
Northwest	0.58	US Air	4.12	Southwest	2.02	Northwest	0.16
Continetal	0.62	Northwest	5.08	United	3.60	US Air	0.17
AirTran	0.84	Alaska	5.20	US Air	3.70	Southwest	0.18
Southwest	0.84	Delta	7.19	Northwest	4.51	United	0.23
Alaska	1.22	AirTran	8.83	Alaska	4.62	Continetal	0.27
US Air	1.30	Southwest	8.85	Delta	5.79	Delta	0.27
Delta	1.34	United	9.28	Jet Blue	6.65	AirTran	0.28
Jet Blue	1.46	Jet Blue	9.82	American	7.15	Jet Blue	0.28
United	1.90	American	10.99	Continetal	7.82	American	0.37
American	2.28	Continetal	11.11	AirTran	7.86	Alaska	0.39

Source: U.S. Bureau of Transportation Statistics: January, 2009 through June, 2009.

2. From a newspaper, identify three measurement cases from news stories and discuss the measurement quality and the measurement error.

3. From a newspaper or magazine, identify three measurement cases from the advertisements and discuss the measurement quality and the measurement error.

4. Develop a rating criteria for a "Best Town to Reside" ranking and apply it to 10 towns. Compare your results with those obtained by other groups.

5. Construct three examples for each of the Six Laws of Measurement discussed in the chapter.

Case Study

This case study provides a legal application of laws #4 and #5 (every measurement carries the potential to change the system and the human is an integral part of the measurement process).

Ledbetter v. Goodyear Tire & Rubber Co., Inc., 550 U.S. 618 (2007)

Upon retiring in November of 1998, Lilly Ledbetter filed suit against her former employer, Goodyear Tire & Rubber Co., Inc. (Goodyear), asserting, among other things, a sex discrimination claim under Title VII of the Civil Rights Act of 1964. She alleged that several supervisors had *in the past* given her poor evaluations because of her gender; that as a result her pay had not increased as much as it would have if she had been evaluated fairly; that those past decisions affected the amount of her pay *throughout* her employment; and that, by the end of her employment, she was earning significantly less than her male colleagues were. The jury agreed with Ledbetter and awarded her back pay and damages.

The above facts illustrate an example of error or bias in measurements. Upon retirement, Lilly Ledbetter was notified by an anonymous note that her salary was significantly lower than that of her male colleagues with similar experience. This was due to the ripple effect of several biased measurements of her performance throughout her career. She was never aware of such discrimination upon occurrence because of Goodyear's policy of keeping employee salaries confidential.

On appeal, Goodyear noted that sex discrimination claims under Title VII must be brought within 180 days of the discriminating act. Accordingly, Goodyear contended that the pay discrimination claim was time barred with regard to all pay decisions made before September 26, 1997, or 180 days before Ledbetter submitted a questionnaire to the Equal Employment Opportunity Commission (EEOC). Additionally, Goodyear claimed that no discriminatory act relating to Ledbetter's pay had occurred after September 26, 1997. The Court of Appeals for the Eleventh Circuit reversed the jury award and held that "because the later affects of past discrimination do not restart the clock for filing EEOC charge, Ledbetter's claim is untimely." The U.S. Supreme Court affirmed the Eleventh Circuit, barring Ledbetter from any recovery.

In the context of equal employment opportunities, the laws impose a rigid 180-day limitation on sex discrimination claims. This limitation is essentially a scale for determining the allowable discrimination claims. As a result of this 180-day scale, an employee who was actually discriminated against more than 180 days ago, and did not bring the claim within the 180-day period, was not legally discriminated against. The 180-day limitations period reflects Congress's strong desire for prompt resolution of employment discrimination matters. The 180-day period appears especially unfair with respect to pay discrimination claims, which naturally take longer to discover.

On January 29, 2009, President Obama signed the Lilly Ledbetter Fair Pay Act of 2009 amending the Civil Rights Act of 1964 to state that the 180-day limitations period starts anew with each discriminatory act.

Case Study Questions

1. From this case study, identify examples or effects of the measurement laws discussed in this chapter.
2. How can employment performance records be checked for measurement bias in practice while maintaining confidentiality?

ENDNOTES

1 Daniel Kleppner, "A Milestone in Time Keeping," *Science*, Vol. 319, March 28, 2008, pp. 1768–1769.
2 S. R. Jefferts et al., "NIST Cesium Fountains—Current Status and Future Prospects," NIST—Time and Frequency Division, Proc. of SPIE Vol. 6673, 667309.
3 USDA Food Safety and Inspection Service Fact Sheet: Food Labeling. September 2006.
4 http://aviation-safety.net/database/record.php?id=19830723-0 (accessed November 8, 2009).

5 Mars Climate Orbiter Mishap Investigation Board Phase I Report, November 1999.

6 Garrnett P. Williams, *Chaos Theory Tamed*. London: Taylor & Francis, 1997, p. 241.

7 Neil J. Dorans, "The Recentering of SAT® Scales and Its Effects on Score Distributions and Score Interpretations," College Board Research Report No. 2002–11, ETS RR-02-04.

8 Jack Gillum and Greg Toppo, "Failure Rate for AP Tests Climbing," *USA Today*, February 4, 2010.

9 Advanced Placement Report to the Nation 2006, The College Board.

10 Scott Jaschik, "College Board to End Penalty for Guessing on AP Tests," *USA Today*, August 10, 2010.

11 "Toward a More Accurate Measure of the Cost of Living: Final Report to the Senate Finance Committee from the Advisory Commission to Study the Consumer Price Index," December 4, 1996.

12 Robert J. Gordon, "The Boskin Commission Report: A Retrospective One Decade Later," NBER Working Paper No. 12311, June 2006.

13 Owen J. Shoemaker, "Variance Estimates for Price Changes in the Consumer Price Index", January–December 2008.

14 *Popular Mechanics*, March 31, 2008.

15 Mel Poretz and Barry Sinrod, "The First Really Important Survey of American Habits," Price Stern Sloan, Inc., 1989.

16 Fumiaki Taguchia, Song Guofua, Zhang Guanglei, and Seibutsu-kogaku Kaishi, "Microbial Treatment of Kitchen Refuse with Enzyme-Producing Thermophilic Bacteria from Giant Panda Feces," *Seibutsu-kogaku Kaishi,* Vol. 79, No. 12. 2001, pp. 463–469.

17 Javier Morales, Miguel Apatiga, and Victor M. Castano, "Growth of Diamond Films from Tequila," 2008, Cornell University Library, arXiv:0806.1485.

18 Donald L. Unger, "Does Knuckle Cracking Lead to Arthritis of the Fingers?" *Arthritis and Rheumatism*, Vol. 41, No. 5, 1998, pp. 949–950.

19 Stephan A. Bolliger, Steffen Ross, Lars Oesterhelweg, Michael J. Thali, and Beat P. Kneubuehl, "Are Full or Empty Beer Bottles Sturdier and Does Their Fracture-Threshold Suffice to Break the Human Skull?" *Journal of Forensic and Legal Medicine*, Vol. 16, No. 3, April 2009, pp. 138–42.

20 Gideon Gono, *Zimbabwe's Casino Economy—Extraordinary Measures for Extraordinary Challenges*. Harare: ZPH Publishers, 2008.

21 Katherine K. Whitcome, Liza J. Shapiro, and Daniel E. Lieberman, "Fetal Load and the Evolution of Lumbar Lordosis in Bipedal Hominins," *Nature*, Vol. 450, December 13, 2007, pp. 1075–1078.

22 U.S. patent 7,255,627, granted August 14, 2007.

23 Catherine Bertenshaw [Douglas] and Peter Rowlinson, "Exploring Stock Managers' Perceptions of the Human-Animal Relationship on Dairy Farms and an Association with Milk Production," *Anthrozoos*, Vol. 22, No. 1, March 2009, pp. 59–69.

24 "A *Noble Side* to the Ig Nobels," *The National*, September 26, 2009.

25 http://neo.jpl.nasa.gov/news/2008tc3.html (accessed January 3, 2009).

26 *Popular Science*, October 2009, pp. 56–57.

27 W. Heisenberg, Über den anschaulichen Inhalt der quantentheoretischen Kinematik und Mechanik. *Zeitschrift für Physik*, vol. 43, 1927, S. 172–198.

28 "Application and Testing of Safety-Related Diesel Generators in Nuclear Power Plants," U.S. NRC Regulatory Guide 1.9 Rev 4, March 2004.

29 http://EzineArticles.com/?expert=Aubrey_Allen_Smith.

30 R. B. Jones, *Risk-Based Management: A Reliability-Centered Approach*. Gulf Publishing, 1995.

31 R. B. Jones, *20% Chance of Rain: A Layman's Guide to Risk*. Bethany, CT: Amity Publishing, 1998

32. Fred R. Shapiro, *Collected Papers on Legal Citation Analysis*. Fred B. Rothman Publications, 2001.

33 http://www.library.hbs.edu/hc/hawthorne/anewvision.html#e.

34 Tomas Philipson, J. Posner, and A. Richard, *Private Choice and Public Health: The AIDS Epidemic in an Economic Perspective*, Harvard University Press, 1993.

35 Office of Inspector General Audit Report, U.S. Dept. of Transportation, Report No. FE-1998-103, March 30, 1998.

36 Bill McGee, "Think Flight Times Are Being Padded? They Are," *USA Today*, June 29, 2009.

37 Daniel E. Moerman, *Meaning, Medicine, and the "Placebo Effect."* New York: Cambridge University Press, 2002, p. 10.
38 Ibid., p. 14.
39 Ibid., p. 11.
40 National Institute for Occupational Safety and Health, "Workplace Use of Back Belts," NIOSH 94–122.
41 H. Brody, "The Lie That Heals: The Ethics of Giving Placebos," *Annals of Internal Medicine*, Vol. 97, Issue 1, July 1982, pp. 112–118.

Chapter 3

Statistics: Numbers Looking for an Argument

Not everything that counts can be counted, and not everything that can be counted, counts.

—Sign hanging in Einstein's Princeton office

Measurement in one form or another is integral to life's routine decisions. And statistics, one form of measurement, are becoming the "language of the informed." Debates and discussions seem to be judged on who has the best (or the most) numbers. Whether you speak the language or not, you can still understand statistics: their strengths, their limitations, and how they help us understand risk in its many forms.

Statistics compress data into facts, but there are many so-called facts that data can support. One phrase I've heard to describe this is: "Statistics means never having to say you're certain." You've probably heard others, like "liars figure and figures lie" and "there are lies, damned lies, and statistics." Statistics are used as ready ammunition in arguments; in many cases, opposing sides use the same data to develop statistically correct arguments in support of their side. To set the stage for the chapter let's take a look at this list of 2009 salaries for the New York Yankees baseball team [1].

I'm not trying to start an argument about whether or not professional athletes are overpaid. This table is used strictly for its value in comparing statistical interpretations of data. First let's calculate the arithmetic mean, or average salary for the team members. By adding up all of the salaries in this list and dividing the total by the number of players, we compute the "average" player is making $8,021,117 a year.

Let's look at the median salary, or the one halfway down the list. There are 25 players listed so the middle player is number 13, Damaso Marte, at a salary of

20% Chance of Rain: Exploring the Concept of Risk, First Edition. Richard B. Jones.
© 2012 John Wiley & Sons, Inc. Published 2012 by John Wiley & Sons, Inc.

Table 3.1   New York Yankees Salaries: 2009 Season

Alex Rodriguez	33,000,000	Jose Molina	2,125,000
Derek Jeter	21,600,000	Jerry Hairston Jr.	2,000,000
Mark Teixeira	20,625,000	Eric Hinske	1,500,000
A.J. Burnett	16,500,000	Melky Cabrera	1,400,000
CC Sabathia	15,285,714	Brian Bruney	1,250,000
Mariano Rivera	15,000,000	Joba Chamberlain	432,575
Jorge Posada	13,100,000	Brett Gardner	414,000
Johnny Damon	13,000,000	Phil Hughes	407,650
Hideki Matsui	13,000,000	David Robertson	406,825
Robinson Cano	6,000,000	Alfredo Aceves	406,750
Andy Pettitte	5,500,000	Phil Coke	403,300
Nick Swisher	5,400,000	Ramiro Pena	400,000
Damaso Marte	3,750,000		

$4,575,000—that's $3,446,117 less than the average salary. Half of the players make more than this amount and half make less.

Both the average and the median are valid statistical measures. Both represent the "middle" of a set of numbers. But as you can see in this case, the word "middle" has two interpretations. Both are correct but they cannot be used interchangeably. This is just one way data can support differing statistical conclusions. You can see how two parties can use the same data to reinforce different positions and both be technically correct. And by the way, neither the average nor the median is going to make a bit of difference to the wallets of the players at the bottom of the list!

Sometimes the use of statistics goes far beyond even the needs of the consumer. My wife purchased a broom-type vacuum cleaner and when she read a certain section of the booklet, she was surprised that the manufacturer would put such technical content in the same book that tells people how to unpack the thing from the box. Here is the exact text. Do you think this information would help you use the vacuum cleaner?

Cleaning Effectiveness Per Amp Rating

This is not *an Amp rating. Amps do* not *measure dirt removal, only the amount of electricity used. Cleaning Effectiveness Per Amp is determined by dividing this model's Cleaning Effectiveness* by its Amps.*

*Cleaning Effectiveness is the percentage value obtained from dividing:

(a) *the geometric mean of the amount of embedded carpet dirt removed in testing under ASTM Test Method F608-89, by*

(b) *the value 29 (the geometric mean of the midpoints of the % dirt pickup scales selected by* [brand name] *as a reference for its rating system).*

This illustration shows why even the word "statistics" often causes pain to some people. It does have its own special jargon. Statistics examines the behavior of a population of things or events. It does not and cannot be used to accurately predict singular, individual events. Nevertheless, the legal profession does use it in court-rooms with surprising results. Some examples have been compiled in a book of case studies, called *Statistics for Lawyers* [2], describing how statistics can be used in trial law. It makes interesting, imaginative, and sometimes sad reading. Just like if you were on a jury listening to evidence, the message throughout this chapter is "caveat emptor" or "buyer beware" of exactly what is shown by the statistics you're hearing and how they are derived.

Statistics have become integrated into many facets of our lives. They are no longer the esoteric language of scientists, technicians, and engineers. One powerful example comes from today's world of "rightsizing." Statistical tests are performed by corporations to ensure that layoffs are done in nondiscriminatory patterns. Before a layoff occurs, an analysis is done of the workers to be laid off to see if the distri-bution of the elderly, minorities, women, and others is fair or excessive. These studies sometimes become part of the company's legal defense in the event of dis-criminatory lawsuits [3].

Statistical interpretations and applications can be volatile subjects in certain situations. Why? One answer is often "money," of course. But more broadly, the use of statistics is important to our understanding of our world, its hazards, and the risks we take living in it. Let's begin exploring this subject in more detail.

MORE STATISTICS YIELDS MORE _____?

Consider the following partially completed sequence of statements:

Technology yields more data.

More data yield more statistics.

More statistics yield more __.

The cynic would complete this sequence with "opportunity to prove anything you want." And there is some truth in this. But for now, let's take the high road and explore some other possible answers.

Some people would say that the word that best completes the passage is "knowl-edge." They would argue that, as a society, we are converting data into information at an ever-increasing rate. Information is being continuously processed to improve our technology, finances, health, and quality of life. I would agree that to some extent this is true, but let's now hear from another point of view.

Other people would end the statement with the word "confusion." You might wonder how anyone who lives today could believe that technological evolution is leading us down the pathway of confusion and possibly conflict. In response, I chal-lenge you to pick up any newspaper and see if the statistics reported on any subject really convince you of anything. The numbers may give you an illusion of precision, true, but do they provide you a genuine package of knowledge? The title of this book is a perfect example. As I explained in the preface, a "20% chance of rain" gives you a very precise, numeric weather forecast that is adequate for a general area, but for pinpointing what will happen right where you're standing, it means almost nothing! Doppler Radar, color graphic display technologies, and computer-based weather models give us tremendous amounts of meteorological data, but the weather forecast is still a "20% chance of rain." The sequence "Technology leads to data which lead to statistics which lead to..." certainly could be ended with the word "confusion." Concerning the weather forecast, all you know is that it might rain today. Weather forecasting is still an art, regardless of the technology, and with a forecast style of "20% chance of rain," meteorologists are never wrong!

Medical science has combined technology and measurement skills, but with mixed results. And sometimes, conflicting results do little to instill public confidence in medical studies. For example, one report says caffeine is okay, while the next says that caffeine causes problems. Some say that butter is bad for you and margarine is okay; others say the opposite. The studies are not at fault. It is the interpretation of the statistics. This brings us back to the "statistics lead to..." sequence. The news media plays a role in this communication gap. There's often insufficient time or space in sound bites and headlines to give the audience the complete story. It's also true that sensational news sells better than does a lesson in survey statistics. We'll revisit this issue once we have discussed some basic facts about the strengths and limitations of statistics.

FACTS OR STATISTICS?

Here's a list of statistical statements that have varying degrees of truth. I've made some comments after each statement to get you thinking about their accuracy.

1. **You are about as likely to get struck by lightning as you are to win the lottery.**
    So what? The three states in the United States that have the most lightning deaths are Florida, North Carolina, and Texas. Do more people play the lottery in these states? Is there a correlation between people who play the lottery and those struck by lightning? If I play the lottery, am I more likely to be struck by lightning? Or, if I get struck by lightning, am I more likely to win the lottery? This statement is a classic example of pseudoprecision.

2. **If you play Russian roulette with a six-shooter once, your chances of shooting yourself are 1 in 6.**
    Suppose you take a revolver, empty the cylinder, put one bullet back in, spin the cylinder, put the gun to your head, squeeze the trigger, and are subsequently killed by the bullet that just happened to be in the right (or wrong) place. What was your probability of shooting yourself? It was one—not one out of six for this singular event now in the past. Based on theory, the probability of having the bullet in the right (wrong) cylinder may be one out of six, however, this statistic does not apply to individual behavior. You can consider statistics in making decisions, but the personal outcome of the uncertain event is real, not theory.

3. **About one in every 123 people in the US will die each year.**
   How does this apply to you?

4. **Cigarettes are not habit-forming.**
   Who do you think would have funded this study?

5. **You can drown in a lake with an average depth of 1 inch.**
   There is nothing wrong with this statement. It's perfectly valid. Think of this example the next time you hear or read the word "average" in some context. Without more information, the statements surrounding the statistics' use can be misleading.

When you read these statements you can see that they could be true sometimes but that does not make them facts. Facts are true all of the time. But statements like these are often used together, in concert, sometimes interspersed with real facts to get people to believe in a certain point of view. If you remember anything from this chapter remember this: statistics do not, and cannot, prove anything. The field of statistics is incapable of proving anything. Statistics can provide information to help us make decisions, but the decisions are still ours to make.

Let's take a "pure" situation where there is only one answer and then generalize it to a circumstance where statistics apply. Suppose you want to measure the "length" of an object. If the object is square, then "length" has a unique numerical answer. However, if the object has a ragged, irregular edge, such as that possessed by a splintered board of wood, there is no single value for length. Statistics redefines "length" here as a variable, such as the average length, median length, or some other quantity. The result is a statistical value representing all of the individual values of the splintered board. Since this one value is designed to be the most representative length from all of the other values, it must be accompanied by information on how representative this special value is relative to the whole family of values from which it was derived. This information can be in the form of confidence limits on the average, quartile or percentile values, the maximum and minimum values or other stuff. The point is that since statistical results (in this case for length) are representative, single for a family of values, they need to be stated with some additional information that indicates the overall variability in the data.

Statistics are applied in many ways but we'll restrict ourselves here to the area called inference, or trying to deduce information from data. This is the area that's most visible to the general public, usually through the news media. Inference is a method of using data to support or not support ideas. It gets its name from the base word "infer," meaning "to derive a conclusion from facts or premises." This is the first definition from the dictionary. The second meaning says the same thing but in a much more straightforward way. The second definition is "guess, surmise." Take your choice. Either way, the process of inference is the process of guessing, but in a way that utilizes information in a structured manner.

Here's another term that has special meaning in statistics: hypothesis. You usually don't see the phrase "statistical hypothesis" in the newspapers, but you read and hear statistical hypotheses almost every day in the news. A hypothesis is a statement suggesting some relationship in a process. Hypotheses can be very simple or very esoteric, depending upon what's being studied. Here are some typical examples:

1. Hypothesis: In one out of every seven weddings, either the bride or groom or both are getting married for the third time.
2. Hypothesis: A person is more likely to die in his or her bathtub than be killed by lightning.
3. Hypothesis: Sixty-five out of every 100 men between the ages of 21 and 65 will suffer baldness or excessive hair loss.
4. Hypothesis: Flying in commercial airlines is safer than driving your car.

At least one study in each case has shown that they are statistically true. However, if taken as potential mathematical theorems, they would all be false. Why? For a mathematical theorem to exist, the statement must to be true all of the time. In each of the hypotheses listed above, a mathematician could prove each statement false by finding just one exception. Even though these statements are not mathematical theorems, some of them are generally true since the results were observed for some overall group. It's this area of bringing quantitative information to practical applications where statistics plays a major role. The only drawback is that the results cannot be directly applied to a specific member of the group. When we look at risk, we are partially looking at the probability of an event happening within a group. If you are concerned about the behavior of groups and not about the behavior of individual items, then statistics works. If your concern is oriented towards the behavior of an individual then statistical results may be of little to no value.

INSIGHTS FROM STATISTICAL INFORMATION

We exist in a world in which we are exposed to great amounts of information in our everyday lives. Much of it is accompanied by "the numbers" that support the presenter's point of view. Together with charts, graphs, and "an attitude," statistics provide a seemingly compelling mechanism to support or reject certain facts. There is a general perception that the more quantitative or "scientific" the measurement tool, the more likely the result is correct. When people use numbers in a convincing manner, it is easy to suffer the illusion of accuracy and precision. The mystique of mathematics, the technical complexity of statistics, and the "science of analysis" lend themselves very well to abuse. The general population is poorly equipped to argue or even question "the numbers."

When it comes to the proper use of statistics, there are serious pitfalls that many people, including the press and regulators, encounter. The use of measurements and conclusions based upon an application of statistics represents a powerful and convincing medium. However, the benefits are accompanied by a responsibility to communicate effectively, clearly, and completely.

Let's take some examples using what is probably the most used (and abused) statistic in the world: the arithmetic average. As we described in the example using the salary schedule for the New York Yankees, you compute the average of a set of values by taking their sum and dividing it by the number of terms in the sum. There is a saying that if your head is on fire and your feet are freezing, on average, you're doing just fine. The average is a statistic that "measures the middle" of a process. If the process contains wide variations, then the average, by itself, doesn't mean much, as in the Yankees example. You'll hear the term "average" used a lot on the news and in TV or radio commercials. The associated claims may be statistically correct, but they may not really describe the activity adequately.

In 2007 the U.S. "average death rate" for motor vehicle accidents was 14.3 deaths per billion vehicle-miles [4], and about half of the deaths occurred during the night. This could lead you to conclude that there is no difference in risk driving or riding in cars during the night versus during the day. But here's the missing piece of information. Even though the number of deaths from each time period is about the same, the number of people on the roads in the day and night is vastly different. Including this fact into the calculations shows the chance of death from car accidents is actually nearly *three times greater* at night than it is in the day. Driver age is another major factor. For teenagers between 15 and 19, the statistic is approximately 25 deaths per billion vehicle-miles. Factor in time of day, day of week, and day of year, and you'll find even more reasons to be concerned about teenagers and cars. This illustrates how useless and potentially misleading "average" statistics can be for your daily, nocturnal, weekend, and holiday roadway risk management decisions.

Here's a case from nature. There is a growing amount of evidence that the genes for intelligence are located on the X-chromosome [5]. Men, genetically described as "XY," inherit the X-chromosome only from their mothers, receiving the Y-chromosome from their fathers. If the evidence about the placement of genes for intelligence is valid, then men receive their intelligence from their mothers. Women, genetically XX, receive an X-chromosome from each parent, so they can inherit intelligence genes from both their father and mother. Although men and women tend to have the same average IQ scores, there are more males who are either geniuses or mentally retarded. Researchers argue that because a man has only one X-chromosome, he experiences the full effect of the intelligence genes it carries. Since women have X-chromosomes from both parents, they receive genetic intelligence information from each, buffering the expression of a single characteristic. So even though "on the average" men and women have IQs that are about the same, the distribution across each of the populations differs. The risk management lesson from this section is that if men want smart sons, they had better marry smart women!

There are more subtle uses of statistics that can make an uncertain, scientific result sound like a sure thing. Consider the ongoing case of talcum powder cancer risks. Talcum powder is made by grinding the mineral talc or hydrated magnesium silicate into a fine powder. It is similar in structure to asbestos and both substances are found in similar mining locations. In the 1970s, talcum powder was required by law not to contain asbestos, but the similar nature of talc and asbestos has prompted several studies analyzing its suspected carcinogenic properties.

Talcum powder is commonly used today on babies to absorb moisture and odors and also by women in hygiene sprays. The concern is that the small fibers in the powder, when applied to the genitals, might travel to the ovaries and trigger a process of inflammation that enables cancer cell formation.

In the 1990s there were clinical studies performed to examine suspected ovarian cancer connections with talcum powder [6, 7]. The studies reportedly identified a correlation, but they also admitted that the results were inconclusive due to the way the data were collected and analyzed. In 2000, another research project considered 121,700 female registered nurses in the United States who were aged 30–55 years at enrollment in 1976. Talc use was ascertained in 1982 by use of a self-administered questionnaire. After exclusions, 78,630 women formed the cohort for analysis. From this relatively large sample, scientists found that talc use in the genital area did not increase overall ovarian cancer risk [8]. This study did not stop the research, however. Additional studies with smaller sample populations under different conditions found a 30% increase in ovarian cancer for women who were regular talc users and in 2008, an Australian study also confirmed the connection without specifying a definitive percentage increase [9].

First of all, the incidence rate of ovarian cancer in the overall female population is about 12 out of every 100,000 women [10]. It is the eighth most frequent type of cancer. To put this into perspective, the number one cancer type for women, breast cancer, occurs in about 118 out of every 100,000 women. The number two type, lung and bronchus cancer, is 55 out of 100,000.

With ovarian cancer frequency just 12 in 100,000, statistically it is difficult to perform conclusive studies given the practical realities of acquiring a statistically ideal testing sample. Add on other risk factors such as age, obesity, reproductive history, gynecologic surgery, fertility drugs, diet, family history, androgens, estrogen and hormone surgery, personal history, analgesics, smoking, and alcohol use, you can see why it is tremendously difficult to isolate the effect of one risk factor over another or the effect of risk factor combinations. Adding talc powder to this list gives 14 risk factors, representing over 87 billion risk factor combinations. With a 12 per 100,000 ovarian cancer incidence rate, the number of women who can practically be involved in the research is small, so the talc powder effect statistical results are going to be stated with considerable uncertainty ranges. Nevertheless, the research is good science and you shouldn't criticize the results or the researchers for developing conclusions that appear to be contradictory. The studies represent how scientists use statistics to advance knowledge in their art. Each study provides not precision, but guidance and more insights to understand causality and the basic mechanisms of this horrible disease.

In this example, the term "incidence rate" was used; notice I did define it. Differences in term definitions can cause people to misinterpret statistical conclusions. What makes the awareness of this dichotomy even more subtle is that terms can appear intuitive (as incidence rate does) yet the difference between intuition and actual definition can be significant.

Interpreting statistical terminology is a basic requirement for the proper application of statistical results. This subject is large enough to fill another book, but

I'll include two examples here because they are used in many situations and the concepts have precise definitions. The terms are "odds ratio" and "relative risk."

The odds ratio is exactly what it says. It is the ratio of the odds for two events. Let's start by looking at what is meant by the term "odds." Odds is the ratio or comparison of the number of winning possibilities and the number of other possibilities. For example, the odds of rolling a 1 on a single throw of a die is 1:5, or 1 to 5. This is not the same as the probability of rolling a 1, which is 1/6.

Consider perhaps a more relevant example in reviewing the odds of dying from various hazards shown in Table 3.2 [11].

The lifetime odds are interpreted as "1 in X" chance of dying by the type of accident. For example, the odds of dying by dog bite or dog strike are 1 in 117,127. These numbers are statistical averages over the U.S. population so they do not necessarily apply to you. You may not own a dog, live around dogs, or play with fireworks.

The numbers in the table are not easily understood since they are all fairly large and the larger the odds, the smaller your chances. The odds ratio is more intuitive since it compares two hazards to show how more likely one is relative to the other. It is computed just as it says: by taking the ratio of the odds. Table 3.3 shows the odds ratios for the accident types given in Table 3.2.

The table is read like this: the row is X times more likely to occur than is the column. For example, death by lightning is 1.5 times more likely to occur than death from a dog bite or dog strike. These comparisons are a lot easier to understand than the large numbers in Table 3.2. That's why they are used.

Both odds and odds ratio values compare the number of occurrences of one event to another. Relative risk compares the probability or percent chance of

Table 3.2   Odds of Death Due to Injury United States, 2005

Type of accident or manner of injury	Lifetime odds
Electric transmission lines	39,042
Lightning	79,746
Bitten or struck by dog	117,127
Fireworks discharge	340,733

Table 3.3   Odds Ratios for Accident Types

	Electric transmission lines	Lightning	Bitten or struck by dog	Fireworks discharge
Electric transmission lines	1.0	2.0	3.0	8.7
Lightning		1.0	1.5	4.3
Bitten or struck by dog			1.0	2.9
Fireworks discharge				1.0

Table 3.4 Rollover Percentages and Relative Risk

Driver age	Rollover percentage Vehicle type SUV	Non SUV	Relative risk
All	5.19	2.00	2.60
16–24	7.43	3.14	2.37
+25	4.42	1.55	2.85

occurrence of two events. This is the difference between relative risk and the odds ratio. For an example, in 2003 a research note was published that discussed the safety issues with young drivers and sport utility vehicles. In this report [12], the percentage of SUV rollovers was studied as a function of age, gender, and vehicle type. Table 3.4 shows some of the results.

Table 3.4 shows the percentage of all vehicle crashes that involved a rollover. For example, in the 16–24 driver age group, 7.43% of all crashes involved rollovers in SUVs, and 3.14% involved rollovers in all other types of cars. The relative risk for this age group of rollovers in SUVs was (7.43/3.14) = 2.37. This is how relative risk is used. Yes, there are more details involved with these statistics, but this discussion is intended to provide you at least with the awareness that when you read or hear statistics being used, you should make sure you understand the exact meaning of the terms.

Statistics in many ways are like colors and textures to a painter. They are mixed together and applied to present a picture or portrait of a process, activity, or situation. "Numerical artists" use statistics as tools or instruments and different combinations produce the desired textures and colors. However, in some cases, what you see is not what you get.

There is no better example of how good, basic statistics have painted an inaccurate picture than with the issue of divorce, especially in the United States. The common "statistic" often quoted is: "50% of all marriages in the United States end in divorce."

This conclusion is simply and completely wrong. But where did this 50% divorce rate "fact" originate? No one knows for sure, but there is speculation that it was born from an improper conclusion drawn from valid statistics. Here's the story.

In 1969, California under the leadership of Governor Ronald Reagan passed the first "no-fault" divorce laws. Up to this point, divorce was a legal option only if the grounds included adultery, physical abuse, mental cruelty, desertion, imprisonment, alcohol or drug addiction, or insanity. This was the first law of its type that allowed divorce without the labeling of a "bad guy (or girl)" in a relationship. Other states quickly followed by either replacing the old system or by adding the no-fault legal divorce option.

In the 1970s, when "no-fault" was implemented across the United States, there was an increase in the number of divorces that brought the issue to the attention of several professional groups. The divorce rate increased to its peak in 1981. The reported statistics indicated [13]:

Marriage rate: 10.6 per 1,000 total population

Divorce rate: 5.3 per 1,000 population

It is speculated that someone misinterpreted these facts and saw the divorce number was 50% of the marriage number and concluded (incorrectly of course) that 50% of all marriages end in divorce. The correct way to interpret the above statistics is to say that there was one divorce for every two marriages over the year. For comparison, the 2009 (provisional) data on the annual number of marriages and divorces show about the same 'two for one' result [14].

Marriage rate: 7.1 per 1,000 total population

Divorce rate: 3.45 per 1,000 population (44 reporting states and D.C.)

Let's look at the data published by U.S. Census Bureau for 2004 [15] to see what the real numbers are. The number of men at or over the age of 15 was 109,830,000. Out of this group, 68.8% (75,563,040) were ever married and 20.7% (22,734,810) were ever divorced. Given that 75,563,040 men were married at least once, and 22,734,810 men were ever divorced, a more accurate representation of the male divorce rate is the ratio of these figures: 22,734,810/75,563,040, or 30.1%— not 50%.

Now the women: In 2004 there were approximately 117,677,000 at or over the age of 15. The number of women that were married at least once is 87,316,334 (74.2%) and the number ever divorced, 26,948,033 (22.9%). The female "divorce rate" estimate is then the ratio of these numbers, or 30.8%—again, not 50%.

In fact, there is no precise answer to this question. Referring back to the 50% divorce rate statement, the language may appear precise, but in mathematical terms, it's vague. For example, does the statement refer to only first marriages for people or for the second, third, or further marriages? Does the time interval of the "50% statistic" apply over the lifetime of a marriage or for 2, 5, 10, 15, or another amount of married years? This subject is a current topic of research, so depending on how you analyze this complex issue, you will end up with different answers. The single point I am making here is that the basic premise for the "50% statistic" appears to have been an improper interpretation of technically correct statistics.

HOW STATISTICS SHOULD BE REPORTED: RULES OF THUMB

So far I've made a big deal of the fact that people can be mistaken, misinformed, and misguided by inaccurate, imprecise, or incomplete reporting of statistics. So much for cursing the darkness. Let's move on to how you can identify when the information you read is based on responsible statistical reporting and when it's more akin to "media jive." We'll start with basic four points and add some others as we go on. Every statistic reported should contain elements from these essential areas.

I.   **Sample Size:** How many items were counted in forming the statistic?

II.   **Event:** What type of item was counted? How was it defined?

III.   **Time Period:** Over what amount of time was data collected?

IV.   **Geographical or Other Limitations:** What region, state, and general physical space, or other restrictions, were applied?

Let's take an example from Chapter 13: An Incomplete List of the Risks of Life, and see how you should interpret the statistic. Here it is:

"One out of every 123 Americans will die this year."

Here are the guidelines for the correct reporting of statistics from the perspective of this example.

Sample:	Population of U.S.
Time Period:	One year, 2008
Event:	Death
Geographical Limitations or other Limitations:	U.S.

This statistic passes the quality test, except it has very little value for each of us. There are gender, age, behavioral, and several other factors that need to be considered before such a statistic can be useful. Chapter 13 has some of them, just to show how such factors can influence mortality.

If you apply these guidelines to commercial advertising or news reporting, you'll be surprised at the extent to which statistical abuse exists. There are a multitude of claims full of sound and fury, yet signifying little or nothing. Here are some examples of claims with little stated substance.

There is a general type of strategy I've seen in advertisements that can be elusive. Do you recall advertisement ploys that say things like: "4 out of 5 doctors recommend ..."? What you don't know from this type of advertisement relates to the type of doctors surveyed. To use an example, suppose the issue is baby food and the company claims 4 out of 5 doctors recommend its brand. We don't know if the "4 out of 5" is across all doctors, or just those who recommend a particular brand of baby food. Not all doctors recommend products. Pay careful attention the next time you're presented with such information.

I was in a limousine one rainy afternoon on the way to O'Hare Airport in Chicago. The driver asked me if I would like to read the newspaper. I usually can't read in a moving car without getting a headache, but for some reason I accepted his offer and began reading. As I paged through the newspaper, one story stood out from all of the rest of the news. Here it is:

HEADLINE: Dirty Air Killing 3,400 Yearly

Despite a significant drop in pollution, dirty air still kills 3,479 people each year in metropolitan Chicago, a new study estimates.

What got my attention from the article's first sentence is its accuracy. This study says exactly 3,479 people are killed, not 3,480 or 3,478. How did they know this? I don't think "air pollution" is a cause of death written on many toe-tags. And any prediction with such accuracy is absolutely nothing short of a miracle! This amazing article went on to say that there are 64,000 pollution-related deaths in some 239 metropolitan areas. (At least they didn't say 63,999 deaths.) Finally, in the seventh paragraph of the nine-paragraph article, it did confess that estimating pollution deaths is an inexact science (What a surprise!). Critics of these figures say that even if pollution does not kill, lives are shortened by just a few days. Saying this statement is like saying that the majority of the health expense incurred by most Americans takes place in the last two weeks of life. We could cut healthcare tremendously if we just cut back costs in those last two weeks. The trouble with this is that no one knows exactly what day anyone will die, so it's impossible to know when to start the clock.

Consider another headline example [16]: "California Air Pollution Kills More People Than Car Crashes, Study Shows." According to a study performed by researchers at California State-Fullerton, improving the air quality of Southern California and the San Joaquin Valley would save more lives each year than the number killed in motor vehicle crashes in the two regions. The study goes on to state that 2,521 vehicle deaths were recorded in 2006 compared to 3,812 deaths attributed to air pollution causes in the two regions. In this example there are no computer models. There are real people behind these statistics in both cases. Yet the sudden and unanticipated nature of motor vehicle fatalities does not necessarily compare to people dying of lung cancer, bronchitis, or pneumonia. This study goes on to conclude that if pollution levels were improved just to federal standards, then there would be 3,860 fewer premature deaths, 3,780 fewer nonfatal heart attacks, 470,000 fewer worker sick days, 1.2 million fewer sick days for school children, a savings of $112 million in caregiver costs, and more than 2 million fewer cases of upper respiratory problems. That is quite a list of conclusions.

The analogy with auto deaths brings the air pollution health effects into perspective for the readers and no one is ever going to argue that the air quality in these regions is good for you or that air pollution is not a major health factor. However, the different time-dependent nature of the deaths diminishes the comparison's relevance.

SCIENTIFIC STUDIES: *CAVEAT EMPTOR*

In situations of uncertainty in which statistics are used, we are eager to believe so-called "scientific studies" because we have a simple desire to learn and sort out facts in our information-overloaded world. All of us search for the truth in complex issues and when science gives us information important to us, it's easy to skip over the details, leading us to misguided or misleading conclusions. Sometimes, however, it's not our fault. Scientific and statistical results can be distorted by inaccurate reporting. Here are some other examples.

One study concluded that left-handed people die an average of nine years earlier than those who are right-handed [17, 18]. This study included 987 people from two Southern California counties. Another study was performed with 2,992 people, and found the mortality rates between right- and left-handers just about equal, if you take into account the fact that left-handers over 60 years of age were often forced to change their dominant hand as children.

A final example in this section has to do with the most watched athletic event in the world: the Super Bowl. It was claimed and largely believed that domestic violence increased on Super Bowl Sunday. The TV networks went so far as to run public service announcements before the football game to help quash the impending wave of violence. Afterward, however, investigations found little evidence to support this claim. What the investigators did find in one instance were details of a press conference where antidomestic violence advocates cited a study performed in Virginia which showed a 40% increase in hospital emergency room admissions of women after football games won by the Washington Redskins. When the study author was contacted about the applicability of the work to the Super Bowl, the author denied the 40% figure and said that the study findings had been misinterpreted. Of course, even an admission of bad facts couldn't kill a good story. The morning of one Super Bowl found a Connecticut television reporter adamantly asserting that the Super Bowl led to the beating of women. He even challenged skeptical viewers to check emergency room numbers… on the following day. Interestingly enough, we must have missed the follow-up report he must have done to verify his own assertions … Hmmm … I wonder if his Super Bowl predictive powers were as good.

In 2003 researchers at Indiana University studied the connection between domestic violence and the Super Bowl and reported they did find a specific correlation [19]. However, the study also reported that the correlation results were similar to the rise in domestic violence seen for other holidays in the year. Their conclusion was the correlation might be due to the "holiday" rather than to the Super Bowl itself.

This is a fairly typical example of how you must be careful when interpreting statistics, regardless of how they are quoted. Think for yourself and be careful of accepting or rejecting any statistics, unless you believe and trust in the study's methods or its authors. Realistically, even if we wanted to, we have neither the time nor the knowledge to sift through all of the details in order to judge statistical research. How, then, can we ever really apply any statistics that are designed to help us better manage our own lives? My suggestion is that most of what you read and hear has at least a thread of truth to it. Use your own judgment to decide how much credibility you want to put in the information.

If you remember back in Chapter 2, Southwest Airlines' on-time performance was held suspect because the flight crew gave the arrival times manually. Other airlines, whose track record was not as good, used an automatic system that transmitted arrival times to a central computer when the front cabin door was opened. Even though everyone knew Southwest's performance record was outstanding, the mere possibility that the published arrival statistics could be fudged caused a lack of confidence in the airline's performance. In Southwest's case, there was no impropriety but their competition's suspicions are noteworthy. The subtle fact is that statistics may be accurately reported, but the foundation on which they are based may contain powerful, hidden biases. This brings us some more issues to consider when judging the veracity of statistical results.

Issue #1: Who funded the study?

You might wonder how reputable research studies can be biased. Well, if the research is measuring the glacial ice distribution on Mars or the radiation profile of a distant star cluster then there's a pretty good chance that you can take the results at face value. However, let's look at an example that is a lot closer to our experience, one involving the universe of nutritional choices we make every day.

In 2001, Senator Patrick Leahy (D-VT), introduced a bill to restrict sales of soft drinks in schools called "The Better Nutrition for School Children Act of 2001." The legislation was intended to act on the statistical finding relating children's soft drink consumption to obesity. At this time there was an "independent" study that demonstrated a strong link between these factors, yet another study that showed no connection. Sound confusing? Well, there is more. The study that showed the connection was funded by the neutral federal agencies: the Centers for Disease Control and Prevention and the National Institute of Diabetes and Digestive and Kidney Diseases. Both of these agencies have encouraged children to avoid soft drinks. Further investigation of the research authors show they have a history in researching childhood obesity and have also correlated obesity to diet and TV-watching habits. The study that showed no connection between obesity and soft drinks was based on two national surveys: The U.S. Department of Health and Human Services' National Health and Nutrition Examination Survey and the U.S. Department of Agriculture's Continuing Survey of Food intake. From the analysis, the conclusions were:

- No relationship between carbonated soft drink consumption among 12 to 16 year olds and obesity
- Soft drinks did not reduce calcium consumption among 2- to 20-year-olds
- Teens who consumed more soft drinks were as physically active as those who consumed fewer soft drinks
- Soft drink consumption did not harm diet quality among children and teens as measured by the USDA's Healthy Eating Index.

Now, guess who funded this study: The National Soft Drink Association. Even though the funding source had virtually no control over the outcome of the study, simply the mere possibility of bias can discredit conclusions where statistics play a major part.

I am not implying a lack of integrity on the part of the players here, just the opposite. The point is studies using statistics can develop different conclusions for the same problem—the art is not exact. The teenage obesity–soft drink example is only one element of the larger issue. You could say that this example is an outlier and that funding source does not correlate generally to study results. You could—but read further ...

A research project was undertaken at the Children's Hospital in Boston to examine the relationship between funding source and conclusions among nutrition-related scientific articles. The particular area of nutrition was limited to nonalcoholic drinks including soft drinks, juices, and milk. Between 1999 and 2003, they reviewed 206 published articles, the conclusions drawn, and the funding sources. They concluded that articles sponsored exclusively by food and drink companies were four to eight times more likely to have conclusions favorable to the financial interests of the sponsoring source than articles that were not sponsored by these types of companies [20]. And by the way, this study was sponsored by a grant from the Charles H. Hood Foundation and by discretionary funds from the Children's Hospital. The report also notes that one of the investigators is an author of a book on childhood obesity.

I am not implying or suggesting that researchers are doing anything that is unethical. How can this apparent contradiction be resolved? The reason all of these different and obviously biased assumptions can be made is that no one really knows what the answers are. The types of measurements suggested by these assumptions are not the kind that can be answered by classical science. Scientists can measure the mass of the earth, the distance between stars, the characteristics of the atomic nucleus with unquestionable accuracy and error limits. It is ironic that in some ways we know more about the physical nuances of our universe than we do about the connection between adolescent soft drink consumption and obesity.

Permit me to elaborate on this point a little bit. We all have opinions and beliefs that direct our lives and yet we can't really prove them. I can't speak for you, but here are some of my own opinions. They pretty much fall into the same category as the soft drink influence on teenage obesity.

(Some) Beliefs of Rick Jones

- The likelihood, if you approach two glass doors to any building, that one will be locked is 90%.
- At fast-food restaurants there is always at least one french fry in the bottom of the bag (given that you ordered french fries).
- Murphy's Laws work about 80% of the time.
- When driving, you always encounter more red traffic lights when you are in a hurry.
- You cannot go on vacation without forgetting at least one thing.

I could go on, but you get the idea. Now, I could also organize a project team to design a testing and evaluation program to verify my beliefs or I could just use "expert opinion." Expert opinion means the beliefs of somebody else who has hopefully more experience in the subject area. Frankly, for the above laws, I think I'm an expert and I am sure you could come up with your own list for which you're an expert. If you were funding a scientific study, wouldn't you look around for ethical scientists who had the same beliefs as you? From a research firm's viewpoint, if someone gives you a wheelbarrow full of money to research an issue, what do you think your chances for a renewed contract are if your results are contrary to the beliefs of the company that gave you the money? Generally, the company tries to find a research firm that mirrors its beliefs. The influence of the sponsoring company is subliminal and indirect, but it is there and reinforced by their presence throughout the course of the work. (And by the way, what are the odds that a study with results adverse to the needs of its sponsor would be released?)

This brings us to conclude that there is more to know about statistics than the analytics. Since statistics are used to obtain information on unsolvable problems, how the tools are applied can have a major bearing on the veracity of the results. Sponsorship and intent are issues you must address before you accept a study's findings.

Issue #2: How were the questions worded?

When it's foggy, ships use radar to avoid collisions. Airplanes use radar to avoid severe weather. And in the foggy, sometimes murky world of public opinion, surveys are the radar of the commercial marketers, political media, and others. Let's face it. There are too many people to just go ask everyone what they think, so some sort of statistical sampling with a structured set of questions is an efficient way to get feedback on specific issues.

On the surface it may seem simple to write up a set of questions and go ask a bunch of people and then compile the results. But the choice of question wording can strongly influence the results. Sometimes this is done intentionally if the interested parties are seeking certain answers, though other times it is an unforeseen result of the survey's construction.

There is no better place to go for examples than politics. In March 2004 an Associated Press-Ipsos poll used these two questions, asked to random samples of adults [21].

- *If you had to choose, would you prefer balancing the budget or cutting taxes?*
- *If you had to choose, would you prefer balancing the budget or spending more on education, healthcare, and economic development?*

The results showed 61% or 36% were in favor of balancing the budget depending on which question. You can guess which question solicited each response.

Ideally, measurement by polling should not affect the process, (Law #4, Chapter 2) yet there are polls that have been (and probably will be) designed to do just the opposite—influence the process under the camouflage of unbiased polling. These are called "push polls" [22] and you hear about them mainly in political polling.

A push poll attempts to feed false or damaging information about a candidate under the guise of conducting a survey of voter preferences. I am not going to discuss this subject here, just be aware of this subtle type of manipulation disguised as campaigning. This shady technique could be used for any poll so if you receive a call about a poll regarding a certain issue and the questions appear prefaced with a lot of negative or one-sided statements, the best thing you can do is just hang up.

However, not all mistakes are by politicians nor are they on purpose. In April of 1993, a poll commissioned by the American Jewish Committee, developed by the well-known pollster Burns W. Roper, drew widespread, shocked reaction from the Jewish community as well as emotional commentary from journalists [23, 24]. The problem was the question—not the real views of the polled population. The question construction, in attempting to stimulate a thoughtful response, obfuscated its real meaning with a double negative. The flawed question read:

Does it seem possible or does it seem impossible to you that the Nazi
extermination of the Jews never happened?

The survey results indicated that 22% of the adults and 20% of the high school students surveyed said they thought it was possible that the Nazi extermination of the Jews never happened. An additional 12% of the adults said that they did not know if it was possible or impossible. When the Gallup pollsters tested this same question they found a full 33% of the people thought that it was possible that the Nazi extermination of the Jews never happened.

The numbers changed, however, when the following, more direct question was used:

Do you doubt that the Holocaust actually happened or not?

Only 9% responded that they doubted it. When the question was put even more clearly, less than 0.5% responded that the Holocaust "definitively" did not happen, 2% said it "probably" did not happen, 79% said it "definitely" happened, and 17% thought it "probably" happened.

In today's society, polls have become almost unquestioned sources of exactly what the public is thinking. And it is true that poll results do give an indication as to what's on people's minds. Yet the devil is in the details. It's statistically possible to sample a small number of people and make conclusions about a population. However, there are so many technical details we cannot always validate to our satisfaction that a survey is free of error or bias. Questions about how people were selected and the exact wording of the questions are seldom mentioned along with the sound bite or headline conclusions.

We see today that our politicians use polls to measure public opinion regarding key news issues. The founding fathers of our constitution used the House of Representatives to ensure that public interests were being followed. I suspect that's why the election period for these public servants is every two years. If someone is not adequately representing their district, the people can quickly elect someone else. Polls are taking on a power and role that was not defined by our government's design. Sure, it's true that polls were not around in 1776 and the polling process may just be another "technology" we need to understand. We have a long way to go. Why? Here's one example from recent history. In 1998 an ABC poll asked people if they wanted "Congress to censure President Clinton." A 61% to 33% ratio said they did. What's wrong here? This assumes that the people knew what is meant by the term "censure." Another poll indicated that 41% of those who favor censure admitted they "don't know" what the term actually means.

What are the lessons here? The next time you fill out a questionnaire or answer polling questions, read or listen very, very carefully. And the next time you hear the results of a poll, ask yourself exactly who might have found the time to answer the pollster's questions.

Issue #3: How was the problem or quantity being measured or defined?

An example of this issue is found by looking at the U.S. government's poverty level definition. It shows a political side of statistics misuse that has persisted for over 30 years. Here's the story.

Every year the Commerce Department in Washington, D.C. announces the number of Americans living in poverty. Along with this depressing statistic are other indicators such as household incomes and conclusions about the rich getting richer and the poor getting poorer. The stakes are high. Bureaucrats use these figures to set eligibility criteria for 27 federal programs including food stamps, Head Start, and school lunches. There's nothing wrong with this process, but there is one little problem. The definition of the poverty level is flawed and is not representative of today's society. This fact is readily acknowledged in the halls of Congress and the Commerce Department. A committee of 13 academics was commissioned under the 1988 Family Support Act [25] to make recommendations on how to update the methodology used to define poverty in today's economy. One million dollars later, a 500-page report was presented to Congress. So far nothing has happened. The reason it hasn't changed for such a long time is that Congressional allocations have become routine and no one wants to upset the taxpayers with another redistribution, even if it means continuing with actions based on an inaccurate metric. You see, the Office of Management and Budget uses census figures to determine poverty standards and to distribute state funding. Congressional districts that have become accustomed to having this cash would not understand having it taken away. To displease your constituents like this is not a good way to be re-elected. So Congress ducks when the logic (or lack thereof) is questioned and these so-called poverty statistics are announced every year. To get an idea of how this all started, we must journey back to the Great Society of the Johnson Administration.

In 1963, a civil servant statistician in the Department of Health and Human Services named Mollie Orshansky [26] developed federal poverty thresholds. President Johnson, looking for a national measure for poverty, appropriated it in a context that neither Ms. Orshansky nor anyone else had intended. Johnson subsequently used it in his speeches and in his policy-making.

Let's take a quick look to see what's wrong with this measure. Orshanksy's poverty level is based on an annual Agriculture Department estimate of the money required for food. This number is multiplied by three to account for all other expenses, and then is seasonally "adjusted" to reflect family size. This is pretty vague, especially the word adjusted in the last sentence. Regardless of these details, here's what's wrong with the poverty level statistic.

The poverty line measures pretax income only, not including benefits like food stamps and Earned Income tax credits. Including items like these would decrease poverty significantly. No allowance is made for childcare and transportation expenses, hence understating the number of working poor. If you have ever compared living in Birmingham, Alabama with living in New York City, you noticed a vast cost of living difference. Yet no regional effects are considered in Orshansky's work. These errors cause the current poverty measure to overstate poverty in some areas and simultaneously deprive truly needy people of benefits in others [27].

Some new methods have been tested. Their results contradict trends observed using the old poverty level. Using the new measures of poverty, the number of people at and below the poverty level is dropping and the gap between the so-called rich and poor is actually shrinking, not growing. The controversy isn't over on this political science metric. And this is just one example. Do you think there could be other government indicators with the same type of history?

After reading all of these ways in which we are led to believe that certain facts are true or false, you could conclude that if someone wants to trick you with numbers they probably can. The sad and scary thing is I believe that this is true, and, worse still, it occurs more than we either know or are willing to admit.

The guidelines I've discussed here will help, but they are not a perfect defense against statistics misuse. There are other, more subtle ways that are almost impossible to stop. You see, one way to make your "stats" look good is to change the math, another is to change the data, but a third, more elegant approach for statistics misuse is to just not record or report events that go against your goals. An example of this is crime under-reporting. For example, I read in the newspaper that the crime rate in Tallahassee, Florida, for theft, assault, and rape is higher than in New York City. From the actual numbers, this is true. However, if you look more closely, you'll find that many, many crimes in New York City were simply not reported because people believed that nothing would be done and reporting all of the crimes will probably just make their insurance premiums go up. In fact, the NYPD appeared to implicitly discourage reporting of the low-level crimes simply because there are so many. In Tallahassee, it's a completely different story. Because crime is not so widespread, every injustice is deemed significant and is reported. And this is really why the stats just don't seem to add up. There are examples of crime under-reporting in Great Britain [28] (to name one country), in U.S. schools [29], and in U.S. cities [30]. Most likely this type of data censoring will be a growing problem as public opinion, politicians' careers, and other self-centered objectives are "measured" by data. In the growing digital world, the human is still the weakest link. Being a skeptic about believing anyone's statistics should most likely become a "best practice."

NEWS MEDIA STATISTICS

Statistics printed in newspapers, stated in news reports, and presented in other media are always subject to interpretation. Without qualification, these numbers are absolutely useless. The quoting agency, however, is seldom held to the standard of proper reporting of numbers. This neglect can create fear and mistrust where none should exist.

Here are two examples of this media incompetence. Statistic: "One out of eight women will get breast cancer." Before I start discussing this topic, let me state that breast cancer is a serious concern for women. I am not trying to minimize the health hazard. Periodic breast exams and mammograms are good preventive medicine. Nothing I am about to say here is designed to change your mind regarding their importance. They are worth the time and money without a doubt. I merely disagree with the partial "truth" in the reporting of the statistic. Let me explain.

Suppose you are a 35-year-old woman and you find a lump in your breast. The statistic goes through your head. "Am I the one in eight?" My wife and several of our closest friends and our families went through the mental agony and worry over the cancer threat when lumps were identified, removed, and biopsied. The irresponsible propagation of the "one in eight" has caused unwarranted pain all because someone decided not to communicate the whole story.

And what is the whole story? It is not as dramatic as you might think. In fact, the little bit that is left out makes the "one in eight" statistic much less frightening. What the statistic does NOT tell you is that "one in eight" applies over a lifetime. We are all going to die of something. The ultimate cause of death for one out of eight women is breast cancer. This is a lot different that the media's implication that one in every eight women gets breast cancer each year. In fact, most women who die from breast cancer are relatively old. In Table 3.5 is a more descriptive statement of the odds of contracting (but not dying from) breast cancer [31]:

A full description of the actual statistics is too long to be included in this text [32], but here are some points worth mentioning. Even though the incidence of breast cancer has increased, the likelihood that it will be fatal has stayed about the same for American women. The median age (50th percentile) age for women to be

Table 3.5   Invasive Female Breast Cancer

Current age	Probability of developing breast cancer in next 10 yrs (%)	1 in:
20	0.06	1,760
30	0.44	229
40	1.44	69
50	2.39	42
60	3.40	29
70	3.73	27
Lifetime risk	12.08	8

diagnosed with breast cancer is 64. The odds of a woman in her forties dying of breast cancer within the next ten years are approximately 0.4%. This is about 1 in 250. The actual death rate for breast cancer was about constant at 32 per 100,000 to about 1990. Since then it has declined to where it is now: the data says 24.5 out of every 100,000 American women died of breast cancer each year, not one in eight.

Finally, for the record, the National Cancer Institute does tell the whole story on this. For American women:

- 12.7% of women born today will be diagnosed with breast cancer at some time in their lives.
- Because rates of breast cancer increase with age, estimates of risk at specific ages are more meaningful than estimates of lifetime risk.
- An estimated risk represents the average risk for all women in the United States as a group. This estimate does not indicate the risk for an individual woman because of individual differences in age, family history, reproductive history, race/ethnicity, and other factors.

While the discussion here shows the disparity within how breast cancer statistics are reported, there is no getting around the fact that this disease is a very serious concern for women. According to the American Cancer Society, 40,170 women in the United States will die from breast cancer in 2009 [33]. This is nearly the same as the number of people killed by motor vehicles in 2008: 43,313.

Breast cancer is real and frightening, but sometimes the press can create news that wasn't there, and, more shamefully still, play on the racial fears of Americans to sell their product. Here is a classic case of what risk analysts call "media amplification."

In the summer of 1996, responding to the reported increase in arson fires at black churches in the south, President Clinton in a radio address stated, "racial hostility is the driving force." He further stated, "I want to ask every citizen in America to say ... we are not slipping back to those dark days." However, the real fire and arson statistics indicated there was no substantial evidence of an epidemic in black church arson [35]. It appears we were all misled by another incorrect use of statistics. Activist groups, manipulating the emotions of all Americans, played up the sad illusion to further their own cause.

Let's have a look at the real evidence. The newspaper, which will go unnamed in this book, compiled what they termed "the most comprehensive statistics available on church arsons in the South in the 1990s." This may be true, but they didn't say they were accurate! Their chart showed a rapid rise in the number of black church arsons starting in 1994 and continuing. One problem with the statistics is that two states didn't start reporting until 1993 and a third not until 1995 [36]. It's not surprising that the numbers went up when states were added. In short, the shape of the curve represents real data, but the number of states contributing also increased during the same time period. This is like saying your food budget increased without your mentioning you had added more people to your family. When you adjust the data for this, there was no increase in arsons and the epidemic went away. Furthermore, data from federal agencies doesn't add any evidence to the purported epidemic either. The largest private group that compiles data on church fires is the National Fire Protection Association. While its data does not identify black churches, it does state that church arsons have decreased from 1,420 in 1980 to 520 in 1994 and this trend appears to be continuing.

You might be wondering where this black church arson epidemic came from anyway. The main sources were the National Council of Churches (NCC) and the activist group Center for Democratic Renewal (CDR). The stated mission of the CDR is to assist activists and organizations in building a movement to counter right-wing rhetoric and public policy initiatives. The CDR was originally called the National Anti-Klan Network, but changed its named after the Klan generally disorganized in the 1980s. In a March 1996 news conference, this group began to claim a tremendous surge of black church arsons. With the intense media coverage, the "domestic terrorism" theme fell on listening ears. A data search after the conference picked up over 2,200 articles on the subject. Hence the "epidemic" was born.

In June of 1997, the National Council Arson Task Force confirmed what the data said all along. There was no conspiracy to burn black churches. The task force noted that a third of all suspected church arsonists arrested since January 1995 were black. This is more than twice the proportion of blacks in the general population. In 2003, a study at Mississippi State University [37] confirmed this conclusion, although their work does note two spikes of violence that, along with the other aspects of this racially sensitive issue, fueled the epidemic's flames.

There is a political component added to the statistical equation. That's why in this equation, 1 + 1 does not equal 2. The data says no conspiracy, yet this is not what we heard from our leaders at this time. Both former President Clinton and former Vice President Gore, through their public statements, seemed to want us to believe the black church conspiracy did exist. We can only conjecture as to why.

There is a tragic aspect of this media amplification. After the press moves on to new territory, the fear created by the distortion of the facts is still in the hearts of many people, both black and white. Worse yet, without resolution, the fear grows. And like other horrendous crimes, publicity breeds mimicry, making the fears more real still.

In this case, the so-called journalistic press is one irresponsible culprit, although, as you can see, there are others. In any case, journalism is the medium. Reporting the news and providing solid information, and not rhetoric, should be their objective, and most of the time they do a reasonable job.

There is one way in which you can feel confident that statistical conclusions are based on good, sound scientific principles. Check to see if a professional peer group reviewed the study. Scientific and professional disciplines have developed a very good quality control procedure to ensure that the public is not exposed to poor research work or conclusions based on biased results. Many journals that publish research findings employ a small peer group to review and check papers before they are published. More often than not, the referees either find mistakes or suggest improvements that necessitate rewriting or at least editing the work before publication. If a journalist refers to a publication such as the Journal of the American

Medical Association (JAMA), chances are the results will have truth in them. In this way, science's self-discipline ensures as much as possible that numbers, logic, and statistics are used correctly and without bias.

The point of this lesson in economic history, crime statistics, and various other aspects of living in the second decade of a new millennium is that the definitions and procedures used in collecting data and in developing statistics makes a big difference. Whether it is the government, your local newspaper, or a private research firm reporting statistics, I have one simple suggestion when it comes to conclusions based on statistical arguments:

> *The pure and simple truth is rarely pure and never simple.*
>
> —Oscar Wilde

DISCUSSION QUESTIONS

1. Choose an advertisement that uses statistics in its message and examine its accuracy using the content of this chapter. How would you rewrite it to be more precise?

2. Choose a news article that uses statistics in its content and examine its accuracy. Are there any relevant external factors the article has left out?

3. Crime statistics are often used by groups either to show some program is working or to document the need for changes. For a city or region of your choice, investigate the potential for crime under-reporting and estimate a range for what some actual statistics could be. Be sure to document all of your assumptions.

4. Identify and explain three reasonably current examples of media amplification using statistics.

5. Develop ten poll questions for any hazard, exposure, or social issue using the scale: Strongly Agree, Agree, Disagree, Strongly Disagree, Don't Know. Now examine the questions for bias and ranges of interpretations.

Case Study: Probability and Statistics: A Legal (Mis-)Application

The People v. Malcolm Ricardo Collins, *68 Cal. 2d 319*

In a court of law, the burden of proof lies with the accuser to show "beyond a reasonable doubt" that the defendant committed the crime. Statistical arguments are often used by both plaintiff and defendant attorneys providing evidence to convince the judge or jury that proof "beyond a reasonable doubt" has or has not been achieved. This case study represents a classic example of how not to use statistics and probability in legal proceedings. The case study has some interesting lessons.

This case stemmed from a woman being pushed to the ground by someone who stole her purse. She saw a young blond woman running from the scene. A man watering his lawn witnessed a blond woman with a ponytail entering a yellow car. The witness observed that a black man with a beard and mustache was driving the car.

Another witness testified that the defendant picked up her housemaid at about 11:30 A.M. that day. She further testified that the housemaid was wearing a ponytail and that the car that the defendant was driving was yellow. There was evidence that would allow the inference that there was plenty of time to drive from this witness's house to participate in the robbery.

The suspects were arrested and taken into custody. At trial, the prosecution experienced difficulty in gaining a positive identification of the perpetrators of the crime. The victim had not seen the driver and could not identify the housemaid. The man watering his lawn could not positively identify either perpetrator.

The prosecutor called an instructor of mathematics who testified that the "probability of joint occurrence of a number of mutually independent events is equal to the product of the individual probabilities that each of the events will occur." For example, the probability of rolling one die and coming up with a 2 is 1/6; that is, any one of the six faces of a die has a one chance in six of landing face up on any particular roll. The probability of rolling two 2s in succession is 1/6 x 1/6, or 1/36.

The prosecutor insisted that the factors he used were only for illustrative purposes—to demonstrate how the probability of the occurrence of mutually independent factors affected the probability that they would occur together. He nevertheless attempted to use factors which he personally related to the distinctive characteristics of the defendants.

The prosecutor then asked the professor to assume probability factors for each of the characteristics of the perpetrators. In his argument to the jury, he invited the jurors to apply their own factors, and asked defense counsel to suggest what the latter would deem as reasonable. The prosecutor himself proposed the individual probabilities set out in the list below. Although the transcript of the examination of the mathematics instructor and the information volunteered by the prosecutor at that time create some uncertainty as to precisely which of the characteristics the prosecutor assigned to the individual probabilities, he restated in his argument to the jury that the probabilities should be as follows:

Characteristic Individual Probability

A. Partly yellow automobile	1/10
B. Man with mustache	1/4
C. Girl with ponytail	1/10
D. Girl with blond hair	1/3
E. Negro man with beard	1/10
F. Interracial couple in car	1/1000

According to his "illustrative argument" the probability of an interracial couple, with a black man, with a beard and mustache, girl with blond hair and a ponytail, in partly yellow automobile is:

$$A \times B \times C \times D \times E \times F = 8.333 \times 10^{-8} \text{ or 1 in 12,000,000}$$

Upon the witness's testimony it was to be "inferred that there could be but one chance in 12 million that defendants were innocent and that another equally distinctive couple actually committed the robbery." After testimony was completed, the jury convicted the couple of second-degree robbery.

On review, the Supreme Court of California found that the mathematical testimony was prejudicial and should not have been admitted. "The prosecution produced no evidence whatsoever showing, or from which it could be in any way inferred, that only one out of every ten cars which might have been at the scene of the robbery was partly yellow, that only one out of every four men who might have been there wore a mustache, that only one out of every ten girls who might have been there wore a ponytail, or that any of the other individual probability factors listed were even roughly accurate." Another problem that the court found was the fact that the prosecution did not have adequate proof of the statistical independence of the six factors (i.e. that wearing a mustache is truly independent of wearing a beard). The court further expressed that "few juries could resist the temptation to accord disproportionate weight to that index; only an exceptional juror, and indeed only a defense attorney schooled in mathematics, could successfully keep in mind the fact that the probability computed by the prosecution can represent, at best, the likelihood that a random couple would share the characteristics testified to by the People's witnesses—not necessarily the characteristics of the actually guilty couple."

The court found this admission of evidence to be a key error and reversed the jury's guilty verdict. "Undoubtedly the jurors were unduly impressed by the mystique of the mathematical demonstration but were unable to assess its relevancy or value ... we think that under the circumstances the 'trial by mathematics' so distorted the role of the jury and so disadvantaged counsel for the defense, as to constitute in itself a miscarriage of justice."

In a criminal case, it was prejudicial error to allow the prosecution to offer, through an expert, a formula in statistical probability, logically irrelevant and evidentially inadequate, from which the jurors were invited to infer that the odds against defendants' innocence were one in 12,000,000, where the circumstantial nature of the evidence and length of the jury deliberation showed that the case was a close one, and where, under the circumstances, the "trial by mathematics," with which the jurors were not technically equipped to cope, so distorted their role and so disadvantaged defense counsel as to constitute a miscarriage of justice.

The prosecution's misuse of mathematical probability statistics was prejudicial where the testimony lacked an adequate foundation in evidence and in statistical theory, where it was used to distract the jury and encourage them to rely on a logically irrelevant expert demonstration, where it foreclosed the possibility of an effective defense by an attorney unschooled in mathematical refinements, and where it placed the jurors and defense counsel at a disadvantage in sifting relevant fact from inapplicable theory.

In this robbery prosecution, the prosecutor misused mathematical probability and statistics in an attempt to show that defendants were the robbers when there was no evidence relating to any of the six individual probability factors assigned to the distinct characteristics of defendants. No proof was presented that the characteristics selected were mutually independent, even though the mathematical expert witness acknowledged that such condition was essential to proper application of the statistical probability rules that he used in his testimony.

Case Study Questions

1. Knowing the outcome of this case on appeal, how could the defense attorney have improved his/her argument strategy to counter the reversal decision argument?

2. Independent of the strategy issues, what would be the correct way to compute this probability?

3. The use of statistics, or in this case probability, is always accompanied by an inherent error or uncertainty of occurrence that can have implications for both sides. Comment on the legal application of "reasonable doubt" using statistics. For example, if the chances the defendant was guilty were 1 in 10, 1 in 100, or 1 in 12,000,000, how could a jury decide which level constitutes reasonable doubt?

ENDNOTES

1 http://espn.go.com/mlb/teams/salaries?team=nyy (accessed November 10, 2009).

2 Michael O. Finkelstein and Bruce Levin, *Statistics for Lawyers*. New York: Springer-Verlag, 1990.

3 Harriet Zellner, "Reducing the Risk When Reducing the Force—It's a Question of Impact and Treatment," *Business Law Today*, November/December 1998.

4 National Safety Council, *Injury Facts*®, 2009, Itasca, IL.

5 Gillian Turner, "Intelligence and the X-Chromosome," *Lancet*, Vol. 347, 1996, pp. 1814–1815.

6 Alice S. Whittemore et al., "Personal and Environment Characteristics Related to Epithelial Ovarian Cancer," *American Journal of Epidemiology*, Vol. 128, No. 6, 1988, pp. 1228–1240.

7 B. L. Harlow, "A Review of Perinal Talc Exposure and Risk of Ovarian Cancer," *Regulatory Toxicology and Pharmacology*, Vol. 21, Issue 2, 1995, pp. 254–260.

8 D. M. Gertig et al., "Prospective Study of Talc Use and Ovarian Cancer," *Journal of the National Cancer Institute*, Vol. 92, No. 3, 2000, pp. 249–252.

9 M. A. Merritt et al., "Talcum Powder, Chronic Pelvic Inflammation and NSAIDs in Relation to Risk of Epithelial Ovarian Cancer," *International Journal of Cancer*, Vol. 122, No. 1, January 1, 2008, pp. 170–176.

10 U.S. Cancer Statistics Working Group, *United States Cancer Statistics: 1999–2005 Incidence and Mortality Web-based Report*. Atlanta, GA: Department of Health and Human Services, Centers for Disease Control and Prevention, and National Cancer Institute, 2009.

11 National Safety Council estimates based on data from National Center for Health Statistics and U.S. Census Bureau. Deaths are classified on the basis of the Tenth Revision of the World Health Organization's "The International Classification of Diseases" (ICD). Numbers following titles refer to External Cause of Morbidity and Mortality classifications in ICD-10. One-year odds are approximated by dividing the 2003 population (290,850,005) by the number of deaths. Lifetime odds are approximated by dividing the one-year odds by the life expectancy of a person born in 2003 (77.6 years).

12 John Kindelberger and A. M. Eigen, "Younger Drivers and Sport Utility Vehicles," *Traffic Safety Facts*, Research Note, September 2003, DOT HS 809 636.

13 Vital Statistics of the United States, 1981, Volume III—Marriage and Divorce, U.S. Department of Health and Human Services, Public Health Services, National Center for Health, 1985 (PHS) 85-1 121.

14 National Vital Statistics Reports, CDC, Vol. 58, No. 3, January 2009.

15 *U.S. Census Bureau, Population Division, Fertility and Family Statistics Branch*, "Number, Timing, and Duration of Marriages and Divorces: 2004," Last Revised August 27, 2008.

16 http://www.huffingtonpost.com/2008/11/13/california-air-pollution-_n_143521.html (accessed November 21, 2009).

17 D. F. Halpern and S. Coren, "Do Right-Handers Live Longer?" *Nature*, vol. 333, 1988, p. 213.

18 D. F. Halpern and S. Coren, "Handedness and Life Span," *New England Journal of Medicine*, Vol. 324, 1991, p. 998.

19 http://newsinfo.iu.edu/news/page/normal/977.html (accessed November 22, 2009).

20 L. I. Lesser, C. B. Ebbeling, M. Goozner, D. Wypij, and D. S. Ludwig, "Relationship between Funding Source and Conclusion among Nutrition-Related Scientific Articles," *PLoS Medicine*, Vol. 4, No. 1, 2007, p. e5.

21 David S. Moore and William I. Notz, *Statistics Concepts and Controversies*, 6th edition. New York: W.H. Freeman, 2006, pp. 65–70.

22 Douglas N. Walton, *Media Argumentation: Dialectic, Persuasion, and Rhetoric*. Cambridge: Cambridge University Press, 2007, pp. 229–230.

23 "1 in 5 Polled Voices Doubt on Holocaust," *New York Times*, April 20, 1993.

24 "Poll Finds 1 Out of 3 Americans Open to Doubt There Was a Holocaust," *Los Angeles Times*, April 20, 1993.

25 N. Gregory Mankiw, *Principles of Microeconomics*. Mason, OH: Southwestern, 2009, p. 429.

26 http://www.ssa.gov/history/orshansky.html (accessed November 27, 2009).

27 Rebecca M. Blank, "How to Improve Poverty Measurements in the United States," University of Michigan and Brookings Institution, November 2007.

28 "Metropolitan Police amongst 18 Forces to Have 'Under-reported Violent Crime for 10 Years,'" http://www.telegraph.co.uk/, January 22, 2009.

29 http://www.schoolsecurity.org/trends/school_crime_reporting.html (accessed November 27, 2009).

30 http://www.10news.com/news/5340605/detail.html (accessed November 27, 2009).

31 L. A. G. Ries, D. Harkins, M. Krapcho, et al., *SEER Cancer Statistics Review, 1975–2003*. Bethesda, MD: National Cancer Institute, 2006.

32 M. J. Horner, L. A. G. Ries, M. Krapcho, N. Neyman, R. Aminou, N. Howlader, S. F. Altekruse, E. J. Feuer, L. Huang, A. Mariotto, B. A. Miller, D. R. Lewis, M. P. Eisner, D. G. Stinchcomb, and B. K. Edwards (eds.), *SEER Cancer Statistics Review, 1975–2006*. Bethesda, MD: National Cancer Institute. http://seer.cancer.gov/csr/1975_2006/, based on November 2008 SEER data submission, posted to the SEER web site, 2009.

33 National Cancer Institute, "Breast Cancer Facts and Figures, 2009–2010."

34 http://www.cancer.org/downloads/stt/CFF2009_EstDSt_5.pdf (accessed November 28, 2009).

35 David G. Savage, "Probe Finds No Conspiracy in Church Arsons," *Los Angeles Times*, June 9, 1997.

36 Christopher B. Strain, *Burning Faith: Church Arson in the American South*. Gainesville: University Press of Florida, 2008, p. 159.

37 "MSU Study Offers New View of Past Decade Church Burnings," Mississippi State University Press Release, March 4, 2003.

Chapter 4

The Value of Life and Limb, Statistically Speaking

In the book of life, the answers aren't in the back.

—Charlie Brown

There is no more difficult application of measurement and statistics than in the valuation of life and limb. At the individual level, life is valued as priceless. But when someone suffers accidental death or injury, the legal system quantifies the value of the life that has been lost or injured. The person's age, family status, income level, and a host of other factors are used to produce a settlement cost, effectively giving a "value of life or limb" on an individual basis. The amount, regardless of its size, is not intended to replace or heal. It is an imperfect and unemotional measurement that substitutes money for life following legal protocol and precedent. In spite of this harsh reality, there is no finite, explicit monetary value for individual human life.

At a societal level, however, the value of life is finite, although it is not a precise number. Common sense tells us that there are simply insufficient resources to prevent all accidents, illnesses, and diseases, but there's no divine formula to determine where we should draw the ethical line when it comes to putting a dollar value on life. Some events may be out of the control of anyone or anything. For example, genetic composition plays a role in human longevity, but that is not what we're talking about here. There are clear situations where personal and government decisions are the primary or initiating events of a sequence that ends in human fatality or injury. When these events happen, we ask what could have been done to prevent such tragedies. This question can be answered. In hindsight, there is no uncertainty, no risk, and consequently little or no doubt about what could have been done on a case-by-case basis. Yet for every second in the future there is a countless number of outcomes, with only one being expressed. Here lies the rub. How do you guard against a seemingly infinite set of outcomes with a finite set of resources?

20% Chance of Rain: Exploring the Concept of Risk, First Edition. Richard B. Jones.
© 2012 John Wiley & Sons, Inc. Published 2012 by John Wiley & Sons, Inc.

This is not a rhetorical question. It is routinely answered on a societal or group level with risk- or cost-benefit analyses that usually accompany any program or project requiring financial support. In this chapter we'll explore some methods used in statistical life valuation, review their important characteristics, and illustrate several complex technical and ethical issues as examples where they are applied.

Not all life or injury valuations are ominous and ethically compromising tasks. Let's start with speeding, a situation familiar to almost all car and truck drivers, and examine the risk-benefit characteristics. What does speeding provide? In urban situations, it usually gets the driver to the next stoplight faster and in rural or interstate driving, the higher speeds can translate into slightly shorter travel times.

Let's compute the representative time saved in an urban and a rural scenario. For the urban environment, let's take a 10-mile cross-town trip with an average posted speed limit of 30 mph (48 kph). Assuming the speed limit is maintained for the entire trip, the nominal trip time is 20 minutes. If we travel at 40 mph (64 kph), the trip time is reduced to 15 minutes. So a 33% increase in speed produces a 25%, or 5 minute, decrease in time.

For the rural or interstate example, let's take a nominal 180 mile, 2.8 hr, drive with a posted speed limit of 65 mph (105 kph). If you drive at 80 mph (129 kph), assuming you have no interruptions, the trip takes about 2.3 hours. So by increasing your speed about 23%, you gain 30 minutes—about a 19% reduction in travel time.

There are several factors that together promote speeding as a temporary or constant driving behavior. If you know anyone in the law enforcement profession, they can tell you a long list of reasons drivers give them while they are filling out their traffic tickets. Speeding does have consequences, maybe not on your next trip but in the long run, that are not worth any amount of travel time saved. Aside from the considerably reduced fuel efficiency and vehicle wear, speeding, a voluntary behavior, is a factor in about 1/3 of all fatal accidents in the United States. For example, in 2008, speeding was a factor in 31% of all motor vehicle crashes [1, 2] in the United States, killing 11,674 people [3]. We will further address the issue of speeding in Chapter 12 but for now it suffices to say that speeding has a definite consequence: the increased likelihood of fatality or serious injury.

Both of the calculations above show a mediocre time savings return for the percentage increase in speed (and fatality risk). From a return on investment (ROI) point of view, the percentage of time saved is less than the percentage increase in speed, so the ROI is less than one, signifying a bad deal. No one in his or her right mind would undertake an activity where the quantified benefit is less than the investment. Another way of stating the return on investment is to say that you gain 30 seconds for each mph of speed increase for the urban scenario and for you gain 2 minutes for each mph of speed increase the rural example. In my opinion for both examples, if you want to arrive 5 minutes earlier, leave five minutes earlier.

Logic and math aside, speeding is common, and in some places more the norm than the exception. Drivers are willing to accept the increased likelihood of fatality and injury to enjoy the expectation of shorter travel times. The concept of the "willingness to accept" risk, sometimes abbreviated in the literature as WTA, is a basic element of the risk-benefit decision-making process. What makes people willing to accept (or reject) an increase in risk, however small, is a very complex subject. Sometimes, as with speeding, people make the decisions for themselves, while other times, the decisions are made for us, as we'll see later in this chapter.

Along with the increase in probability of death and injury that accompanies speeding, there are the costs associated with speed-related deaths. At this point we leave the individual point of view to analyze the behavior on a national or societal level. According to the National Highway Traffic Safety Administration, the annual economic costs of speed-related accidents in the United States are more than $40 billion [4]. Speeding is a behavior that drivers engage in voluntarily. Drivers are willing to accept the risk and gamble (yes, "gamble" is the right word) for the shorter travel time benefit. Yet hospitals, insurance companies, insurance buyers, law enforcement agencies, and, most importantly, over 11,000 families and friends pay the price each year. And that number shows just the fatalities.

Now we have the ingredients of what may seem a macabre statistic but is actually an important tool in the valuation of risk management alternatives at societal levels. The statistic is called the "value of a statistical life" (VSL). The word "statistical," as used in the phrase, means we are not valuing individual lives but are computing valuations based on aggregated data. Here is one example of the VSL calculation: If 11,674 people are killed in speed-related accidents, with the associated economic costs estimated at approximately $40 billion, the VSL ($/death) estimate is about $40 billion/11,674 or close to $3.4 million per life lost.

This amount reflects the societal costs of speeding-related deaths. The main application of these seemingly morbid VSL statistics, in this case speeding-related deaths, is to value the cost savings of programs designed to improve safety. It's hard to quantify the value of safety or improvement programs because averted deaths or injuries can't be included quantitatively in the direct accounting of multimillion dollar program costs. This is where VSL estimates are applied. For a given improvement program, the product of the value of a statistical life and the forecasted number of deaths and injuries averted provides a quantitative estimate [5] of the society-level monetary savings. These types of calculations are often used in prioritization of projects during budget planning and prioritization.

For example, let's assume that there is a program plan in Congress to reduce the national speed limit to a maximum of 55 mph (89 kph). The plan also calls for new and aggressive law enforcement activities costing a total of $15 billion with the reduction in speeding estimated to save 5,000 lives a year. Using the VSL estimate for speeding deaths, saving 5,000 lives translates into (3.4 Million $/life × 5,000 lives) or $17 billion in averted costs. So from an economic perspective, the program shows a positive return and looks like a good effort to pursue. From a purely mathematical perspective this is true, but there is another issue to be included in the decision-making process. Are the recipients of this program, who would enjoy as a group the life-saving results of the program, willing to pay the individual increase in travel time? The "willingness to pay" or "WTP" issues are often the key variables in the implementation or nonimplementation of risk management actions. In the case of speeding, the well-documented life and cost saving attributes of lower speed limits [6] are not enough for the public to pay the price of the increase in travel time. Thus even though the program shows a positive financial return, it may not be passed by Congress if the elected officials want to be re-elected.

Not all willingness-to-pay decisions are this ambiguous. Consider, for example, your car insurance premium. This amount is within your willingness-to-pay threshold, because you perceive the transfer of risk to the insurance company an equal or greater benefit than the dollar amount of the premium. Or from a less esoteric perspective, you purchase the insurance to stay out of court, as having basic car insurance is required by law in the United States. If your premium was enormously high, above your willingness to pay, you might decide on an alternative method of risk reduction such as posting your own bond, not driving, or, as almost 14% of U.S. drivers [7] nationwide do, just not buying car insurance.

The public's willingness-to-pay criteria are crucial parts of risk management because they affect people in the most sensitive spots: their wallets and their behavior. The willingness to pay measures the public's perceived change in wellbeing that results from reducing the risk associated with a certain activity or object. By stating the value in terms of money, it measures how much money a person is willing to divert from buying other goods and services in order to receive a perceived reduction in risk. Totaling this amount over an entire population is another way to estimate the statistical value of a life or an injury, depending on what risk is being reduced.

There are other variables that can make the valuation more specific, but the increased precision leads to more complexity. For example, if different age groups are exposed to the risk, then the statistical value of a life should be given in terms of life-years to differentiate between the older and younger populations being exposed. Also, since the incomes of the populations exposed to different activities vary, perhaps the incomes of the groups should be addressed. Safety enhancements or risk reduction projects that are expected to save lives in the future can also be valued by discounting the future savings to present value dollars. From a financial perspective, one could discount the statistical value of a life to account for the time delay between when the money is spent and when it is saved. All of these techniques have been applied to life valuation studies but here we will just look at how age can influence life valuation results. Society generally values the lives of young people, especially children, more highly than the lives of adults [8], so it makes sense that statistical life valuations should take this into account.

Auto death rates illustrate this variability. The death rate of 16- to -18-year-old drivers is more than twice that of 30- to 55-year-old drivers [9]. Every parent intuitively knows this from the shock of paying his or her auto insurance premium after his or her teenager begins to drive. Some of the cost is due to higher accident frequencies but some portion can be attributed to the higher total cost of death for people in this age group.

Whether it's speeding, buying car insurance, or a host of other activities, there are often controversial aspects accompanying risk management decision making where the math, science, engineering, and societal benefits are simply overshadowed by the subjective willingness of the participants to pay for the individually small decrease in risk. Speeding is common. For car and truck speedsters, assuming the outcome is uneventful, speeding is intentional and the benefit is clear and immediate. Most people are winners when they roll the dice in this gamble. And you can enter the speeding game anytime—even for a short duration. It is also a risk exposure that appears to be more prevalent with (but not unique to) young drivers. Driving risk by age is an example wherein the insurance companies have run the numbers and the net cost is distributed among the insureds. But just to focus on how much age can change the valuation results, let's consider a more illustrative example.

In simple terms, (using round numbers for the sake of easy-to-view calculation) if we assume people live 80 years, saving the life (or averting the death) of a 20-year-old person produces 60 life-years compared to saving the life of a 60-year-old person that saves only 20 life-years. *Thus saving the 20-year-old person has 3 times the life-year value of saving the 60-year-old.* Now you might say that all lives are equally valuable and age shouldn't matter. This is true in a world with unbounded resources. But to manage finite resources, especially in setting public policy, decisions should be made to determine the best, though not a perfect, strategy to pursue. Policies that influence public health undergo considerable ethical scrutiny by all affected parties. When it comes to life valuations, even the way the questions are worded was found to make a major difference in perception. I mention this because these questions often invoke emotional discussions, and the subtle wording of questions attempting to obtain public sentiment on issues can be overlooked. For example, in research where vaccine policies for a flu epidemic allocation were described in terms of lives saved, people judged them on the basis of the number of years gained with no reference to age. However, when the policies were described in terms of lives lost, people considered the age of the policy's beneficiaries, taking into account the number of years lived to prioritize young people for the vaccine [10]. Other studies in different cultures have obtained similar results, indicating that the value of a saved life decreases rapidly with age and that people have strong preferences for saving life-years rather than lives. Overall the research clearly shows the importance of the number of life-years saved in the valuation of life [11].

To illustrate the application of the "life-years" metric and show a comparison to lives saved, let's take a look at a hypothetical flu epidemic that is threatening a nation, region, or city. The current level of epidemic has produced the number of deaths by age group shown in Figure 4.1. The number of deaths in each age group is divided by the number of people in the total population. It appears that the disease affects the younger age groups almost the same amount. The 45–64 age group shows nearly double the number of deaths as compared with the next youngest group, 25–44. And clearly, the elderly, 75+, have experienced the greatest number of lives lost.

In Figure 4.2, however, we divide the number of deaths in each group by the number of people in the age group to produce a death rate (number of deaths per 100,000 in age group population) and we see a different picture. The epidemic is more likely to kill the young and senior citizens (65+), with the highest death rate occurring in the elderly.

Assuming no additional information, from this viewpoint the data clearly suggest that the 65+ groups should be treated first, followed by the very young. The

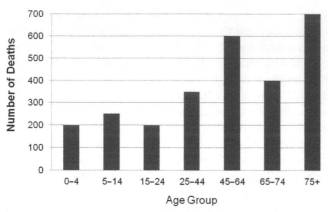

Figure 4.1   Number of epidemic-related deaths by age group.

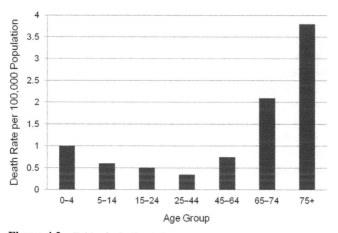

Figure 4.2   Epidemic death rate by age group.

transition of the number of deaths by age group (Fig. 4.1) into death rates taking into account the actual number of people in each group (Fig. 4.2) shows a different picture of how the epidemic affects different age groups. Figure 4.2 focuses on saving lives since we are looking at mitigation actions to reduce the number of deaths. It tells policymakers we should vaccinate the elderly first and then the very young.

Now we go back and start again from the data in Figure 4.1. If we assume that everyone will live to 80 years and compute the life-years lost for the people who died in each group we get a different picture, shown in Figure 4.3.

The vaccine allocation strategy conclusions drawn from each figure are different. Figure 4.3 tells policymakers they should apply resources to save the young.

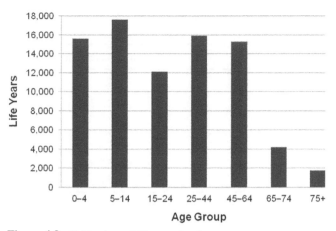

Figure 4.3   Epidemic total life years lost by age group.

The 65+ age groups that received the highest priority using the death rate metric become the lowest priority for vaccine allocation using the life-years-remaining valuation.

This example is designed only to illustrate potential differences in valuation results possible between life-years and lives. It was not taken or derived from any real public policy analysis so there is no implicit message. The point is that the metric, life-years, can change the priority of risk mitigation actions. It depends on what the public values: "lives saved" or "life-years saved." By this example I am not suggesting that risk mitigation decisions are this clear cut. In practice, there is usually a large list of additional issues to consider specific to the threat under analysis, such as the mitigation costs, the expected success of mitigation actions, population demographics, and a host of other factors.

The vaccine allocation example also highlights an important economic issue common to valuation computations. Statistical value of a life calculations done for different age groups are not directly comparable by themselves since the economic value is computed at different times. Just like in annuity calculations, to make the valuations equivalent, the age-dependent values can apply a discount factor recognizing the time value of money. That's right, just like your 401(k), bank accounts, and home mortgage rates, the statistical value of a life (VSL) or a life-year (VSLY) depends on the economy. The specific discount rate applied, 3%, 4%, or another value, enables monetary values to be compared equally in time. You could argue that in a specific case, VSL or VSLY valuation results are inaccurate because of an incorrect discount rate, but you cannot argue with the general methodology that enables the direct comparison of time-based financial information.

Since the notion of paying someone for loss of life and limb is common in the workplace, it shouldn't surprise you to learn that the U.S. government has its own numbers. In fact, it has lots of them depending on where you look. A concise description is found in a U.S. Department of Transportation Memo, "Treatment of Value of

Table 4.1   Value of a Statistical Life Estimates Applied in Cost vs. Benefits of Federal Regulations

U.S. government component	Value of a statistical life ($million)
Office of Management and Budget	5.0
Food & Drug Administration	5.0–6.5
Consumer Product Safety Commission	5.0
Environmental Protection Agency	5.5–7.0
Department of Labor	5.5–7.0
Department of Agriculture	5.0–6.5
Department of Transportation	5.8

Life and Injuries in Preparing Economic Evaluations" [12]. The memo describes the history and updated statistical life valuation for several governmental components. It also clearly states that the values are imprecise results that represent the current state of knowledge and that higher numbers could be used in the future. A summary of some of the government's life valuation results are listed in Table 4.1.

These figures may make it seem like the government is making important life-saving decisions for you that don't necessarily match your views. But remember, we're dealing with life, death, and injury in a population, not at the individual level. At this level of statistical analysis, no one is making decisions about your individual value. The government is doing exactly what the taxpayers are paying it to do: looking out for the common good of the people, not for any single person. For a good compilation of international VSL estimates, see François Bellavance's work [13].

The actual numbers used for the value of a statistical life or cost savings for an averted fatality are not really the issue. As researchers continue to probe the population with surveys and new technologies to make our world safer and more reliable at lower costs, the numbers are going to change. The point is this: life saving and injury saving measures have costs and savings. Yet the question remains: how can we manage our societal risks with the societal benefits?

Each budget year, our government agencies must decide whether or not to fund certain programs. Because some are funded, safety is enhanced and lives may be saved. However, some safety improvements are not funded and as a result, some people may die. These are difficult ethical and political decisions. Here is how the value of a statistical life and the public's WTP fit into this decision-making process.

Suppose you have to review a project proposal for its importance or priority for funding. Let's set our hypothetical total project costs at $60 million. Since the objective is to reduce accident severity and frequency, the project description estimates that the property damage and other direct savings is $10 million with the saving of 10 lives each year. Here are the calculations in table form:

Total Project Costs:	$60 million
Property Damage & other Direct Savings:	$10 million
Net Project Cost:	$50 million
Annual Number of Lives Saved Estimate:	10
Net Cost per Fatality Averted:	$5.0 million project VSL

These figures produce a statistical value of a life of $5.0 million, about the same as the $5.8 million WTP that could be applied as the standard for project valuation. In practice, the most current value would be used. The value is only used as a guide to judge the cost-effectiveness of the intended work.

Here's another way of actually measuring the project benefits, which this time contains the public's WTP inside the calculations.

Benefits:	10 Lives Saved × $5.8 million/statistical life:	$58 million
	Property Damage & Other Direct Savings:	$ 6 million
	Total Project Benefits:	$ 64 million
Costs:	Total Project Costs:	$ 60 million
	Net Project Benefit:	$ 4 million

One of the more subtle aspects of estimating lives saved is that reducing the number of deaths may be coupled with increasing the numbers of injuries. Depending on the situation, injuries can cost more than deaths. Consider the airline industry. Before the last two decades, passengers in most crashes either walked away or were killed. As wide-body jets and numerous other safety-enhancing improvements were incorporated, aircraft accidents became more survivable. Now, instead of accidents killing all of the passengers, passengers are more likely to suffer nonfatal injuries. Since lives saved and the value of a statistical life is computed, so should the increasing number of injuries, the cost of these injuries, and the public's WTP for injury aversion.

On the other hand, any project or program that has health and safety components should be valued on the ability to reduce not just deaths but also injuries. For example, consumer product safety programs are often designed to reduce injuries for situations where accidental death has not occurred. For these situations, injury reduction valuation is a necessity since death may not be the primary or even secondary issue.

The Department of Transportation memo that produced the valuation results for quantifying the cost savings for averted deaths also has a schedule quantifying the cost savings of averted injuries. The injury valuation results are shown in Table 4.2. This schedule (or other schedules with similar information) is important for quantifying the total value of initiatives, activities, programs, and projects that have health and safety implications.

MAIS is an abbreviation for Maximum Abbreviated Injury Scale. It is one of several classification systems used to categorize injuries [14].

Table 4.2 U.S. Department of Transportation Estimates for Valuation of Injury Prevention

MAIS level	Severity	Fraction of VSL	Injury statistical value
1	Minor	0.0020	$ 11,600
2	Moderate	0.0155	$ 89,900
3	Serious	0.0575	$ 333,500
4	Severe	0.1875	$1,087,500
5	Critical	0.7625	$4,422,500
6	Fatal	1.0000	$5,800,000

Does this process of monetizing death and injury seem cold and callous? Maybe it is, depending on the use and user. These types of calculations are not unique to governmental policy. Different commercial sector elements may have their proprietary versions for internal strategy analyses, but let's recognize two absolutely noncontroversial facts.

Fact #1: Cost-benefit analyses are an essential part of any rational system for work prioritization whenever the number of projects exceeds the available financial resources.

Fact #2: Cost-benefit analyses are imperfect. They represent a person or a group's collective wisdom about the future—no more, but no less.

And not all programs need to pass the financial test. There are situations where a project or program's statistical value of a life may be greater than the public's WTP, yet the work should be done. This is the case when, for example, Congress mandates certain actions or the activity is of significant public concern. High-profile accidents and the silent, latent effects of chemical toxins are examples where the government may disregard the mathematics of cost-benefit analyses to implement policies and laws to protect the general public.

The legal profession has been dealing with questions about the value of life and limb for decades. In the case where a worker dies or is injured in the workplace, workers' compensation statutes, depending on the state, define benefits according to a well-documented protocol that includes the past year's earnings, family size, age, and many other factors. If there was negligence involved then a tort action is filed with the lawyers fighting the case, usually for larger sums of money.

There is one interesting aspect about workers' compensation. The law seldom limits the amount paid. Payments are allocated based each case's merits. In this system, each body part has a dollar amount reward in case a worker loses it. I have a friend who lost the partial use of one of her fingers while at work. When the surgery was over and her partial permanent disability was established, her compensation was a check for about $2,000. If you live in the United States and take the time to look at your state's workers' compensation statues, you'll see the dollar values allocated for different injuries. One thing is certain, life and limb do have a statistical worth in the developed world. Figure 4.4 shows some examples of workers' compensation

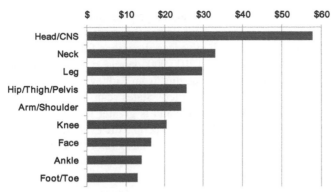

Figure 4.4   Workers' Compensation claims values by part of body (2005–2006) (claim amounts in $1,000).

CNS: Central Nervous System

Source: National Safety Council, *Injury Facts®*, 2008.

values for certain injuries. Keep in mind these are average values taken from National Council on Compensation Insurance files. Specific claim values, depending on the type and characteristics of the accident, could be much higher. These claim amounts include both the medical and indemnity awards. This is how much is paid, on the average, back to the worker for his or her on-the-job injuries. The claim amounts are representative total costs for injuries by body part.

You can replace leg, arms, feet, and certain joints with prosthetics, but what if you need a new kidney, heart, lung, pancreas, or liver, or an intestine splice? For these types of replacements, other valuation models are applied.

In developed countries where legal and administrative services are effective, life and limb valuation exists. In countries without these critical infrastructure elements, the statistical value of a life or limb reverts largely to individual financial wealth or social position. However, in many parts of the globe, there is another type of body part valuation that at least needs to be mentioned here: cadaver and organ transplant donations.

The growth of the medical community, pharmaceutical industry, and medical instrument makers depends on body parts for teaching and research. Medical schools need cadavers and body parts for training and research, the pharmaceutical industry needs human parts to test and research new drug effects, and medical instrument companies need samples to develop and test new products. After all, who would want to be treated by a physician who has no practical knowledge of body parts, take a drug that never has been exposed to human tissue, or be the first live human to have a new type of CAT scan. For all of these reasons and many more, the human body and its parts (fingernails, tissue, organs, limbs, torsos, and heads) are in high demand for transplantation and medical research.

Just as your car would probably cost three or four or more times its purchase price if you were to build it from scratch by buying parts, your body's worth is a lot more than the few dollars for its chemicals if you were to sell it piece by piece, organ by organ, or part by part. There are several reasons for the huge value of "pre-owned" human body parts. For organ transplants, the most obvious is the waiting period duration uncertainty for receiving organ donations for the critically ill. In the United States today, there are over 100,000 people registered in the organized process awaiting organ transplants ranging from kidneys and hearts to intestines [15] (The current totals for the United States can be viewed at www.organdonor.gov). The tragic reality is the demand is greater than the supply. About 130 people are added to the transplant list and 18 people die every day waiting for their turn [16]. You can understand why people in desperate need for organ replacements would be highly motivated, to say the least, to obtain what they need any way they can. International black market dealers have made and probably still are making money paying people to donate body organs and then selling them for over 10 times more [17]. And the market isn't only for organ transplants. Everything from fingernails to tendons is available for a price. There are laws against illegal marketing of body parts but the black market is a powerful international commercial marketplace [18].

The main reason for the black market is simply the demand is much greater than the supply. The tragic aspect of this issue is that there are plenty of organs potentially available but people choose not to donate them before their death. To alleviate this shortage, the United States and other countries have debated some innovative ways to increase organ donations by paying for them. One proposal presented by the American Society for Transplant Surgeons in a Congressional bill would provide a $10,000 tax credit in exchange for donated organs [19]. In a more recent detailed analysis done by Nobel Laureate economist Gary S. Becker, Ph.D. with coauthor Julio Jorge Elias, Ph.D., the authors estimate that a $15,000 payment for kidneys would increase the number of transplants by 44%, to about 20,000 a year. A $40,000 cash (not tax credit) payment to donors would increase the number of liver transplants 67%, to more than 8,600 a year. The analysis was based on the authors' estimates of how much it would take to convince a person earning the median American salary of $35,000 to become a donor, considering the actuarial value of a statistical life, the small risk of death in surgery, quality-of-life effects, and the value of the time spent in recovery [20].

The life-saving benefits of providing financial incentives for organs is offset by the far-reaching ethical issues in some lawmakers' and medical professionals' opinions. Families, hospitals, and others may be motivated to provide a lower standard of care for dying and life-threatened patients in expectation of the life-saving rewards obtained in harvesting the high-value organs. Another complex issue is the fine line between life and death that can exist with patients connected to life-support technology.

In some countries today, however, the financial incentives are working [21]. Iran provides a legal payment system for organ donations. In the case of liver donations, the organ donor receives $1,200 from the government plus one year limited health-care insurance coverage. The program was started in 1988 and literally eliminated kidney organ shortages by 1999. In March 2009, Singapore legalized a plan for paying organ donors but specific payment amounts in the neighborhood of

$35,000 are still being discussed. They also have an interesting nonmonetary incentive. Anyone can opt out of the presumed organ donation consent, but those who do this are assigned a lower priority on the transplant list should they ever need an organ in the future. Israel has a more flexible version of Singapore's model, implemented in 2010. Individuals are given points if they sign an organ donation card, placing them higher on the transplant list should they need an organ in the future. They also get more points for close relatives who also sign their organ donation cards. The United States is considering several methods but as of 2011, it has a long way to go. Consider the most common organ transplant—kidneys. In 2008 there were about 16,500 people who received a kidney. In the same year, nearly 5,000 people died waiting for one.

This type of valuation is not the output of a statistical or analytical analysis. In the dynamic, unconstrained, growing world of organ black market free enterprise, the valuation is simply and precisely what the market will bear. This is a glimpse into an area of life and limb valuation where the end result is good—the supply of needed organs and other body parts—but the means is, at best, highly questionable. In reality, while the lawyers, economists, and policymakers can calculate group-level life and limb valuations, there is almost a parallel world that has its own figures.

Now let's consider some specific and growing risk exposures where applying the life valuation measurements is much more complex than the simple arithmetic you have seen up to this point in the chapter. We will discuss two major issues and explore the risk factors where life valuation metrics apply and leave the math to the economists, policymakers, and ethicists.

Usually, if people are going to assume a risk exposure, they want to enjoy the benefits quickly. Speeding is one example discussed earlier where the risk exposure and benefits occur at the same time. It is a deliberate behavior that can be entered and left quickly by simply adjusting the car's speed. There is another set of human risk exposures that possess similar characteristics, albeit not as quickly as with speeders or as infrequently as with smoke detectors, but with a very high level of personal commitment: clinical drug testing.

Before we get into the details of this very active industry, let's start with a little background information. There is no argument that one of the benefits of our combined scientific and clinical knowledge is in the development of drugs to treat and sometimes cure disease. For pharmaceutical manufacturers, the time and expense to get new drugs to the public generally requires about 8 years and costs several hundred millions of premarket expense dollars. Add in the basic research for drug discoveries and you can understand the tremendous investment these companies have to make before they sell the first dose to the general public. As the global population continues to grow, there is no doubt that these companies have a healthy future.

Part of the product development process is testing new drugs' effects on humans in various ways. These tests are mandated by the U.S. Food and Drug Administration (FDA) and have specific guidelines to ensure the participants' safety. The FDA provides several guidelines but they are the first to point out that no drug is 100%

safe and there can be adverse reactions ranging from minor discomfort to death. The objective of the clinical research trials or "experiments" is to determine a drug's detailed safety and effectiveness profiles in the treatment of the identified disease or condition. FDA officials use this information as a part of their cost-benefit analysis, probably using some of the metrics we discussed in the speeding example, to make public release decisions.

Drug testing is normally performed in three phases prescribed by the FDA for admission into their approval process [22]. Phase 1 drug tests are on healthy subjects who are usually volunteers, although sometimes patients. The objective is to observe the metabolic effects, drug action mechanisms, and, of course, side effects. The testing experiments normally include anywhere from 20 to 80 subjects. In Phase 2 testing, the drug is applied to patients with a specific disease or condition to study side effects and drug effectiveness. These experiments have several hundred patients. Phase 3 tests the drug on several hundred to several thousand patients examining more side effects and detailing effectiveness data to forecast the drug's performance in the general public.

The willingness to accept risk issue for Phases 2 and 3 testing on medical patients is easy to explain. In a Harris poll conducted on cancer patients in 2000, 76% of the respondents participated in the study because they believed the trial offered the best quality of care for their disease [23]. Sure, there is risk in new drug experiments, but here at least the test subjects may receive the benefits.

Phase 1 clinical trials examine the first real application of a new drug in humans. After all, a drug isn't much good if it makes healthy people sick or worse so testing it, and, of course, the dosages, makes sense. However, you might ask, "How do drug companies find perfectly healthy people to become human testers?" The answer? They pay them.

Actually, the money isn't bad for a young, healthy person. To give you some idea how much money you can make, clinical Phase 1 testing compensation varies by the term of the test. The approximate amounts are listed here [24]:

3 days	$500
5 days	$1,000
10 days	$1,500
15 days	$2,250
20 days	$3,000
25 days	$4,000
30 days	$5,000

Phase 1 testing usually requires sequestering the subjects to a controlled environment for the test duration. During this time, clinicians provide a structured diet and living space to closely observe behavior while routinely analyzing blood, urine, fecal samples.

Some people do this for a little extra money and others do it as a profession to a varying degree. After all, what other job requires no special training or education, allows you to travel, to work part-time, and to make between $20,000 and $25,000 a year. The reason you can only work part-time is that a 30-day drug-free period is required to ensure your system is clear of the previous drugs before you begin a new test. Any synergistic reactions would negate the trial validity and, of course, might kill you.

There are several websites where you can find out more information about these studies. For example, www.gpgp.net (guinea pigs get paid) presents a global listing of medical and other types of testing trials. Any standard search using keywords like "list drug trials" will show you how active this industry is.

And it should be an active industry. Drug research represents a tremendous positive effort to improve and extend the quality of life for all people on this planet. For most of us, the research will help us at an unknown time in the future. For the sick, the research represents something immediate and very important: hope.

The fact that people, primarily young people, participate in this activity is an interesting but somewhat obvious characteristic of this age group. Most 20–30 year olds have little understanding of their fragile mortality. Test subjects in Phase 2 and 3 testing are at least patients so they can potentially benefit from the treatment. For Phase 1 testing, the primary benefit is the money. Whether it's $600 for a 3 day study or $30 for a blood sample, people are willing to accept the risk, inconvenience, and sometimes pain for a financial reward. Sure the acceptable level of remuneration varies by person and situation, but it seems like there is always a breaking point (price) at which people are willing to accept risk. However, in the case of medical testing, the test subjects can be exposing themselves to more risk than they know and test subjects have died under conditions in which the tests have been found to be a contributing, if not the primary, factor [25].

I have prefaced this discussion with talking about the FDA safety guidelines but the guidelines and oversight capabilities pale compared to the medical testing industry's size. In September 2007, a report from the Inspector General of the U.S. Department of Health and Human Services concluded that FDA did not have a system for identifying all ongoing clinical trials. The report also noted that the FDA had checked less than 1% of the estimated 350,000 testing sites. This isn't too surprising—the FDA only has about 220 inspectors [26]. The large size, diversity, and dynamic activity of the medical testing industry make it impossible for the FDA or any governmental agency to provide firm oversight.

There are protocols to follow and medical studies generally do a good job in self-monitoring via their internal and external review boards which are responsible for the technical overview. However, quality can vary by site, especially for studies outside of the United States. One of the common problems with these studies is the ability to track the medical test subjects. People can (and do) travel from study to study and there is no current way to check their testing history. Clinicians rely on test subjects to tell the truth in filling out the forms. The system is not robust enough to guard against enterprising professional test subjects bending the 30-day rule or from falsifying other information. By doing this, subjects put the study's quality and their own health at risk. If the study's quality is compromised, we all can pay the price once the drug is released. But in the case of the test subjects, their willingness to accept risk, albeit for a price, is flawed because they do not know how the chemical residues in their bodies will interact with the current study drugs.

The globalization of our economy has seen the migration of many labor-intensive industries to countries with cheaper costs. The pharmaceutical industries are also following this trend. The major countries for medical testing are Brazil, China, Russia, and India. The pharmaceutical companies' clinical trial testing costs in these countries can be less than 10% of the costs in the United States [27].

The increase in medical testing in developing countries was encouraged due to the widespread adoption of best practice guidelines developed from the International Conference on Harmonisation of Technical Requirements for Registration of Pharmaceuticals for Human Use Good Clinical Practice (ICH-GCP) [28].

Using these guidelines, clinical researchers around the world now have the protocols that are the current best practice consensus for testing procedures, analysis methods, and results reporting. From a technical perspective this is good news. The outsourcing of testing to these countries offers the benefits of shorter testing times and lower development costs. Everybody wins—right?

Not quite. There is another side of this business model that's not as encouraging and in fact, is a major bioethics issue associated with medical testing in developing countries. These issues influence test subjects' willingness to accept risk and society's view of putting these people in such compromising situations, even if they agree.

To operate clinical trials in developed and undeveloped countries, pharmaceutical companies employ contract research organizations (CROs). These companies are on the ground with the infrastructure and the contacts required for acquiring medical test subjects quickly depending on the drug test requirements. The financial incentives for the test subjects can be intense drivers. The financial compensation for study participation may exceed a person's normal annual wages and participation in the trial may be the only source of treatment available for the patients [29, 30].

The willingness to accept risk is simple in concept, yet complex in practice. For medical trials in developing countries, the potential to bring new drugs quickly and cheaply to the market is due to countless thousands of people of all ages willing to act as human lab rats. True, sometimes these people participate for money, but often they do because it's the only or best choice of healthcare available. Thanks to the ICH-GCP, there are uniformity guidelines available, yet in practice the risk exposure of the participants may be higher than they know. Aside from the relatively large financial incentive motivation to accept risk, there are other more subtle risk exposures that are involved. For example:

1) Language and reading skill differences can be a barrier for test subjects' complete comprehension of their personal commitment.

2) Legal action against negligence is difficult to employ in developing countries.

3) FDA oversight is minimal.

The ethical and risk management issues of the growing medical testing industry will be with us for a long time to come. At the time of the writing of this chapter, there were 84,228 trials with locations in 171 countries listed on www.clinicaltrials.gov. This website, supported by the U.S. National Institute of Health, actively compiles detailed information on medical testing from around the globe. Check out what the current numbers are.

So you don't get the wrong idea here, not all "testing" involves drugs, or complex ethical issues. NASA's Test Bed Facility paid $17,200 for participants just to stay in bed for 90 days, except during the limited times required for testing. In bed, the test subjects had to lie in a slightly head down, feet up position. Lights were on for 16 hours and off for eight hours each day for 90 days [31].

Also, I'd like to extend my thanks to the 13,329 men and women, aged 45–84 years, who participated in the Copenhagen City Heart Study [32]. During the 16 years of follow-up testing, the doctors have concluded that moderate wine drinking is associated with reduced stroke risk.

The subject of human testing of new medicines illustrates the complexity of life valuation in practice. It highlights the common risk management decision-making ingredients: the test subjects, industry, and society. All share the benefits, risks, and complex ethical decisions. Yet this trinity is not unique to medical human testing and, in fact, is quite common in contemporary life. To give you some idea of how widespread these conflicting risk management issues are, let's look at an industry we absolutely need to support the level of food production required to sustain our growing world population: pesticide production.

The pesticide industry is another important business in which human testing has been and still is controversial. There always seems to be people willing to accept the risk of pesticide exposure testing for financial gain. While it's true they are providing society a benefit for their sacrifice I would bet that the money at the end of the test has a little to do with their commitment to humanity. The human test subjects have an interest in testing safety, everyone has an interest testing validity, and the pesticide company and its shareholders want to make money. You can understand the motivations of all players in this equation. They are common drivers of people, society, and industry. We are now going to look at human pesticide testing to identify the drivers that create the risk management and ethical issues. As you'll see, these concepts are actually inherent to a wide range of industries and situations— more than you might think before you read on. But before we get into that, here's a primer on why the pesticide business is so important.

Pesticides are a diverse group of chemical compounds consisting of insecticides, fungicides, herbicides, and rodenticides. They help farmers dramatically increase crop yields and help limit the spread of disease. As we'll discuss in more detail when we talk about food, without these compounds it would be difficult to produce sufficient food to support the current world population. As the world's population grows, this industry will also grow in monetary value, societal value, and size.

These compounds are designed to kill. Remember the suffix "-cide" is derived from Latin and means "the act of killing." The objective is to develop environmentally friendly chemical compounds that kill their targets while being harmless to humans. That's the theory, but in practice the ideal is rarely achieved. The range of health effects include permanent and partial damage to nervous systems, lungs, reproductive organs, the immune system, and the endocrine (hormone) system [33]. So the simple, irrefutable conclusion is that we need pesticides to support human quantity and quality of life, but they do possess adverse health effects.

As you might expect, in the U.S. government, there is a long list of federal regulations governing pesticide use; we will page down to the Food Quality and Protection Act (FQPA) that was unanimously passed by Congress in 1996. This legislation fundamentally changed how the government regulates the pesticide industry and gave the Environmental Protection Agency (EPA) explicit responsibility to make human health the primary goal of their pesticide-related policies. The law further required the EPA to reassess 9,721 current pesticide residue tolerances by 2006. The difference in residue tolerances for infants and children compared to adults was a major concern. The EPA was charged with the task of studying this issue and publishing testing guidelines for the industry. This work required the collection of better data on food consumption patterns, pesticide residue levels, and pesticide use [34].

Because of the uncertainty in pesticide dose-response data in humans, the law requires a 10 fold additional reduction in tolerances as a safety factor to ensure the safety of infants and children, unless reliable data are available that show that a different factor can be used.

The EPA has historically accepted studies from private companies for pesticide testing and up to 1996, rodent studies were the standard for pesticide dose-response analyses. These tests were done to establish thresholds for symptoms otherwise known as NOELs, or NOAELs, short for No Observed Effect Levels or No Observed Adverse Effect Levels. In precise terms, NOEL is defined as a dose level at which there are no statistically or biologically significant effects attributable to the pesticide under test. Pesticide doses are first tested in rodents to determine the NOEL. Then this level is reduced by a factor of 10 to account for the differences between rodents and humans. That number is divided by a factor of 10 again to account for variations in among humans. This procedure has been in place for several years and is standard practice in the United States. Traditionally, the NOELs determined in pesticide tests using rodents are divided by 100 to determine tolerance standards for humans.

However, after the passage of the FQPA, the law required an additional child-protective safety factor of 10 to the previously established NOELs unless human test data exist that would negate the need for this reduction. This is the industry's motivation for human testing: unless they can prove through tests that their pesticide products are safe at 1/100 or 1% of rodent levels, the EPA will require levels at 1/1000 or 0.1% of rodent NOELs.

The uncertainty of pesticide dose-response analysis is an active biological research issue around the world today. The 1.0% or 0.1% rodent dose reduction represents the current scientific consensus of safety today. This is the best we can do. However, when it comes to testing, the industry has shown that its interests have not always been aligned with those of test subjects or society. The deficiencies (both ethical and technical) in human pesticide testing were made public three years after the FQPA in a landmark report called the English Patients [35]. The report cited several studies that failed to meet scientific and contemporary ethical research standards and it restarted the serious review of human pesticide testing.

From 1999 to 2010, government's reaction on this issue has been a dynamic sequence of risk-based decisions or, in plain language, policy reversals. In an effort to protect society the Clinton administration in place in 1999 directed the EPA to reject any pesticide industry data that included human testing. In November 2001, the first year of the Bush administration, it changed that policy and stated that human pesticide test data would be accepted. And in December 2001, the EPA issued a moratorium on human tests that was allowed to expire in 2003. The controversy has continued [36] ever since.

You can expect to hear more about pesticides and human testing throughout your lifetime. Just as our population grows, so does our scientific and biological knowledge. The wild card in this equation is the ability of pesticide manufacturers to create new products at a faster rate than our ability to understand their effects on humans. So far it appears society's willingness to pay for quality and quantity of life is greater than the value of the statistical lives adversely affected by pesticide exposures. But as you can see from this short discussion of human testing and pesticides, while there is no ideal solution in reality, the VSL, WTP, and WTA characteristics do provide a systematic framework to help make the best decisions for individuals, society, and industry.

I hope all of this hasn't been a remedy for (or the cause of) your insomnia. This discussion should help you understand that risk analysis and the management of risk are much more than mathematical exercises. There is no way for us as individuals, as a society, or as a global community to put all of the ingredients of risk into an equation and compute "the answer." Answers to these questions on the valuation of life and limb are not in the back of any book.

DISCUSSION QUESTIONS

1. The National Fire Protection Association reported in 2008 there were 403,000 home fires in the United States, claiming the lives of 2,755 people (not including firefighters) and injuring another 13,560, not including firefighters [37], at a cost of $7.5 billion each year [38]. Furthermore, 40% of the fire deaths occurred in homes with no smoke detectors [39].
   a. Estimate the statistical value of a life and of a life year.
   b. Discuss the public's willingness to pay for requiring smoke detectors in residential homes.
2. Examine the VSL, WTP, and WTA issues for a risk exposure of your choice other than for auto-related exposures. Include life-years in your analysis.
3. What risk-based alternatives would you suggest to reduce pesticide effects on humans?
4. Discuss the pros and cons of instituting the Iranian system for organ donations from a VSL perspective. How much and in what ways should organ donors be compensated?

5. The content of this chapter has shown that ethical issues are a critical part of the analysis when considering societal risk management issues. Since risk is future's uncertainty, look into the future and identify and describe the top five critical life and injury valuation and ethical issues for the next decade.

Case Study: The Value of Wrongful Birth and Wrongful Life: The Monetary Value of a Life with Defects versus No Life at All

This case study discusses the value of life from a direct, nonstatistical perspective related to some tragic situations that can be associated with birth. The lesson here is that while economists and risk analysts compute statistical life values, they are not the only ones doing this type of math. The legal system is also involved. Our society is based on laws and the value of a life, statistical or specific, is strongly influenced by how the courts treat this extremely difficult issue.

During the past few decades, wrongful birth and wrongful life claims have gained increasing recognition as avenues for imposing civil liability on third parties for undesired births and lives. The plaintiffs in such lawsuits claim life itself, that is, being born, as the injury. Although such claims cover a broad spectrum of factual situations, it's reasonable to broadly categorize them into illegitimacy and malpractice cases. Illegitimacy cases usually involve children born out of wedlock who claim injury by the virtue of the illegitimacy stigma. For example in *Zepeda v. Zepeda* [40] a child sued his father alleging that the father had fraudulently promised to marry the mother, and that the father's misconduct had caused the child to suffer the dishonor of "life as an adulterine bastard" [41]. The court refused to recognize such a claim and did not award any damages.

Courts across the country typically refuse to award damages in illegitimacy cases as it's nearly impossible to put a dollar value on, for example, the pain and suffering associated with being an illegitimate child. The situation is not same with medical malpractice claims. The courts have been more accepting of medical malpractice wrongful birth claims, mainly because of the legal system's evolution and the advancements in medical technologies. Specifically, the Supreme Court's confirmation of parents' constitutional right to prevent the birth of a child (*Roe v. Wade*, 1973) and use contraception (*Griswold v. Connecticut*, 1965) laid the foundation for imposing a legal duty on healthcare providers to disclose all information relevant to the parents' decision to allow or prevent their child's birth. Finally, advancements in medical technologies make it possible to predict the occurrence of genetic defects and to diagnose abnormalities in the unborn fetus, allowing the courts to more clearly define the contours of this duty to inform.

The identity of the plaintiff is the distinguishing factor between wrongful birth and wrongful life claims:

Wrongful birth refers to the claim for relief of *parents* who allege they would have terminated the pregnancy but for the negligence of the physician charged with counseling the parents as to the likelihood of giving birth to a physically or mentally impaired child [42].

Wrongful life, on the other hand, refers to the claim for relief of the *child* who alleges that the physician's breach of the applicable standard of care prevented an informed prenatal decision by the parents to avoid his or her conception or birth [43].

In a wrongful life claim, therefore, the child is essentially alleging that but for the physician's negligence, he or she would not have been born to experience the pain and suffering attributable to their disability.

Many states allow wrongful birth claims brought by a disabled child's parents whereas only a few states (e.g., New Jersey and Washington) allow claims of wrongful life brought by or on behalf of a disabled child. This is important because in many instances, the parents' claim is time-barred by the relevant statute of limitations and the child and, thus, the family, are left without any relief. The reason for this discrepancy is that in case of wrongful life claims, most courts hold that it's impossible to rationally determine that a child was actually damaged by the birth because to do so "would require weighing the relative benefits of an impaired life versus no life at all" [44]. New Jersey courts, which allow a child to bring a wrongful life claim to recover the extraordinary medical expenses associated with the disability, reason that "the right to recover the often crushing burden of extraordinary expenses visited by an act of medical malpractice should not depend on the wholly fortuitous circumstance of whether the parents are available to sue" [45]. The New Jersey Supreme Court, explaining the rationale for its decision to allow wrongful life claims, noted:

> *The foreseeability of injury to members of a family other than one immediately injured by the wrongdoing of another must be viewed in light of the legal relationships among family members. A family is woven of the fibers of life; if one strand is damaged, the whole structure may suffer. . . . When a child requires extraordinary medical care, the financial impact is felt not just by the parents, but also by the injured child. As a practical matter, the impact may extend beyond the injured child to his brothers or sisters. [46]*

As for the numerical value of available damages in wrongful birth claims, most courts allowing a claim of wrongful birth have limited damages to the extraordinary expenses required for the care and treatment of the disabled child [47]. Such damage calculations do not consider expenses associated with the raising of a healthy child and instead include institutional, medical, hospital, and educational expenses necessary to properly manage and treat the child's disorder. Moreover, in states where a claim of negligent infliction of emotional distress is available, parents may also be able to recover damages for the emotional distress associated with raising a disabled child (Texas, for example, does not recognize a claim for negligent infliction of emotional distress). U.S. courts appear unwilling to extend damages to the disabled child's potential lost wages mainly on the basis that such damages cannot be calculated with reasonable certainty.

The wrongful birth or life medical malpractice claims appear to be limited to those arising out of the physician's alleged negligence taking place before the birth of the child. For example, parents and children have brought wrongful birth or life claims arising out of allegedly negligent resuscitation efforts. In *Steward-Graves v. Vaughn* [48], the child was born prematurely and without a heart rate. The attending doctor immediately began resuscitation efforts checking for a heartbeat every two minutes. The physician did not give up. It was after 24 minutes of fighting for this child's life that the heart began to beat. The child survived but suffered severe health problems. The parents sued the doctor for malpractice in an attempt to recover money to help support the child's long-term care. They claim that if the attending doctor would have informed them about the time-critical need to resuscitate and the risk of long-term health problems, they would have chosen to stop the treatment and let the premature, nonbreathing baby die. The court held that the doctor did the right thing:

We recognize the complex issues involved in withholding life-sustaining treatment. In this case we are asked to recognize the right of parents and infants, through their parents, to hold healthcare providers liable in negligence for failing to withhold resuscitative medical treatment from an infant born without a heartbeat. We decline to do so. Rather, we hold that the trial court here properly dismissed the negligence claim by (the) parents because (the physician) had no duty to obtain their consent before resuscitating him. [49]

Looking at the way the U.S. courts treat these claims paints a complex picture of our society's willingness to provide relief for the injured and to impose liability on the conduct causing the injury.

In *Steward-Graves v. Vaughn,* the court's ruling suggests the monetary value of a child's life born with health problems either directly or indirectly caused by 24 minutes of resuscitation efforts is no different from those of a child born without any issues whatsoever, that is, neither parents are entitled to any monetary compensation arising out of the child's birth. The court ruling implies that both types of births are equal and the added healthcare costs in the case of the disabled child are risks parents assume with childbirth.

Birthing risks are a tremendous liability for pediatricians and related healthcare professionals who are in the middle of this ethical, medical, and legal arena. Long-term medical and mental conditions either occurring during pregnancy or discovered at birth can result in extraordinary, extended care expenses. It's not surprising that parents seek financial relief. But in these tragic situations, doctors are not the ethical decision-makers who are going to weigh the parents' financial and emotional distress costs to care for the disabled child against giving a helpless infant a chance to live. What happened in *Steward-Graves v. Vaughn* was tragic and expensive for the parents but a life was saved. The doctors didn't know the extent of brain damage at that time. After an aggressive 24-minute fight for life, they just saw a baby breathe his first breath. Would you want your doctor to do anything else?

The court willingness to award damages in some cases and not others might change with medical technology advancements and evolution of the legal framework in this area. For example, when we better understand the reasons why a child is born prematurely, not breathing, and without a heart rate, then this knowledge may contribute to the courts' willingness to impose a duty on doctors to at least consult the parents about resuscitation efforts in cases in which it's likely that such efforts may be necessary.

Let's suppose these conversations someday do occur and medical advances give parents the ability to foresee these types of situations. As a parent (or possibly soon to be parent) how would you decide—your child's future life versus future expenses of $100,000, $250,000, $500,000, $1,000,000, or $5,000,000? Would your decision change depending on the size of the anticipated costs?

Questions

1. Contrast the beginning-of-life ethical issues with those associated with the end-of-life issues from a risk management perspective.
2. This case study indicates that the location or state of the legal claim can strongly influence the outcome. Examine the regional variation of wrongful death and wrongful life issues in another country of your choice.

ENDNOTES

1 National Safety Council, *Injury Facts*®, 2009, p. 99.
2 A crash is considered speeding-related if the driver was charged with a speeding-related offense or if racing, driving too fast for conditions, or exceeding the posted speed limit was indicated as a contributing factor in the crash.
3 National Highway Traffic Safety Administration, *Traffic Safety Facts, 2008: Speeding*. Washington, DC: U.S. Department of Transportation, 2009.
4 Ibid.
5 Estimate is defined as: "a reasonable, educated, mathematical guess," Dr. J'aime L. Jones, 1996.
6 "Higher Speed Limits Cost Lives, Researchers Find," *Science Daily*. University of Illinois at Chicago, July 18, 2009.
7 Insurance Research Council, "Economic Downturn May Push Percentage of Uninsured Motorists to All-Time High." January 21, 2009.
8 Olof Johansson-Stenman, "Does Age Matter for the Value of Life? Evidence from a Choice Experiment in Rural Bangladesh," Working Papers on Economics, No. 389, Sweden, October 2009.
9 National Safety Council, "Injury Facts®," 2008.
10 Meng Li et al., "How Do People Value Life?" *Psychological Science*, Vol. 20, No. 10, 2009, pp. 1–5.
11 John Hartwick. "The Discount Rate and the Value of Remaining Years of Life," Working Paper 1191, Queen's University, Department of Economics, 2008.
12 U.S. Dept. of Transportation Memo, "Treatment of the Economic Value of a Statistical Life in Departmental Analyses," http://ostpxweb.dot.gov/policy/reports/080205.htm (accessed February 14, 2010).
13 François Bellavance et al., "Canada Research Chair in Risk Management Working paper 06–12, The Value of a Statistical Life: A Meta-Analysis with a Mixed Effects Regression Model," *HEC Montreal*, January 7, 2007, pp. 46–47.
14 Leonard Evans, "Traffic Safety," *Science Serving Society*, 2004, p. 22.
15 http://www.donatelife.net/UnderstandingDonation/Statistics.php (accessed February 21, 2010).
16 United Network for Organ Sharing (UNOS)/Organ Procurement and Transplantation Network (OPTN), as of January 8, 2010.
17 Jeneen Interlandi, "Not Just Urban Legend," *Newsweek*, January 10, 2009.
18 Annie Cheney, *Body Brokers: Inside America's Underground Trade in Human Remains*. Random House, 2007.
19 Deborah Josefson. "United States Starts to Consider Paying Organ Donors," *BMJ*, Vol. 324, No. 7335, 2002, p. 446.
20 Gary S. Becker and Julio Jorge Elías, "Introducing Incentives in the Market for Live and Cadaveric Organ Donations," *Journal of Economic Perspectives*, Vol. 21, No. 3, 2007, pp. 3–24.
21 Alex Tabarrok, "The Meat Market," *Wall Street Journal*, January 8, 2010.
22 http://www.chemcases.com/cisplat/cisplat15.htm (accessed January 16, 2010).
23 http://www.fda.gov/Drugs/ResourcesForYou/Consumers/ucm143531.htm (accessed January 14, 2010).
24 http://www.paidtotakedrugs.com/how-much-money-do-clinical-drug-studies-pay.html (accessed February 19, 2010).
25 http://www.geneticsandsociety.org/article.php?id=495 (accessed February 19, 2010).
26 "Human Lab Rats Get Paid to Test Drugs," http://www.msnbc.msn.com/id/23727874/ (accessed January 17, 2010).
27 J. P. Garnier, "Rebuilding the R&D Engine in Big Pharma," *Harvard Business Review*, Vol. 86, 2008, pp. 68–76.
28 http://ichgcp.net/ (accessed January 18, 2010).
29 G. J. Annas and M. A. Grodin, "Human Rights and Maternal-Fetal HIV Transmission Prevention Trials in Africa." *American Journal Public Health*, Vol. 88, 1998, pp. 560–563.

30 J. L. Hutton, "Ethics of Medical Research in Developing Countries: The Role of International Codes of Conduct," *Statistical Methods in Medical Research*, Vol. 9, 2000, pp. 185–206.

31 http://www.wired.com/wiredscience/2008/05/nasa-bed-rest-l (accessed January 16, 2010).

32 Thomas Truelsen et al., "Intake of Beer, Wine, and Spirits and Risk of Stroke," *Stroke*, Vol. 29, 1998, pp. 2467–2472.

33 Christopher Oleskey et al., "Pesticide Testing in Humans: Ethics and Public Policy," *Environmental Health Perspectives*, Vol. 112, No. 8, June 2004.

34 Environmental Protection Agency, http://www.epa.gov/pesticides/regulating/laws/fqpa/fqpa_implementation.htm (accessed February 6, 2010).

35 Todd Ken, "The English Patients: Human Experiments and Pesticide Policy," *Environmental Working Group*, June 1998.

36 Juliet Eilperin, "EPA Will Test Pesticides' Effect on Endocrine System," *Washington Post*, April 15, 2009.

37 M. J. Karter, *Fire Loss in the United States during 2008*. Quincy, MA: National Fire Protection Association, Fire Analysis and Research Division, 2009.

38 E. A. Finkelstein, P. S. Corso, T. R. Miller, et al., *Incidence and Economic Burden of Injuries in the United States*. New York: Oxford University Press, 2006.

39 M. Ahrens, *Smoke Alarms in U.S. Home Fires*. Quincy, MA: National Fire Protection Association, 2009b.

40 *Zepeda v. Zepeda*, 190 N.E.2d 849 (Ill. App. Ct. 1963).

41 Id. at 851.

42 *Clark v. Children's Mem. Hosp.*, 391 Ill. App. 3d 321, 326 (2009).

43 Id. at 351.

44 *HCA v. Miller*, 36 S.W.3d 187, 196 (Tex.App.–Houston [14th Dist.] 2000).

45 *Procanik v. Cillo*, 97 N.J. 339, 352 (N.J. 1984).

46 Id. at 351.

47 See, e.g., *Clark v. Children's Mem. Hosp.*, 391 Ill. App. 3d 321, 327 (2009). Another issue is whether damages should be calculated according to the child's estimated life span or be limited to the period of the child's minority. Illinois allows recovery beyond the age of majority.

48 *Steward-Graves v. Vaughn*, 170 P.3d 1151 (Wash. 2007).

49 Id. at 1160.

Chapter 5

The Evolution of Risk

Who controls the past controls the future; who controls the present, controls the past.

—George Orwell, *Nineteen Eighty-Four*

The concept of risk is not new. It has been around since humans first began walking the Earth. Even though risk management is getting a considerable amount of attention today, I believe that risk assessment is a very special characteristic of people that hasn't changed much over the millennia. It is a fundamental element of our unique ability to rationalize. No matter how you make your decisions, risk always implies chance or uncertainty. People have always been intrigued by this notion even though they have dealt with it very differently. Let's start at the beginning and trace its evolution by going on a brief tour through the history of risk.

Even in prehistoric times it appears that the early people had some understanding of chance in their lives. There are cave drawings of an irregular piece of bone found in the foot of some vertebrates called the astragalus. This solid, irregularly shaped bone approximates the shape of a die. No one knows exactly how these early people used them, but anthropologists speculate they were part of some amusement [1]. The first documented application of the astragalus was by the Egyptians in 3500 B.C., when they were used commonly as dice for board games. Even though 3500 B.C. may seem like a long time ago, on an evolutionary scale it is not. 2000 years corresponds to 80 generations of 25 years each. The Egyptians were using astralagi only 220 generations ago. From an evolutionary perspective, this is just a blink of the eye. Aside from being a much younger population, these people had the same basic intellectual capacities, physical attributes, needs and desires as we do today. Analysis of their board games demonstrates that the Egyptians had developed the basic laws of physics, geometry, and mathematics. The notion of chance, while recognized, was not an integral part of their societal development. It appears that a very sophisticated, controlling culture, probably not totally unique for this time in history, essentially ignored the mathematical development of chance or risk. They

20% Chance of Rain: Exploring the Concept of Risk, First Edition. Richard B. Jones.
© 2012 John Wiley & Sons, Inc. Published 2012 by John Wiley & Sons, Inc.

had a random number generator, astralagi (or dice), but did not develop the mathematics it represented. To see how this could happen we only have to look at the cultural ideology of the time. The Egyptians, like other early cultures, had a pantheon of gods that governed certain aspects of life. If something unexpected occurred outside their boundaries of logic and order, the event was attributed to a god's will. In this way, while chance or elementary risk analysis was accepted and used for games and amusement, the ancient gods took over when it came to serious risk management and societal issues. I doubt that insurance companies would have had much business there.

Probably the oldest and most read essay on risk is "The Art of War" [2–4], written by Sun Tzu sometime between the 5th and 6th century B.C. The text is 5,600 words in length written in the form of thirteen short chapters, each on a different aspect of tactics. The widespread application of the book comes from the reinterpretation of the text for specific situations. Seldom, if ever, does a set words—let alone a book—have only one interpretation so you might ask why Sun Tzu's book is so different. One answer is the general nature of the writing of Chinese classics. They were intended for application in a wide variety of situations. The writing style is simple, rhythmic, and easily understood, making the content easy to remember. This was especially important since book printing technology wasn't discovered for many centuries.

Sun Tzu was a distinguished army general serving a regional king in a period of time where war and intrigue were prevalent. Scholars have little information about this unique person and most of his life, including exact birth and death dates, remain a mystery. What is clear is that he wrote what is most likely the oldest and most read treatise on strategy and risk management in the world. "The Art of War" was well known in Japan and China but it only entered Western cultures in 1772, when a French Jesuit translated the book into French. From France, translations appeared in Russia in the 1800s. The first English translation from Japanese was completed in 1905. In 1910, the original Chinese text was translated into English and since then, Sun Tzu's 2,500-year-old work has become a "must read" for military leaders, diplomats, politicians, corporate leaders, poker players, and anyone else interested in solving problems with the environment of uncertainty.

Control, law, and the will of the gods, although not necessarily in that order, managed to sustain cultures over centuries. This general viewpoint concerning chance and risk prevailed for a long time, if for no other reason than the fact that it worked. Still, even when the great astronomers discovered relationships between the stars and planets, the notion of probability and its implications were not considered. Although probability studies didn't flourish, the field of mathematics, supported by astronomy, logic, geometry, and other discoveries, did. The story of the great mathematician Pythagoras (530 B.C.) is a particularly interesting example of how people related mathematics to their lives. If you remember your geometry class, you'll recall that he is known for the discovery of a theorem that bears his name and states that the square of the hypotenuse of a right triangle is equal to the sum of the squares of the two other sides. While historians tell us that this relationship was actually discovered by Mesopotamian mathematicians around 1900 B.C. [5], Pythagoras gets the credit because of his stature in the mathematics community of his time and the many valuable contributions he did make to geometry and number theory.

Pythagoras was the head of a group that formed special beliefs based on the number 10. This group believed that whole numbers and fractions made with these numbers were references to the gods. Some people say the group actually was more like a religious cult whose members saw mystical or supernatural powers in numerical relationships. The number 10, though probably not divine, is a special number, beginning with the fact that we have 10 fingers and 10 toes. And $1 + 2 + 3 + 4 = 10$.

The Greeks searched for order in their science. You might think that the cult-like worship of numbers is an extremism, emblematic of our ancient past. Well, this type of behavior is still around today, albeit in a slightly different form. For example,

- Do you choose the numbers randomly when (if) you play the lottery?
- Is it coincidence that the Chinese started the Summer Olympic Games in Beijing in 2008 on August 8 at 8 P.M.?
- Why is it some airliners don't have a row 13 and some hotels don't have a 13th floor? What about Friday the 13th?
- The number 9 is a common choice for a lucky number. After all, September 9th is the 252nd day of the year and $9 = 2 + 5 + 2$.

Most of us seem to have at least one number we call lucky or unlucky. In a sense there's a little Pythagoras in all of us.

You can understand this mindset a little by looking at the kind of relationships Pythagoras found in these numbers. Look at the example of music, which we know can stimulate emotions. Pythagoras discovered special relationships [6] between different sound pitches and the ratios of whole numbers. For example:

perfect third	5 : 4
perfect fourth	4 : 3
perfect fifth	3 : 2
perfect octave	2 : 1

As a side note, you can't use these relationships to tune your piano. We use the octave relationship, 2:1, and the others to a lesser degree or not at all. Ask a piano tuner! Of course, Pythagoras didn't know this because piano-type instruments hadn't been invented yet.

The key point here is that the Greeks had an open society not restricted by the religious sector, and yet they chose again to search the universe for order but leave the laws of probability, chance, and risk management to their set of gods. The mathematicians, scientists, and philosophers searched their respective worlds as well for the fundamental, almost mystical order they believed drove their lives.

In mathematics, the Greeks could not accept the "disorder" associated with irrational numbers like $\sqrt{2}$ or those with an infinite number of nonrepeating digits. Any number or fraction formed by a ratio of whole numbers (such as 1, 2, 3, 4, 5, and 6) was acceptable by the Greeks since even though there may be an infinite number of digits, there is a repeating pattern, bringing order to the sequence.

To some extent, advancing the laws of probability required some new tools. Roman numerals were sufficient for basic arithmetic operations and the things Pythagoras wanted to do, but can you imagine writing $27 = 3^3$ as XXVII = III^{III}? It appears that the major change occurred with one of the many technology transfers of the Crusades. The so-called barbaric Arabic civilization had many things that the "civilized" West found very useful, one being their number system [7]. This analytical structure was more conducive to the kinds of mathematics used in probability theory.

The Crusades also brought the West an influx of people, commerce, and ideas, moving from Constantinople to the Adriatic Sea. With this migration began the Renaissance period that, from a mathematical perspective, started with the basic tools of algebra, arithmetic, and geometry.

Christianity, while providing a structure for the development of the arts, did not promote the development of the mathematics of probability. "God's will be done" became the substitute for chance. If something happened, it was not the result of chance; it was God's will. This attitude was strongly embedded in the early Christian lifestyle and to some extent prevails today, with one small difference. At least today you're not burned at the stake if you deviate from the teachings of the Church. The Roman Catholic Church "suppressed" or, more tactfully, "shaped" scientific thought for over 1,000 years. There was no question that "God was in and science was out." The concepts of uncertainty and probability were not compatible with the ideas of the Church and therefore were not allowed.

This brings us to about A.D. 1200, when Italy was beginning to enjoy the benefits of its central location at the gateway from the East to the West. Italian mathematics started to flourish along with several great works like Leonardo Fibonacci's development of a mathematical sequence of numbers, now called the Fibonacci Sequence. Leonardo Fibonacci, also known as Leonardo de Pisa, was born in Pisa, Italy [8], about A.D. 1170. His father was a prominent, respected businessman in North Africa. It was there that Leonardo received his early education. He traveled extensively as a young man, meeting scholars and learning about the different arithmetic systems used around the Mediterranean. Leonardo Fibonacci became a recognized and respected mathematician whose advice was sought even by the Holy Roman Emperor, Frederick II. In A.D. 1200, this honor was probably higher than that of advisor to any world leader today.

Leonardo became the first person to introduce the Hindu-Arabic number system into Europe [9]. This is the position-based system using the ten digits (0, 1, 2, 3, . . . 9) we use today, with its decimal point and a symbol for zero. In his book, "Liber abaci" ("Book of the Abacus") he discussed arithmetic in the decimal system. It was in this book that he developed his famous number sequence. These numbers are supposed to detail the ways certain types of events are grouped.

Did you ever hear someone say, "things always happen in groups of threes?" Well, three is a Fibonacci number. In case you're interested, here are the first few terms of this famous, infamous, and infinite sequence:

$$0, 1, 1, 2, 3, 5, 8, 13, 21, 34, 55, 89, 144, 233, 377\ldots$$

The successive numbers of the sequence are computed by adding the two previous numbers. This sequence deserves a little discussion here because it illustrates the mindset of the ancient scientists in their search for the logic and order they believed were the central forces of life.

Admittedly, the numbers of the Fibonacci sequence have some very interesting properties. Take, for the first example, ratios of successive numbers in the sequence. Here are a few examples.

$$1/1 = 1$$
$$2/1 = 2$$
$$3/2 = 1.5$$
$$5/3 = 1.666667$$
$$8/5 = 1.6$$
$$13/8 = 1.625$$
$$21/13 = 1.615385$$
$$34/21 = 1.619048$$
$$55/34 = 1.617647$$

The actual number or value to which the sequence converges, 1.61803 . . . , is called by many names, the Golden Ratio, the Golden Mean, or Phi. Phi has applications in solid geometry, and has purportedly been applied in building construction projects from the Parthenon in Athens to the United Nations building in New York [10].

In nature, there are many applications that give some credence to the logical, natural order scenario that seemed to dominate these times. On many plants, the number of petals is a Fibonacci number: buttercups have five petals, some lilies and irises have three petals, corn marigolds have 13, daisies can be found with 34, 55, or even 89 petals. Some flowers always have the same number of petals. For those that have a varying number, the average turns out to be a Fibonacci number. Even some leaf arrangements follow the mathematical sequence. Coincidence? Maybe. But in medieval Europe, this discovery certainly enriched the popular concept that God was in charge of risk management.

The applications of this simple sequence of numbers still seem to have a place in our time. Today there is a publication called "The Fibonacci Quarterly" that is solely devoted to the exploration of Fibonacci numbers and their applications [11].

The first person who actually appears actually to have addressed the concepts of probability and risk was Girolamo Cardano [12]. He was born in Pavia on September 24, 1501. His father was a lawyer, mathematician, and a colleague of Leonardo di Vinci. Cardano was trained as a physician, but his interests were truly those of a Renaissance man. He became a well-known physician, with high-profile patients in nobility and government. He was one of first physicians to document a clinical description of typhoid fever. And in addition to his medical accomplishments, Cardano published three works on mathematics, 10 on astronomy, 50 on physics, and many, many more. In his autobiography he claims to have burned many of his books, leaving us 138 books printed and 92 still in manuscript form. His primary contributions in mathematics were in the algebra of cubic and quartic equations. He also invented the "Cardan Joint" or universal joint commonly used today to connect drive shafts to transmissions for rear drive cars and tractors. But his interest in probability had practical yet destructive applications in money and addiction; Girolamo Cardano loved to gamble.

> *Peradventure in no respect can I be deemed worthy of praise; for so surely as I was inordinately addicted to the chess-board and the dicing table, I know that I must rather be considered deserving of the severest censure, I gambled at both for many years, at chess more than forty years, at dice about twenty five; and not only every year, but—I say it with shame—every day, and with the loss at once of thought, of substance, and of time. [13]*

He made an interesting observation regarding gambling and why he preferred dice over chess. He said that with chess, the opponent's inherent skill and knowledge determined the winner, but in dice gambling, the playing field was level. Only the player's knowledge of the odds would influence the outcome. And as a result, he observed for the first time that the outcomes of dice throws adhered to mathematical ratios. Throughout the millennia, people understood the uncertainty of dice throws but it was Cardano who developed the specific odds calculations. This knowledge gave him an edge and he actually used gambling as a source of income in tough times. He developed his thoughts in his treatise on games of chance, *Liber De Ludo Aleae*, the first major work describing the mathematics of probability. The manuscript was found among others after Cardano's death and published in 1663, 87 years after his death. We can only speculate why he didn't publish such innovative concepts during his lifetime, but perhaps he didn't want to educate his competition.

There is no doubt that Cardano was evidently what we would call today a "free spirit." He had a temper but was viewed as a fair person. However, in spite of his accomplishments and intellect, Cardano did not have a stable life. He went from respected to outcast, rich to poor, and happy to sad. Towards the end of his life, he got into trouble with the Church for casting the horoscope of Jesus Christ. This landed him in jail for a while. He also cast his own horoscope, which indicated that he would live to the age of 75. He was forgiven by the church and lived the remaining days of his life in Rome. He was evidently a true believer in the laws of probability and prediction and died just three days before his 76th birthday, though some claim he committed suicide.

At the end of the 16th century, several other people with both mathematical ability and a propensity for, or at least an interest in, gambling continued to develop and structure the laws of probability. Their work identified and quantified the likelihood of certain future events, indicating different rewards or penalties. True, the original purpose of their labors was most likely applied to gaming applications, but the methodology for risk measurement is the same. These early risk managers were developing a system to manage the frequency and severity of events.

Although the Catholic Church remained central to European life, some of its basic teachings came into question as technological discoveries enabled scientists to learn more about their environment. The most notable example of this was Galileo Galilei's scientific belief that the Earth was not the center of the Universe, but rather a planet revolving around the sun. Galileo's contemporary, Italian priest and scientist Giordano Bruno [14], was burned alive at the stake in 1600, partly for claiming that the Earth moves around the sun and for promoting other beliefs and behaviors contrary to Church authority. To replace Biblical teachings with contemporary scientific discoveries was a threat to the authority and credibility of the Church. Galileo [15] himself was tortured in many ways, and finally, at the very old age of 70 years, the weary scientist acquiesced. In 1633, he signed a long declaration saying he was wrong and the Church was right.

Why didn't they just burn him at the stake, too? Most likely because he was too high-profile a figure to eliminate. The Church rulers were familiar with the power of martyrs. To the Roman Catholic Church, scientific thought wasn't just "out," it could be dangerous. With the rise of the Roman Empire, scientific thought had gone to sleep. With the rise of the Roman Catholic Church, scientific thought had gone into hibernation.

But there is more to this story. The issue here was the Roman Church's risk perception of science in relation to its teachings. It is important to understand the risk drivers from Galileo's and the Church's perspectives because this type of risk management scenario is not just a lesson. It occurs today. Let's explore this in more detail.

The concept of an earth-centered universe was first constructed by Aristotle. He cast the Universe as a series of 55 concentric spheres on which each planet, the sun, and all of the stars moved [16]. The outside ring he called the "Prime Mover." Energy from the Prime Mover was transferred to the stars, sun, planets, and, finally, to the earth. About 400 years later, the Greek astronomer Ptolemy codified the details of the planetary and moon motions in the earth-centered concentric spheres from Aristotle's ideas. This coordinate system helped him take measurements with the naked eye, and provided a mathematical framework to assist in calculations. As to whether he really believed the configuration was actually correct is unknown.

Around A.D. 1200, Aristotle's philosophy was promoted by philosopher and Catholic theologian Thomas Aquinas. His philosophy is based on logic and deduction that was very compatible with the Roman Church's scripture and faith. Aristotle's universe configuration, formally published by Ptolemy, was a good match for the Church and subsequently, around A.D. 1200, it was adopted as the official interpretation. From then on, it became part of the general syllabus of "contemporary" scientific education.

It was Copernicus's combination of mathematics, physics, and cosmology that changed the playing field. He postulated with convincing technical detail that the planets revolved around the sun. Even though the work was probably completed sometime during his life, it wasn't actually published until the year of his death, in 1543. What was missing from the convincing argument was the ability to prove or test the theory. This process, which today we call the scientific method, is a standard protocol which many people can use to repeat experiments and obtain similar, if not the same, results. Without scientific experiments, conjectures (logic and deduction) were just hypotheses—easily dismissed by those who chose not to believe in them. And for astronomy experiments, the key equipment was, and still is, the telescope.

Fast-forward 63 years to 1609. Galileo is a 45-year-old university professor who introduces the application (though not the invention) of the telescope to scientific and religious circles in Italy. (Actually, the inventor of the telescope is not known. Lippeshey, a spectacle maker in the Dutch town Middelburg, applied for a patent on a telescope but it was rejected since it was common knowledge that others had developed similar devices before him [17].)

Galileo improved the telescope of his day. He achieved a magnification of 20 times, whereas the Dutch versions had much less magnification. Galileo's stature in the scientific community helped promote his new "invention," especially with the noted potential for commercial and military applications. He was a true Renaissance man: a writer, painter, scientist, inventor, and musician. However, his real love and talent was in scientific experimentation. Galileo's telescope allowed him to see the stars and planets in detail that no one else had ever seen before, and he had the penchant and training to record what he saw in such a manner that others could repeat his steps and see the same thing—the origin of the scientific method.

As Galileo's planetary observations and methods were published, it didn't take long to connect the dots. This was not just from reading his accounts but from the observations of others who made telescopes and repeated his experiments with similar results. The discoveries all together painted a picture of a sun-centered, not earth-centered, solar system.

Church leaders knew Galileo to be a deeply religious man and initially didn't view his work as a threat. In fact, they were amazed at what they saw through the telescope to the point where they even hosted luxurious parties in Rome to view the stars and celebrate his great achievements.

Yet not everyone was enthralled with Galileo's works. He had enemies. Silent at first, they worked behind the scenes to convince church leaders that Galileo must be stopped. Why? There was more at stake than just the proof of an astronomical theory. The academics had a legacy investment in the Aristotelian philosophy of logic and deduction. For many years they had been teaching these two concepts as the pillars of all science. Now along comes Galileo, the telescope, and the scientific method, and all of a sudden, Aristotelian philosophy is out of date.

The academics chose not to attack Galileo by themselves. Instead they directed their energy towards convincing the Catholic Church leadership that Galileo's science was wrong. Galileo's general rejection of the Aristotelian method of logic and deduction for the scientific method of hypothesis, experimentation, analysis, and conclusion made him an enemy of the legacy university leadership. It was this group that convinced the Church to reject Galileo's work and to force him to state publicly that the Copernican (sun-centered) universe configuration was wrong.

This is how Galileo got into so much trouble. The academic institution and the people he threatened with his non-Aristotelian ideas were against him from the beginning. Galileo's experimental results proved the university professors' teachings were wrong. Anyone with a telescope could see the data for themselves. Galileo was right but there was no resolution in his lifetime. He died under house arrest in 1642.

The Vatican eventually changed its mind in 1822 when the solar system structure was common knowledge. They lifted the ban on his book. Yet it wasn't until 1992, 350 years after his death and after NASA named a Jupiter-bound spacecraft after Galileo, that the Vatican publicly cleared Galileo of any wrongdoing.

The problem that the church leaders had with Galileo and countless other people hasn't gone away. Even today, our understanding of history and the current environment is being challenged as science finds more creative ways to apply the basic laws of physics. When the news reported the tentative discovery that there had been life on Mars, the whole philosophy of how we view the physical universe and religion was poised for change. We are pretty much used to this. The early religious leaders were not.

While Galileo was discussing his Copernican "hypothesis" with church leaders in 1623, Blaise Pascal was born in Clermont-Ferrand France [18]. In spite of losing his mother when he was three, Pascal had a caring and nurturing childhood. He had two sisters who proved to be loyal friends throughout his life and a loving father, Étienne, who was an educated professional determined to provide his son with the best education possible. Perhaps due to his professional work or social interactions Étienne became a knowledgeable mathematician and held a high interest in physics. He designed a systematic education plan for his only son. The plan was to start his training in languages and grammar along with discussions of natural phenomena such as gunpowder. The formal study of Latin and Greek was to begin at age 12. Mathematics was in the syllabus but only after he reached the age of 15 or 16.

However, Pascal's intellect and active curiosity would change Étienne's design. Pascal constantly challenged his father with insightful questions and followed up on subjects with his own research, experiments, and writings. Etienne's lesson plan changed when Pascal asked him about the Euclidean proof related to the sum of the angles of a triangle. His father, stunned as to how he would even conceive such a problem, asked him how he thought of it. Pascal replied that he drew figures out of charcoal on the tile floor and formed his own axioms and definitions as one fact lead to another. His father then recognized the true prodigal genius of his son, changed his educational plan, and the rest is history.

Pascal did have another advantage. His father regularly took him to weekly scientific meetings, sort of like the "Academy of Science," where he met scientists and mathematicians working on current research problems from grinding mirror lenses for Galileo telescopes to the mathematics related to mechanics. These friendships that began while he was young provided a rich intellectual environment throughout his life.

The caring and intellectual early family environment of his youth, combined with his natural genius, allowed Blaise Pascal to grow into one of the major pillars of modern science. By the age of 18, with his life nearly half over, he had written a treatise on conic sections that, when published, was to be well received by the mathematician Gottfried Leibnitz. He also had invented a mechanical machine to perform mathematical operations—what we would today call a computer.

Yet Pascal's interest in mathematics went beyond, physics, geometry and algebra. He became interested in games of chance and their applications outside the immediate entertainment of gambling. Pascal devised an arithmetic arrangement of simple integers in a triangular shape that produced results of complicated formulas related to combinations and permutations. The result today is known as "Pascal's Triangle," even though the method was actually first discovered several centuries earlier [19].

Pascal's Triangle

$n = 1$				1		
$n = 2$			1		1	
$n = 3$		1		2		1
$n = 4$	1		3		3		1
$n = 5$	1	4		6		4	1
$n = 6$	1	5	10		10	5	1

This simple geometrical arrangement has many applications. The initial interpretation was that the numbers at each level turn out to be the coefficients of the binomial expansion of $(1 + x)^n$. There are many others, including a special case that reproduces Fibonacci's famous sequence.

The triangle formation was part of Pascal's work trying to solve a nearly 200-year-old problem stated by Luca Pacioli in 1494 [20] called "the problem of the unfinished game." The problem is as follows: *How should the stakes be divided when a game of several rounds must be abandoned unfinished?* For example, suppose you are rolling dice with a friend and you are betting on who gets the highest value each roll. Now you're ahead, say 2:1, and with two throws left, you need to quit the game. How should the winnings be divided?

Pascal solved this problem and sent his solution to longtime friend and colleague Pierre de Fermat for a peer review. They were astonished to learn that both had been working on the problem and that they both had separate solutions. These two great mathematicians, friends, and colleagues communicated over most of Pascal's adult life. The content of this correspondence laid the foundations of modern probability.

However, Blaise Pascal left us more than this. In addition to his analytical talents, he was also a philosopher and theologian. He never married and, over time, took on the practices of monks—praying, reading, and writing about God and ethics. Toward the end of his short life, just a couple of weeks after a serious carriage accident from which he escaped uninjured, Pascal decided to devote his mind and energy towards the rules and laws of God. In his last days, he wrote several short segments about his relationship with God. He did not live long enough to complete this work on religion, dying at the age of 39. But loyal friends and relatives compiled his notes into what would be Blaise Pascal's final publication 8 years after his death. It is called *Pensées de M. Pascal sur la religion et sur quelques autres sujets, qui ont été trouvées apres sa mort parami ses papiers* (Thoughts on religion and on certain other subjects, by M. Pascal; being writings found among his papers after his death). Today this work is known by the shorter title of *The Pensées*.

In one part of this work Pascal applies logic, philosophical insights, and understanding of probability to discuss the existence of God. The content is known today as "Pascal's Wager." Its structure, content, and implications are still being discussed, debated, and applied today. Here is an excerpt that describes how Pascal used probability and consequence (or risk) to show why one should believe in God. Pascal states [21]:

> *Suppose human reason to be in a state of uncertainty on the question of the existence of God. It is, as it were, a game of pitch and toss. I say that reason must lay the odds for the existence of God.*
>
> *And believe me, we are bound to take these odds one way or another. We have no choice in the matter. We are living; and every one of our actions implies a decision touching our destiny. It is evident that we should act in a different manner according as [to whether] God exists or does not exist. On which side shall we wager? We must wager that God exists.*
>
> *In every wager, there are two things to be considered: the number of the chances, and the importance of their gain or loss. Our reason for choosing this or that side is expressed by the product of these two factors. Now to suppose God is to suppose an infinite good. Let us make chances for the existence of God as small as you please, say for example, equal to 1. The contention that God exists shall be represented by $1 \times \infty$. Now opposite the blessedness which God can bestow upon us let us put the good things of this world, and let us grant them as great as you please. They can only form a finite quantity, which we call "a". Again, let us make as numerous as you please the chances that God is not, and that the world exists by itself. This number is finite, since there is one chance that God exists. The contention that God does not exists will thus be represented by the expression $n \times a$. Now this product is necessarily smaller than the first, into which the infinite enters as a factor. Therefore I must wager that God exists.*

Pascal used his great intellect and mathematical knowledge to teach us how to make decisions in the environment of uncertainty. His writing on this subject is one of the first, if not the first, risk or decision theoretic analysis documented in history. Every person is different and the decision theoretic argument of Pascal will not cement everyone's belief in the divine. But for Pascal, the thought process was concise and complete. His last words were: "May God never forsake me."

The understanding of the concept of probability is fundamental to the understanding to the understanding of risk applications. We have seen that the laws of probability were initially derived from quantifying the possible outcomes of gambling situations, and that science clearly challenged some of the foundations the early church. This should be expected since all of us to some degree test the efficacy and implications of any new discoveries relative to our behaviors and beliefs. It is the very nature of the art concisely stated in the scientific method to prove laws by the repeatability of experiments. So from this perspective, it makes sense that the legacy beliefs of the church, taken on faith, not experiment, would be difficult to supplant.

But there is another part of this story that reaffirms the creative genius and perhaps courage of the developers of modern probability. Today we apply of the term "probability" with a clear meaning, but in the 17th century, the word literally did not exist. There are several versions of the origins of the term "probability" but there is general agreement that before the time of Pascal the word "probable" (Latin, *probabilis*) meant *approvable* or *appropriate*. A probable action was an appropriate action [22]. So in addition to creating new math, the probability pioneers were, if not creating new words, creating new meanings for the words of their language.

A linguist once told me that a language has exactly the number of words in it required by the people who use it. As new concepts, behaviors, and products are created, languages add words when their use becomes prevalent. For example, in June 2009, the English language passed the million-word threshold. Well, from Cardano to Pascal and Fermat, both new math and new words were created.

It may be no chance that the next major player in the arena of probability development was Thomas Bayes. Born in London, England, in 1702, Thomas Bayes was a Presbyterian minister and, you guessed it, a part-time mathematician. His contribution, today known as "Bayes' Theorem," represents a powerful way to apply probability to situations where there are very little or no current data [23]. Here is a general idea of how it works, without going into the math details.

Suppose the New York Yankees have just drafted you, and you want to compute the likelihood that you'll get a hit when you go to bat. Based on standard, classical probability, we would divide the number of successes (hits) by the number of times at bat. In this situation, however, you've made no hits and haven't been at bat, so the ratio is zero over zero. The laws of mathematics tell us this is undefined. As the season progresses, and you have more at bats and start making some hits, we'll be able to compute this figure. Right now, we can't do it. Thomas Bayes gave us a way to bridge the gap between the two situations. His theory was to get some "prior information" and start with that. In this case we could use the average batting average for the Yankees, or the previous batting average for rookie players. On your first time to the plate, we would use this value as your batting average. This is an assumption. Now as the season progresses, you develop your own data. Since this is your first year, there still aren't a lot of data, but we do have some. Bayes developed a way of combining "prior information" with "current data" to produce an updated "posterior" estimate of probability —in this case, a batting average. If we didn't have Bayes' work, we would not have any way of estimating the probability and risk of events that have little or no current data. He gave us a way to use all of the information we have to determine probabilities and risks in a very practical way.

From this time on, the field of probability grew steadily as a distinct application of mathematics and statistics. Predicting the frequency and severity of undesirable incidents was basic to the earliest insurers. Data collection and analysis has become easier as nations, cities, towns, and communities around the world became established and communication technology facilitated the transfer of important information.

Risk was a technical, theoretical subject until it was brought to the world's general attention during World War I. The order and conventional wisdom of the early 20th century was shattered as the world's armies fought to realign their borders. The speed at which the war began, the large scale of the fighting, the suffering, and the duration sent an undeniable message to people everywhere that stability, order, and prosperity were privileged states of existence. There was a great deal more uncertainty in people's lives than they'd previously realized. The risks of instability and war were serious issues to be addressed. Risk management now took on a new meaning as the public's perception of risk became tragically seasoned with practical experience. After World War I, the world felt confident peace could be managed. And for the next two decades, it was. Unfortunately, at the same time, the management of financial risk was disastrous. The stock market crash which started on October 29, 1929, triggered the Great Depression and exposed the general public to the penalties of improper financial risk management. Soon after, the man who came to power by managing the ruined German economy, Adolph Hitler, put the entire world at risk.

Today, communication technology continues shrinking the distances between the listening and watching public and disturbing events. Science still continues to challenge our legacy thinking at a seemingly increasing pace. And the more we learn about our planet, our universe, and ourselves, the more we also learn how much we still don't know. From all of this and more, the public perception of risk likely has grown further in the first decade of the 21st century than in all previous time. The world had experienced earthquakes and other large disasters before, but never in history were the horrific events broadcast in real time, or played back from cellphone cameras and home videos to people sitting in their homes and workplaces all over the world. Risk, once a topic for intellectuals, gamblers, mathematicians, and (let's not leave out the new players!) politicians and day-traders, now has become a household term. This is the evolution of risk.

DISCUSSION QUESTIONS

1. The content of this chapter presents one string of events that shows how the concept of risk was used, understood, defined, and finally quantified. There are many versions which could be specifically written for different countries, regions, or cultures. Choose one and write an evolution of risk for a specific culture, country, or region of your choice. You may use some of the content of this chapter if you choose.

2. Trace the evolution of risk as viewed by a single person's life.

3. Apply Pascal's Wager to construct new versions for contemporary problems.

4. This chapter's content on the evolution of risk identifies Thomas Bayes' work as the last significant mathematical contribution. What additional work, not just in probability, would you add to this for the 19th, 20th and 21st centuries?

5. What was the first type of risk transfer that was used in commercial situations? Where was it applied? How is it used today?

ENDNOTES

1 Garth H. Gilmour, "The Nature and Function of Astragalus Bones from Archaeological Contexts in the Levant and Eastern Mediterranean," *Oxford Journal of Archaeology*, Vol 16, No. 2. 1997, pp. 167–175.

2 R. L. Wing, *The Art of Strategy: A New Translation of the Art of War*. Doubleday, 1988.

3 Sun Tzu, *The Art of War*, special edition. Ed. James H. Ford and Shawn Conners, translation by Lionel Giles. El Paso Norte Press, 2009.

4 Marilyn K. Null and Janice W. Larkin, "Sun Tzu and the Art of Stakeholder Involvement," *Federal Facilities Environmental Journal*, Vol. 12, No. 3, 2001, pp. 83–87.

5 Dick Tersi, *The Ancient Roots of Modern Science: From the Babylonians to the Maya*. Simon & Schuster, 2002, p. 17.

6 Harry Partch, *Genesis of Music*. New York: Da Capo Press, 1974, p.398.

7 David Eugene Smith and Louis Charles Karpinski, *The Hindu-Arabic Numerals*. Boston: Ginn and Co., 1911, pp. 1–12.

8 Ibid, pp. 127–133.

9 D. E. Smith, *History of Mathematics*. Toronto: General Publishing, 1951.

10 Mario Livio, *The Golden Ratio: The Story of Phi, the World's Most Astonishing Number*. New York: Broadway Books, 2002.

11 The Fibonacci Association, *The Fibonacci Quarterly*, http://www.mathstat.dal.ca/FQ/.

12 Henry Morley, *Jerome Cardan: The Life of Giralomo Cardano, of Milan, the Physician*. London: Chapman and Hall, 1854.

13 Girolomo Cardon, *The Book of My Life (De vita proria liber)*. Translated from Latin by Jean Stoner. New York: E.P. Dutton, 1930, p. 73.

14 Giordano Bruno, *Cause, Principle and Unity and Essays on Magic*, Ed. Richard J. Blackwekk and Alfonso Ingegno. Cambridge University Press, 1998.

15 Jerome J. Langford, *Galileo, Science and the Church*. Ann Arbor: University of Michigan Press, 1992.

16 Aristotle, *The Works of Aristotle VIII*, ed. J. A. Smith. London: Oxford University Press, 1908, p. 1073.

17 Rebecca Stefoff, *Microscopes and Telescopes*, Tarrytown, NY: Marshall Cavendish, 2007, pp. 15–16.

18 Émile Boutroux, *Pascal*. Translated by Ellen Margaret Creak. Manchester: Sherratt and Hughes, 1902.

19 A. W. F. Edwards, *Pascal's Arithmetical Triangle: The Story of a Mathematical Idea*. London: Charles Griffin, 2002.

20 Keith Devlin, *The Unfinished Game: Pascal, Fermat and the Seventeenth-Century Letter That Made the World Modern*. Basic Books, 2008.

21 Boutroux, *Pascal*, pp.183–184.

22 R. C. Jeffrey, *Probability and the Art of Judgment*. Cambridge University Press, 1992, pp. 54–55.

23 Thomas Bayes, *Philosophical Transactions of the Royal Society*. 1763, pp. 370–418.

Chapter 6

Frequency and Severity: Weighing the Risk

Often the difference between a successful man and a failure is not one's better abilities or ideas, but the courage that one has to bet on his ideas, to take a calculated risk, and to act.

—Maxwell Maltz, M.D.

From the lessons learned in Chapter 5, it's clear that the knowledge, use, or belief in mathematics was not the primary barrier to the acceptance of risk. The main reason it took so long for risk to be recognized was the continued belief in divine control of the future's uncertainty. Regardless of your beliefs about the future, we're now going to look at a framework that helps match some new ideas about risk with the subtle, implicit calculations we all make about risk in our daily lives. We'll start with what we can measure: the frequency and severity of various events. Then we'll add the nonmeasurable, subjective factors of experience, intuition, and perception. Together, these groups are blended to form a framework that may help us recognize the edges of life's cliffs before we go over the precipices. In other words, our framework can help us recognize when we are headed for disaster before we get there. The method will not eliminate uncertainty in your future, but it just might give you a little more knowledge you wouldn't have had otherwise. After reading this chapter you should have a richer mental picture of frequency and severity to help you better understand, interpret, and weigh the risks in your universe.

The word "risk" is used a lot in our society. Here are some recent applications of "risk" I took from news and other media publications. Do these phrases sound familiar?

risk of heart attack . . .	risk of vaccine . . .
risk of suicide . . .	risk of steroids . . .
risk of cancer . . .	risk of HIV infection . . .
risk of gun ownership . . .	risk of losing medical benefits . . .

20% Chance of Rain: Exploring the Concept of Risk, First Edition. Richard B. Jones.
© 2012 John Wiley & Sons, Inc. Published 2012 by John Wiley & Sons, Inc.

The list could become a book in and of itself. There's no question that the media has become one of the most frequent users of the "r" word. They are not alone in their increased use of the term, however. Virtually any group, whether it's a government spokesperson, a healthcare provider, a Wall Streeter, or anyone with a cause, seems to float the term into their vocabulary. The dictionary generally defines the word "risk" as some sort of hazard. Since it appears that the number, or at least our knowledge, of incidents of violence, disease, and tragedies is increasing, it makes sense that we are hearing "risk" mentioned more often.

While equating risk to hazard is normally correct, risk is generally understood to have two components. The first is a measure of how likely it is that an event occurs. That's called its frequency. The second term describes the occurrence's effects, or its severity. The statistical and mathematical sciences put these terms into what I call the *risk equation:*

$$RISK = FREQUENCY \times SEVERITY$$

To determine an event's risk, we multiply its severity, in whatever units that make sense, by its frequency of occurrence.

It's time to present an application of this equation but first, the background information. The Cassini mission, launched by NASA on October 15, 1997, is a deep space probe to Saturn and its largest moon, Titan. With Saturn's distance from the sun being about nine times that of Earth, there is insufficient solar energy to power the spacecraft's onboard electronics for its long journey. NASA developed a nuclear technology called radioisotope thermoelectric generators, or RTGs. These devices have no moving parts and use the high-energy alpha particle decay of non-weapons-grade plutonium-238 (half-life, 88 years) as a heat source. Thermocouples transform the heat to electricity. They are very reliable and highly efficient from an energy output-per-kilogram perspective. We know today that the mission [1] has been successful, performing beyond design expectations. However, back in the prelaunch days of 1997, there were considerable concerns and protests regarding the environmental and health risks related to an accidental release of the plutonium-238 during launch, ascent, or the Earth swing-by scheduled for 9 months after liftoff. The Cassini mission had the largest amount of Plutonium 238 ever put into a space rocket: approximately 72 pounds of plutonium dioxide. This fact triggered a series of health effect claims that NASA took very seriously. Their calculations of Cassini-induced health risks were placed in the context of other risks everyone on Earth shares every day. The events and their frequency, severity, and risks are listed in Table 6.1. Before you read on, ask yourself the following question: Does your homeowner's insurance policy cover asteroid strikes?

Table 6.1 shows the highest-risk scenario had neither the highest likelihood nor the highest severity [2, 3]. The 1.5 km diameter asteroid hitting the earth has the largest frequency-severity product. In this manner, the highest risk scenario balances the two factors so neither one dominates the result. The product version also helps people to understand what really is practically important in a systematic way that does not rely on intuition, anecdote, or subjective judgment. This is one major benefit of using risk in decisionmaking. The qualitative tools are important, but the risk

Table 6.1   Frequency, Severity, and Risk Comparison of the Cassini Mission to Other
Natural Exposures

Potential event	Frequency of occurrence	Estimated fatalities	Risk
1.5 kilometer (1 mile) diameter asteroid hitting the Earth	2 in 1 million yrs	1.5 billion	3,000
10 kilometer (6 mile) diameter, or greater, asteroid hitting the Earth	1 in 100 million yrs	5 billion	50
50–300 meter (150–1000 feet) diameter asteroid hitting the Earth	1 in 250 yrs	3,000	20
Cassini inadvertent Earth swing-by re-entry with plutonium dioxide release	1 in 1 million	120	0.0001
Cassini early launch accident	1 in 1,400	0.1	0.00007
Cassini late launch or re-entry from Earth orbit accident with plutonium-238 release	1 in 476	0.04	0.00008

analysis methodology demonstrated in Table 6.1 provides a more quantitative result
to be considered in the decisionmaking process.

As an interesting update to this table data, in January 2010, astronomers discovered a new, tiny asteroid that looked like it was heading directly for Earth [4].
The object had an orbit period of 365 days and was between 10 to 15 meters
long. These conditions caused people to speculate it was a depleted upper stage
booster from an European Space Agency rocket. As it turned out, it just missed
earth by only 80,000 miles (<1/3 the distance to the moon) and the incident
motivated the National Academy of Sciences to release a report recommending
that Congress expand its near earth orbit objects (NEO) to include objects between
30 to 50 meters in diameter. A NEO with a diameter of 50 meters can create a
blast equivalent to one megaton of TNT [5]. This is over 50 times larger than the
bomb dropped on Nagasaki in World War II. If you're interesting in updating Table
6.1, all of the several hundred NEOs currently being tracked are listed in detail
at http://www.jpl.nasa.gov. I wasn't kidding about the homeowners' asteroid
coverage!

Risk, in Table 6.1, is computed in terms of the number of deaths per year. This
is a statistical term, as accurate records of asteroid-related deaths are hard to verify
internationally. For these types of low-frequency events, risk analyses often use
scientific calculations to estimate event characteristics given the sparse information
available on asteroid collisions here on earth. The use of estimation techniques is
common, as it is impossible to gather actual data on the future's uncertainty since
by definition, the future is never achieved.

Now along with this snippet of philosophy, consider the following question: If we choose not to multiply frequency and severity together, is there still any risk? The answer? Of course there is risk, regardless of whether we do the arithmetic or not. This point highlights a basic principle you need to keep in mind any time you come across reports where risk is described by any equation. Risk is bigger than any equation. The equation represents an approximation of a process; the math does not capture the full complexity as some devoted scientists, politicians, and others would like you to believe. While the risk equation represents probably the most widely accepted model, it has its limitations. It's only an equation—a model, not reality. It is often tempting to accept the mathematical precision of sophisticated risk analyses without understanding that the basic equation used in the analysis is itself just a model of reality. The results can be used to help us make decisions, but remember, all of the science in the world is not a surrogate for our common sense. With the weaknesses of the risk equation exposed, let's discuss its strengths.

Describing risk in terms of hazards, as in the phrases used at the beginning of this chapter, is generally consistent with the risk equation. For example, let's examine the phrases "risk of heart attack" and "risk of suicide." The severity of these events is fixed. It is part of the statement. In these cases, risk becomes directly related to the hazard frequency. When severity is defined as part of a statement, risk becomes the same as frequency. Any time you can replace the word "risk" by "chance," likelihood," or "probability," you have this type of situation. Now you may ask, "so what?" The simple point is that the two ways of referencing risk are really consistent.

I doubt that you think of yourself as a risk expert, but you are—unless there is someone else in your life who tells you what to do, when to do it and how to do it. You make countless decisions daily, weighing all of the competing variables and then taking action or not. Crossing the street, wearing a seat belt (or not), the speed you drive, what you eat for breakfast—you get the idea.

Risk management requires some form of risk quantification. You need some data or information to do this activity. Then how do you "measure" risk? You can measure pressure or temperature with a barometer or thermometer, weight with scales, speed with laser beams, distances with light, but you can't measure risk. Risk is in a separate category when you talk about measurement. What can you measure? The components of risk: event frequency and severity. To obtain risk estimates you must compute the product of these two quantities. We do this all of the time, but in an automatic way without ever separating the calculation into its constituent parts, frequency and severity. Think about this the next time you put on your seat belt, slow down a little bit on the freeway, or eat a carrot.

The analytical methods of calculating risk are much simpler than any method of trying to measure risk perception. But even though the math is easier, the incongruity here is that in many cases, psychologists or sociologists can get a more precise measurement of public risk perception than scientists can in measuring actual risks. Even though questionnaires are biased to some degree based on how the questions are asked, the results can give a pretty good indication of the selected group's collective opinions of the various topics polled. The results are statistical, but in a different way from the statistics presented in a scientific analysis of risk.

From a scientific perspective, the risk of a specific event is equal to its frequency or probability of occurrence multiplied by the event's severity or consequence. That's all there is to calculating risk. I could end this chapter here . . . but as you can see I didn't and for good reason. Even though the equation is simple, its ramifications, interpretations, and limitations warrant a little more discussion. Not all risks are created equal, and the simplicity of the risk equation is deceiving.

The risk equation (risk = frequency × severity) is exact. It is a pure, perfect representation of what statisticians call "expected loss." Here's an example of what this term means. If you flip a coin 1,000 times and keep track of the number of times it comes up tails, the number would be around 500. The "expected" number of tails is 1,000 multiplied by (1/2), the probability of tails. Even though you always don't get the expected value, the math says that if enough people performed this experiment or you flipped the coin an infinite number of times, the number of tails would approach half the total number of trials. In risk analysis, the expected loss is usually expressed as some function of time in terms such as dollars/year or deaths/year. This expected loss is analogous to the coin flip case. We don't expect to have exactly that loss value each year, but it is the statistical average we would achieve over a long period of time, given the same set of conditions.

The risk equation is a fixed statement of relationships. The problem is not the math, it's the data and relationships we use for its terms. If events occur often, then event frequency can be easily determined from data. With plenty of data, the severity scenarios are also measured with relatively high accuracy. In this case, the risk equation makes intuitive sense.

Now let's take a situation like an earthquake scenario. Its frequency of occurrence may be once every 10,000 years. Remember, on the geologic time scale, this amount of time is just the blink of an eye. The consequences of such an event are known to be large, but how large? No one knows. Insurance companies and some other large, global businesses model such scenarios, but there isn't enough data around to know how bad a severe earthquake would be today. We could keep the severity constant by saying that a large earthquake causing $500 billion in devastation could happen. This certainly true, but the trouble then is that no one knows the frequency for this specific earthquake scenario. The same argument could be used if you fixed the frequency. No one knows what the event's severity would be.

The lesson of this discussion is that the risk equation, in its purest form, is statistically based. You can't do statistics without data. The more data you have, the more accurate the results.

What about the many real applications of risk where you are dealing with low frequency–high severity events? These are exactly the types of situations where government policymakers and industry leaders are using risk management principles. As with many aspects of life, there are no exact answers to these issues. The risk equation alone is not the magic bullet. However, I speculate that as our frequency and severity measurement systems become more sophisticated, risk management's role in our society and our daily lives will increase. This won't result from solving equations; in fact, just the opposite. Risk, frequency, and severity provide a cohesive framework to help people make decisions in an environment of uncertainty. This is exactly where the concept of risk and its underlying structure have the most value. Our experience, intuition, and judgment are the factors that take up the slack where the math leaves off. After all, if risk management was as simple as plugging numbers into an equation, someone would have computerized the whole subject a long time ago.

With such a simple equation you might think that nothing of significance could be obtained from studying it any further. If you recall Albert Einstein's equation that has become a symbol of his theory of relativity, $E = mc^2$, while certainly simple, has changed the way we look at our universe. I suggest to you that nature has given us, with the risk equation, a more fundamental, more articulate, more powerful, and immensely simpler tool to help us understand, interpret, and survive our political, social, and natural environment.

Now, with that audacious preface, let's take a look at some of the more esoteric ramifications of the risk equation. We'll do this with the use of several graphics. The pictures portray in a succinct way the salient characteristics of risk.

We start by plotting frequency on the horizontal axis and severity on the vertical axis of a common graph, shown in Figure 6.1. Since event frequencies and severities can vary in size, we use a logarithmic scale in Figure 6.1A where every major tick mark is power of ten. The diagonal lines depict constant risk where the product of frequency and severity has the same value. Two points (A and B) are examples of this relationship. Point A has the values of frequency = 0.001 and severity = 1,000. Multiply them and you get 1. Now take point B. The frequency is 0.1 and the severity is 10. Their product is also 1, and the product of the coordinates of every point on this line has the same value. The same process is true for Figure 6.1B, except that the lines of constant risk are curves since the axes are linear.

Let's generalize the types of events symbolized by these two points. Hurricanes, floods, and earthquakes are examples of events that would be A-type of events: low frequency–high severity. Slips and falls, leaking roofs, cuts and bruises, and electrical fuse or breaker failures are B-type events: high frequency–low severity. The power of looking at these events through the lens of risk is that these seemingly diverse situations can be equivalent from a risk perspective. For example, over some period of time you may have spent $100. You may have spent it $1 at a time or all once. These are two very different scenarios, but with the same $100 result. If you instead want to save that $100, you'll have two very different alternatives for how to go about doing that, depending on the way you would have spent it. This is the same as our calculation of risk. A small problem can occur often or a large problem can occur infrequently. In either case, the risk can be the same. Let's take an important issue in contemporary society and examine it in terms of both frequency and severity.

The number of older workers in the United States has been growing steadily since the 1990s. The population of senior citizens is over 36 million [6] with approximately 5.9 million [7] active workers, about 4% of the total workforce. An extensive

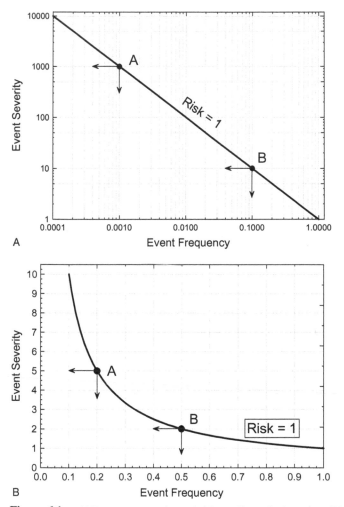

Figure 6.1 (A) Frequency, severity, and risk, nonlinear (log) version. (B) Frequency, severity, and risk, linear version.

report in 2006 on the relationships between work injuries and age focused primarily on workers between the ages of 20 and 64 [8]. The report assumed that people reaching the age of 65 would leave the work force for retirement. However, with the stock market plunge and large decreases in housing values that started in 2008, more senior citizens are postponing retirement. The increase in senior citizens either joining or remaining in the work force prompted a new study to ascertain how this rapidly growing worker segment influences the frequency and severity of workers' compensation claims [9].

Advocates of older workers highlight the fact that workers over 65 years of age have a lower accident incidence rate than their younger, and perhaps more reckless,

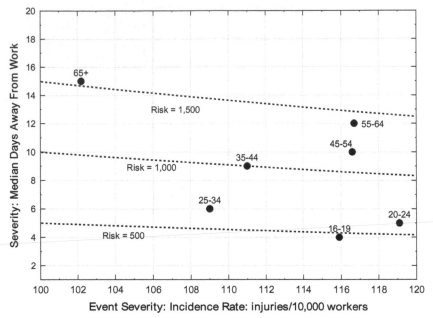

Figure 6.2   Worker injury frequency, severity, and risk by age group.
Source: U.S. Bureau of Labor, 2008.

colleagues. This is the lower frequency argument. But what about the severity dimension of the issue?

Let's take a look at workers' compensation from a frequency, severity, and risk perspective. The pertinent data from the Bureau of Labor Statistics [10] is plotted in Figure 6.2 using the linear version of the risk plots shown in Figure 6.1B.

Frequency is measured by worker incidence rate defined as the number of injuries per 10,000 workers × 20,000,000. The number 20,000,000 is derived from 10,000 workers working 40 hours/week for 50 weeks. Severity is taken as the median (50th percentile) number of days away from work. Risk is measured as the lost productivity due to worker injuries with the units of number of days away from work per 10,000 workers.

Figure 6.2 shows that the highest risk group is the 65+ senior citizens. Even though the group has lowest incidence rate, it has the highest the loss per event: enough to produce a risk value greater than 1,500. Thus the low worker injury incidence rate is only part of the total risk. The plot helps lead you to this conclusion.

This simple format of showing risk as a function of event frequency and severity is a valuable communication tool in practice. The information presented in table format provides the same results as does the plot, yet the visual presentation is generally more appealing to people. It makes data observations easier to comprehend. For example, in Figure 6.2, notice that workers' compensation risks are divided into roughly three equivalent classes. Separate constant risk lines could be drawn

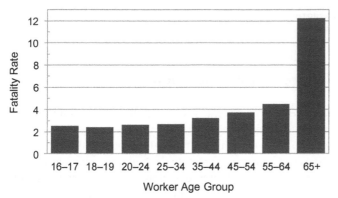

Figure 6.3 Worker fatality frequency and risk by age.
Source: U.S. Bureau of Labor Statistics, 2008.

though the groupings: (16–19, 20–24, 25–34), (34–44, 45–54), and (55–64, 65+). This observation is not as obvious if you just look at the numbers in a table.

Now let's extend this senior citizen risk discussion and examine occupational fatality risk by age. In this situation, severity is fixed at death, so risk is measured by changes in frequency. The fatality frequency (and risk) results are presented in Figure 6.3. Fatality rate is measured as the number of deaths per 100,000 hours of full time equivalent workers [11]. The fatality rate for the 65+ age group is nearly three times the 55–64 age group. This result concludes that the 65+ is clearly the highest risk age group for work-related death. Overall it is the highest group for both for injury and fatality risks.

If you are studying the aging workforce in more detail then there are several additional factors that need to be included before any realistic conclusions can be made. For example:

- This analysis measures injury risk in terms of days away from work; another way to measure it could be in terms of indemnity claims.
- Fatality risk is the highest for 65+ worker age group, but if severity is measured in terms of claim amounts, then the life valuation methods discussed in Chapter 4 could alter this conclusion.
- Worker risk varies by industry—general results for the entire group are not necessarily relevant for any specific application.

On one side of the issue we have a growing number of people over 65 years of age who want and now need to work, and age discrimination laws may restrict an employer's ability to reject qualified older workers. On the other side we have indisputable evidence that the increase in 65+ worker accident severity more than compensates for their lower accident frequency which, from a risk perspective, makes elderly workers a higher risk than their younger coworkers. The lesson here is that if you examine the issue from a frequency perspective, you get an incomplete view. The severity dimension, essential for weighing risks, must also be explored.

Now let's move on to explore some other interesting aspects of frequency, severity, and risk.

Point #1: All risks are not created equal

Knowing the risk associated with certain events is insufficient information by which to make risk-based decisions. You need to understand the frequency and severity factors that determine the risk. Even though their risk values may be the same, their risk management alternatives can be very different. In the high frequency–low severity case, you have a lot of data with which to figure out how to improve the situation. For example, in the home environment, if slips and falls are occurring, you can install nonslip carpets, better handrails, or better lighting. For low frequency–high severity events like floods, hurricanes, or tornadoes, what can you do? These events occur so rarely that there is really little or nothing you can do. Moving out of the hazard zone usually is not a viable choice, and in most cases, all you would do here is trade one set of risks for another, so we're not going to consider the option of risk elimination. For fires, it is true that smoke alarms are good warning devices, but they don't put the fire out. They just tell you to get out of the house.

For these situations your only cost-effective alternative is to transfer the risk. You pay someone else a premium to assume to risk, and this person or entity will reimburse you for the loss if it occurs, per the contractual agreement or policy.

Here lies the notion of insurance—though not necessarily insurance companies. To explain, let's consider flood risk. Only people in flood plains buy flood insurance. In all, the U.S. National Flood Insurance Program had 5.6 million policies in force as of December 31, 2009, with a total insured value of $1.2 trillion and total annual premiums of $3.1 billion [12].

The risk-bearing population has a high potential for experiencing a flood. In order for an insurance company to be profitable, it would require the exposed population to pay for at least all of the flood claims plus company expenses and overhead. Yet looking at the history of flood claims and associated premiums, this doesn't work. The reason it doesn't work is identified by the term "adverse selection." Only those homeowners/renters who feel threatened by flood and live in flood areas will buy the insurance. Thus there is a relatively small population that gets flooded and the payout per insured is large, outside the economics of most people's budgets. Private insurance companies can't find enough people to buy the insurance at a price where they can make any money. For flood and some other special types of insurance, the federal government is the insurer, and we all pay the premiums in our taxes.

Flood insurance is an interesting type of low frequency–high severity event to discuss further. People generally perceive flood risks as lower than fire risk. A survey of victims of floods induced by hurricanes in the southeast United States showed that most of the homeowners who did not have flood insurance did have fire insurance. This is not as dramatic as it sounds. Standard homeowners insurance covers most fire exposures but excludes floods. You need to buy a separate flood insurance policy to get this coverage and most people don't spend the additional time, effort, and money required to do this. Nevertheless, the tragic part of this situation is these folks who elected to transfer their fire risks and assume the flood risks were deceived by their own intuition. Loss statistics show that these people are, in fact, four times more likely to be devastated by flood than by fire.

Here's my speculation about this counterintuitive situation. Fire risks are more real to homeowners, even to those who live in flood plain areas, than flood losses. People are constantly reminded about the fire loss potential. Every city and town has its own fire department with their polished, shiny fire engines that parade down Main Street in 4th of July or Memorial Day parades. There is no "Flood Department" with red boats or helicopters in the parades. The issue here is risk communication. People need to be informed in a manner that will allow them to judge the facts better for themselves. The government has attempted to convince the flood-prone public to buy the insurance through a series of TV commercials. The media coverage of the effects of hurricanes Rita and Katrina also helped convey the message that the risk is real. And don't conclude from all of this talk about flood disasters that flood insurance is expensive. The average premium is about $563 per year [13]. That's less than what many people pay for coffee over the same amount of time. Flood insurance is inexpensive for the policyholder but remember, the U.S. taxpayers are subsidizing the insurance.

The answer to this huge risk management problem is not the U.S. government or, for that matter, the taxpayers. People who build or buy homes in natural hazard zones should be required to adhere to strict building codes to mitigate or reduce losses and to pay the transfer of risk premium as part of their mortgage. It should not be an elective. Too many people elect to gamble by not buying the insurance and then, when a disaster strikes, taxpayers again have to bail out these folks with Federal Emergency Management Agency (FEMA) funds. The insurance is available to everyone. It's not like other insurance products, which limit participation based on certain prequalifications. The issue here is becoming so commonplace that it's no wonder we can't reduce the size of government. People are not being held accountable for their intentional actions.

But you can't blame people for taking advantage of the laws. Here's what I mean. Congress passed its disaster legislation in 1968 and didn't give FEMA the power to raise the premiums of people who continued to be hit by floods or the ability to refuse insuring poor construction. In fact, just the opposite is true. FEMA is required to rebuild infrastructure once it has been built. No thought can be given about how stupid it may have been to build in the location in the first place. If you choose not to buy flood insurance and one occurs, you gambled and lost. Today, the taxpayer will cover your bet. Tomorrow, who knows?

Concluding the discussion of Point #1, knowing the risk associated with certain events is not enough information for action. It's only one part of the risk-based decision process. If you're beginning to question the utility of using risk for any practical situation, the second point is going cast even more doubt. As you'll see however, it's not the concept of risk that is bad. It's how we apply it. Practical risk analysis is much more complicated than multiplying frequency and severity together.

An effective way to manage risk is to combine intuition and experience, mixed with just enough science and statistics. The recipe may vary depending on circumstances, but the ingredients remain the same. Leave out one of them and the result is analogous to gambling without knowing the odds. Now let's move on to the next discussion.

Point #2: Risk reduction is not enough. You need to know the direction in which risk has changed

Suppose you review two businesses' or people's experiences for a year and their resultant overall risk and frequency and severity factors turn out to be exactly the same. Both have the same point in the risk coordinate system as shown in Figure 6.4, where I've set a risk value of 10. Another year goes by and a similar review is performed. This time, results indicate that both firms or people have reduced risk by an equal amount. Both now have value of 1. But, as you can see from the chart, the frequency and severity components are very different. Firm (person) A has experienced increased event severity but at a significantly lower frequency. Firm (or person) B has reduced risk by decreasing severity more than frequency has increased. For the business side of this example, here a few questions: Which one would you like to own? Which one would you like to insure? In which one would you want to invest? Which one would you like to be?

Since I posed the questions you are probably thinking that I know the answers. My answer is: "It depends."

Let's discuss Type B first. There has been a significant increase in the number of failure events. If you read the graph, Type B's failure frequency has actually almost doubled over the past year. The good news is that the severity has been

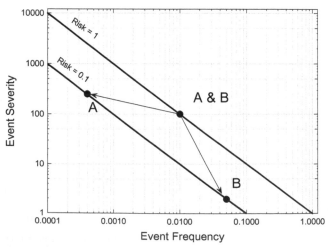

Figure 6.4   Frequency, severity, and risk.

reduced from about 100 to 2, a factor of 50. Type B now has more information for analysis. There is more information to fix and improve what's wrong. Measurements also are more accurate because there's more data to determine the statistics. There is no single general cause for this type of performance. Here's one possible explanation I have seen in industrial situations. A new leader becomes in charge with the goal of improving the company's finances. Record-keeping, documentation, and information systems are improved. Failure events with low severity, even near misses, are now being documented. The old system may have only captured failures if the severity was deemed important enough to document, usually by various criteria depending on who was in charge at the time. In essence, the definition of what is meant by "a failure event" now has been defined and is being enforced. Actually, the business, before the new leadership, was probably measured imprecisely because many failures may have been occurring that were never captured, especially the low-severity events which could go unnoticed.

Is there a downside to Type B? It's hard to say in general terms. Here's one perspective that deserves to be mentioned. Every failure is the result of two probability events. The first probability variable is the time until the next failure. This time varies and refers to the reliability of the system. The second variable, many people neglect to recognize. It determines the severity of the event. In essence you have two bins of numbers from which to draw. You draw a number from the frequency bucket that tells you the time of the next failure. Then you must go to the second bucket and pick a number that tells you how severe it will be. That usually means how much it will cost. As you have more and more failures, you are taking more and more chances with severity. Each time you "draw a number" you have a chance of getting a very, very severe event. This is why the best way to reduce risk is to reduce both the frequency and the severity.

Let's now take a look at Type A's situation. We can characterize its frequency and severity performance by saying that failures are occurring less frequent, they are becoming more severe, and our ability to measure is deteriorating as we're getting less data. This statement is more than a good news/bad news phrase. It illustrates the measurement problems associated with infrequent events. Without sufficient information to analyze the hazards, it's tough to make improvements. Or perhaps someone made the decision not to improve because of the expense involved, or decided to transfer the risk to an insurance company, or felt lucky enough to simply accept it. For example, if you knew that every 10 years, a windstorm would cause $10,000 in damage to your house, you could protect yourself by putting $1,000 in the bank each year. For those of you who are financiers, please excuse the financial simplicity of the calculation. I am making a point about risk here, not the mathematics of net present value calculations.

After this discussion of both situations, which would you rather be—Type A or Type B? The Type A company or person may have filed a claim if the severity was above the deductible. Their insurance company probably does not know how Type B is doing, only that it has not paid a claim. The Type A company may be doing all of the right things and may have just experienced a very infrequent natural peril like a tornado or freak storm that is not emblematic of the normal situation.

Now let's factor perception into the equation by mentioning an issue we'll discuss more in the next chapter. Type A behavior can be perceived as a higher risk than actual while Type B situations may seem to have less than their actual risk. If you're considering public opinion in your decisionmaking, the perception factors can be of significant importance. Factoring perception into the analysis says you want to avoid Type A behavior. Even though the public usually has a short attention span for the low frequency events, they seem to remember high severity events very well. The chemical, nuclear, refining, and oil drilling industries are especially good examples. The disasters at Chernobyl, Bhopal, Three Mile Island, BP's Texas City fire, and BP's Deep Water Horizon oil spill—need I say more?

The last discussion really identifies what is almost a paradox. Type A and B performances, even though they have the same risk reduction, show you the need to consider other issues before you decide which one is best. The graphic in Figure 6.4 helps give you a mental picture of a situation wherein risk and risk reductions are the same, yet the two situations are very different. This is the real power of the picture.

The direction in which risk is changing is another useful piece of information. It is an aspect of risk management that is not part of the risk equation. It can be measured (computed), plotted, and used to understand risk. This is a new aspect in the study of risk regardless of whether it's at a personal, business, or government level.

We'll use Figure 6.5 as a visual aid in our discussion of the direction aspect of risk changes. In this figure, the four major directions are categorized into quadrants, and we'll discuss them in order, from the worst to the best.

Movements in Quadrant I and some of IV directions correspond to increasing risk, driven by frequency or severity respectively. It is impossible to move in the direction of Quadrant I without increasing frequency and severity. It's also impossible to move in the direction of Quadrant IV unless severity is consistently increasing and frequency is decreasing. Notice that there are situations on the plot where risk is increasing even though either frequency or severity is decreasing. Let's look at each quadrant separately.

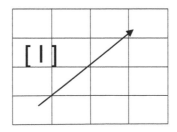

Quadrant I indicates perhaps the most obvious bad news that can be expressed. This direction indicates more events are occurring and their consequences are getting worse. The increased frequency is providing more data to measure the increasing consequences with enhanced accuracy. This is a classic bad news/bad news scenario.

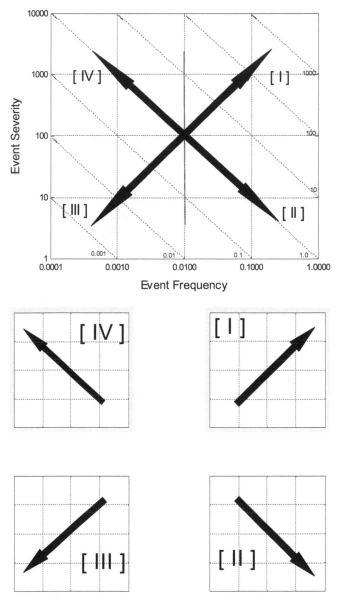

Figure 6.5   Frequency, severity, and risk quadrants.

In real life, these events get the highest priority to fix whatever is wrong. There is no doubt that this direction is the worst case.

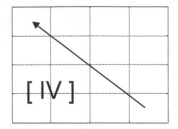

The choice for the next worse direction requires a judgment call. My opinion is Quadrant IV. Severity is increasing; the decreasing frequency may be good in some ways, yet bad from another point of view. This direction implies that less data are being collected, and the ability to measure is diminishing. If the situation under analysis is equipment- or-process related, the lower event frequency can be seen as a reliability improvement even though event severity is increasing. Here you can have situations where reliability and risk are increasing at the same time.

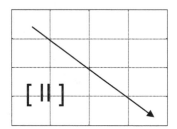

Next in line is Quadrant II, which says that event frequency is increasing while event severity is decreasing. With event frequency rising, reliability is decreasing but the additional events provide more data for analysis. The lower event severity can compensate for the frequency increase to the point where both reliability and risk can be decreasing at the same time. This type of pattern can be seen after the implementation of improvement programs where analysts are capturing more small failures that otherwise would be ignored. And by catching the small failures, the hope is that the high severity events are prevented.

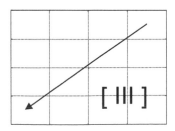

Of course, the best way to reduce risk is to decrease both event frequency and severity. This direction is shown in Quadrant III.

Often, the frequency and severity positions in these types of plots are dictated by the nature of the events themselves. Big equipment costs more than small equipment and car accidents have a different risk profile (frequency and severity) than most home-related accidents. That's the way it is. However, there are two indisputable facts about risk that we all know:

1. Risk is a function of time.

2. Risk is in the future.

Trending changes in risk are good, but it is more important to examine how the risk changes. Herein lies the power of the risk coordinate system picture. The direction of movement, regardless of the position in the picture, can suggest where risk is going. The key to the proper interpretation of risk directional changes is measurement. Are you tracking and documenting the right indicators? Once they are known, the picture of how risk is changing in the risk coordinate system shows which events need more study and, just as important, which events don't require a detailed examination. You must identify what to measure, what criteria to use to define a failure event, and how events should be combined to describe reality in the most direct way. This is our challenge—not the random failure of nature, fate, or so-called fortuitous events.

To demonstrate the application of the previous quadrant discussion, four examples are presented in Figure 6.6. Five points are plotted in each case. The examples are provided to show you how different types of trends can be identified from

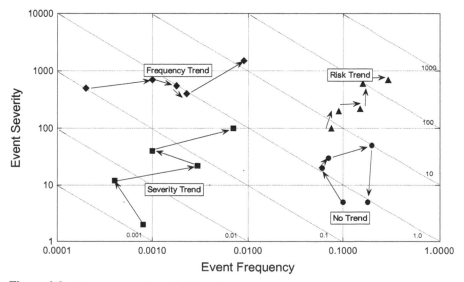

Figure 6.6   Frequency, severity, and risk trends.

frequency–severity plots. The data are fictitious but emblematic of actual situations.

In the frequency trend, notice that the point always moves to the right even though goes severity moves up and down. An analogous pattern is seen for severity in the severity trend. For the risk trend, the next point is always in the same quadrant—in this case, Quadrant 1. The quadrant is relative to the current, rather than first point. The last example shows a situation where no trend exists.

An interesting property of this graphic method is that trends in frequency, severity, and risk can be inferred at a much higher confidence than the certainty associated with the actual position of individual points. Each point is defined as coordinates that are expected or average values. Severity is usually defined as the average loss per event and frequency is a proportion describing the number of events per unit time. If we were interested in the precise position of each point, we could express our knowledge of each point's location as contained within rectangles whose sides are defined by the statistical uncertainties in frequency and severity at the level of confidence desired. The uncertainties can be sufficiently large that statistically significant changes in frequency, severity, or risk are often not seen between time period points. But to identify trends we focus on the direction the point moves between periods.

Here's how we can do this. For each coordinate, there is a 50–50 chance the value will be larger or smaller for the next time period. In other words, assuming the points are independent and movements are random, the probability that a point will move higher or lower in either severity or frequency is 1/2. This means that if the direction changes the same way, up–down or left–right, during each time period, the probability of an actual trend in the process increases. In equation form, the probability of a trend in frequency or severity after moving in the same direction for n consecutive time periods is computed as:

$$P_n = 1 - \left(\tfrac{1}{2}\right)^{(n-1)}$$

Trends in risk are identified by consecutive point movement in the same quadrant. The probability of moving in the same quadrant for n consecutive time periods is:

$$P_n = 1 - \left(\tfrac{1}{4}\right)^{(n-1)}$$

Referring to Figure 6.6, notice that the frequency and severity examples both show a consecutive point movement pattern. Using the above formulas with $n = 5$, the probability of frequency and severity trends is 0.9375. The risk trend probability is 0.9961. Thus with only five data points given by the patterns shown in Figure 6.6, we have a high confidence of trends in the direction of movement. In risk analysis, where virtually all data represents history, we are trying to see into the future. This methodology is not a "silver bullet" but it does represent a useful risk analysis and risk communication tool which will help you make decisions about the future.

Here is one last point of speculation for you to consider. Can we predict the apparent sudden, catastrophic events that polarize us occasionally? Does the phrase "It's an accident waiting to happen" have any connection to this discussion? I believe there is no such thing as a "random failure." I equate belief in random failures to the belief that the Greek god, Zeus, manages the world's thunder and lightning production and delivery. Nature has given us a highly ordered world. If we can't figure it out, that's our deficiency; it is not just randomness. You could say that an apparent random event occurs because we have applied insufficient or inappropriate measurements to describe the phenomena. I believe that a system (use whatever definition you like) trips before it falls. If you can measure the trips, maybe, just maybe, you can prevent the fall. True, the introduction of chaos theory has showed us that our universe is not as simple as we would like it to be. There are situations in which small changes in conditions produce large changes in the process. The dimensional framework discussed here doesn't answer all questions about why seemingly "random" events occur. It can, however, give you a mental picture on which to bet on your ideas, to take the calculated risk, and to act.

DISCUSSION QUESTIONS

1. Research a current topic of your choice and develop a risk-based importance ranking of a set of possible events similar to the ranking shown in Table 6.1 comparing Cassini Mission risks to other common exposures. What assumptions have you made regarding your conclusions?

2. Risk is described in terms of two independent variables: event frequency and severity. What additional variables could you add?

3. Discuss examples of Type A and B events as shown in Figure 6.1 that have the same quantitative risk.

4. Describe a contemporary issue similar to the workers' compensation effects of the +65 age group, using a similar methodology.

5. Using the graphical frequency, severity, and risk trending methodology shown in Figures 6.4 and 6.6 for several time based risk exposures: Can you identify any potential trends? If so document your conclusions and continue to plot new data as it becomes available. Are your conclusions valid? What assumptions have you made?

Case Study

Even though risk is in the future, our ability to "measure" it using historical data has its limitations. Risk management represents, in essence, a leap of faith regarding our assumptions about the future. Here is a case from negligence law in which crime frequency and severity is fundamental to the ultimate verdict. The outcome shows how, in practice, risk management can be treated by the courts.

In a tort claim of negligence, a plaintiff must prove that the defendant owed him or her a duty of care, that the defendant breached the duty of care, that the defendant's breach caused the plaintiff's injury, and that the plaintiff suffered damages. In this case the duty of care was in connection with parking lot security.

As shown on a video recording of activity at the entranceway of a large food store in a shopping center, the plaintiff is seen leaving with her groceries at 4:13 P.M. on Sunday, May 16, 2004 [14]. After putting her purchases into the back of her van, she opted to forego the 25-cent deposit for the shopping cart she had used and decided to just leave it by her car. She got into the driver's seat and put on her seatbelt. A man approached the car and, assuming that he wanted her shopping cart, she opened her window and said he could have it. He thanked her. Then while she turned her attention away, the man opened the car door and grabbed her by the neck. She struggled, called for help, and honked the horn of her car. A passer-by shouted and he fled. The man was arrested a short time later. From the scuffle in the car, both the assailant and the plaintiff had minor injuries that were visible. After the incident, the plaintiff found that rags and a knife that did not belong to her had been placed in a back seat pocket of the van. It became apparent the man had entered the car while she was shopping and by the time of the trial he pled guilty to attempted kidnapping.

The plaintiff sued the shopping center owner and operating company to recover damages for injuries sustained as a consequence of their negligence to provide adequate security. She hired a security expert who performed a security risk assessment of the shopping plaza. The food store was the shopping center's largest business but there were several other commercial retail establishments that shared the parking lot. The expert compiled a list of criminal and security related activities that occurred during the past 18 months. There were incidents of intoxication, thief, threats, fights, shoplifting, and fraud. Based on the frequency and severity of criminal activity at the plaza and other crimes that have occurred at similar facilities around the country, he concluded that the shopping plaza owner and operator were negligent in providing adequate parking security.

The jury awarded the plaintiff damages in the total amount of $275,000. On appeal, the decision was reversed because the appeals judge believed the plaintiff's evidence was insufficient to establish the defendants' negligence. To establish a claim for negligent failure to provide reasonable security, the plaintiff must prove that the commission of criminal acts by third parties was foreseeable under the totality of circumstances and that the owner of the parking facility failed to take reasonable measures to guard against the risk. The court held that the risk assessment information provided during the court proceedings did not show kidnapping as an exposure from the data. Stated differently, the court found that, based on the frequency and severity of all the crimes which had previously occurred in the shopping center and the area, the defendants could not reasonably foresee that a kidnapping would take place in the shopping center. This frequency–severity consideration is often an integral part of the courts' analytical framework while ruling on civil negligence cases. Negligence law is essentially a study of how much risk is acceptable in a given society.

Question

1. If you were the security risk assessment expert working for the plaintiff, how would you argue the case of negligence to provide a reasonable duty of care against the defendants?

2. Assume the role of a juror. Would you vote for or against awarding damages? The jury in the first verdict did award damages most likely because the plaintiff was

injured and could not recover any money from the kidnapper. This common phenomenon usually leads juries to find that unacceptable level of risk taking by the defendant indeed existed, because the alternative is sending an injured plaintiff home emptyhanded.

3. Why do you think the appeals court judge, one person, reversed a decision made by 12 people?

ENDNOTES

1 http://saturn.jpl.nasa.gov/ (accessed March 15, 2010).
2 National Aeronautics and Space Administration (NASA), Draft Supplemental Environmental Impact Statement for the Cassini Mission. Washington, DC, April 1997.
3 Clark R. Chapman and David Morrison, "Impacts on the Earth by Asteroids and Comets: Assessing the Hazard," *Nature*, Vol. 367, January 6, 1994.
4 http://news.discovery.com/space/the-2010-al30-an-asteroid-or-man-made-object.html (accessed March 30, 2010).
5 http://epod.usra.edu/blog/2010/03/nearearth-object-2010-al30.html (accessed March 30, 2010).
6 U.S. Census Bureau, Current Population Survey, Annual Social and Economic Supplement, 2008. Internet release date, June 2009.
7 Bureau of Labor Statistics, Table A6: Employment status of the civilian population by sex, age, and disability status, not seasonally adjusted. Last modified March 10, 2010.
8 Tanya Restrepo et al., "Age as a Driver of Frequency and Severity," NCCI Research Brief, December 2006.
9 Martin Wolf, "Claims Characteristics of Workers Aged 65 and Older," NCCI Research Brief, January 2010.
10 U.S. Department of Labor, "Nonfatal Occupational Injuries and Illnesses Requiring Days Away from Work, 2008," Occupational Injuries and Illnesses by Selected Characteristics News Release, USDL-09-1454.
11 U.S. Bureau of Labor Statistics, U.S. Department of Labor, 2008 Census of Fatal Occupational Injuries (preliminary data).
12 http://bsa.nfipstat.com/reports/1011.htm (accessed March 20, 2010).
13 FEMA, "Flood Safety Awareness Week March 15–19, 2010," release date, March 12, 2010, release number R10-10-014.
14 *Pisciotti v. PF Pasbjerg Development Co.*, NJ Superior Court, Appellate Div. 2009.

Chapter 7

What's in a Risk? Perception Creates Risk Reality

If you think it, want it, dream it—then it's real. You are what you feel . . .

—from *Joseph and the Amazing Technicolor Dreamcoat*

During a wintry Friday morning rush hour in the Washington, D.C., metro system, a young musician dressed in nondescript jeans and a baseball cap walked over to an open area, took out his violin, put its case in front of him for tips and began to play. For the next 43 minutes he played some of the most elegant and beautiful classical music ever written for the instrument. Over 1000 people passed him on their way to work. A handful paused momentarily but then continued walking to make sure they were not late for work. Twenty-seven people gave him money. He made about $32. Doing the math, this rolls up to the revenue generation rate of about $45/hr. This may not be that bad for most street musicians, but this performer was not the norm. Nor was he playing your typical violin. In his element, the symphony hall, he generates revenue at about $1000/minute. His instrument was handcrafted by Antonio Stradivari in 1713 and valued today at over $3.5 million.

The street performer was Joshua Bell, one of the most accomplished classical violinists of our time. Even the nosebleed seats at his concerts typically cost more than $100 each and yet this exceptional musician, playing exceptional music at a Washington Metro Station, was valued by only about 2.5% of the people who heard this music, and even then only for $32 in total. Actually, one person out of all who passed recognized Bell and gave him a $20 bill.

Mr. Bell did not do this as a lesson in humility or for crude adventure. He did it as part of a well-planned and -documented experiment by the *Washington Post* [1] to examine the public's reaction to context, perception, and priorities.

In this case, perhaps it's not possible to separate the context from the perception variables. You could argue that people who value classical violin virtuosity are

20% Chance of Rain: Exploring the Concept of Risk, First Edition. Richard B. Jones.
© 2012 John Wiley & Sons, Inc. Published 2012 by John Wiley & Sons, Inc.

willing to pay for the experience at the symphony and many of the folks passing by may not have had those characteristics. Nevertheless, do you think their behavior would have been different if they knew the caliber of the musician and his instrument, and knew that they were not up on a stage hundreds of feet in the distance, but performing only 3 feet away and at eye level?

Measuring public perception is like touching a cloud—you can tell when you're in it, and you can tell when you're out of it, but it's just about impossible to define the exact edge. Measurement, analysis, and an understanding of physical or statistical phenomena usually adhere to a fixed set of rules. But when measuring perception, all bets are off, for we must factor the moods and moments of the human mind into the equations. What was once logical, rational, and invariant becomes alive with new patterns of behavior outside the physical laws of the universe. The human mind is the "logic engine" that drives perception, making perception more subject to personal experience and the laws of psychology than to those of mathematics and physics. In math and physics, a law is a law. In the absence of new discoveries, these laws don't change over time. Risk perception disregards this type of structure. Perceptions of risk vary by age, sex, vocation, race, culture, and ideology, not to mention the time of day, day of the week, and any other individual idiosyncrasy you can think of. Risk perceptions are internal, sometimes invariant, sometimes fleeting value judgments made by each person. Roll these up to a community, county, regional, state, or national level and you have one of the most powerful forces on Earth: the general public's perception of risk.

Understanding the many dimensions of risk perception is difficult. The challenge is to measure how different people with different beliefs, opinions, and value systems respond to a circumstance or situation, without influencing their views. It is easy to ask people for whom they would vote in an election. The choices are simple and mutually exclusive. As with multiple choice questions, there are no shades of gray.

The public's perception of risk is not so simple. It's a complicated moving target, and its measurement brings unique challenges. When TWA flight 800 crashed into the Long Island Sound in July 1997, the New Haven, Connecticut, area press reported that among the lost cargo items were eight vials of HIV-infected blood, destined for a laboratory in France. The day after this announcement, sales of fish in a large supermarket in nearby Hamden, Connecticut, dropped dramatically. Overall sales receipts were normal—only the seafood sales were unusual. The apparent cause? The perception that the AIDS virus might now be lurking in the ocean [2].

Now if you asked people about the likelihood of acquiring AIDS from eating fish that might have come from the same very large body of water as flight 800's minuscule blood cargo, they'd probably be quite rational and deny the possibility. Yet left to their own unconscious judgments and powers of association, fear will beat rationale and "tainted" fish will remain unsold. Of course, seafood purchases returned to normal levels once the initial shock of the flight 800 tragedy passed.

Flying is not the only transportation mode subject to dramatic fluctuations in associated risk perception. Consider long tunnels, for example. If you have ever been in a long underground tunnel you can understand the committed, almost claustro-phobic feeling of confinement. Tunnels can be dangerous places. But when they are designed, built, and operated with safety as a primary objective, the actual risk can be far less than that perceived by its users. An example of this case is that of the England-to-France tunnel, also called the "Chunnel."

On November 18, 1996, a truck aboard a freight train in the Chunnel caught fire. Injuries were minor, but news headlines like "Panic Under the Channel," "Tunnel of Blunders," and "It was like being in a tomb," created tremendous public worry about safety in the 31.4-mile tunnel embedded in the rock beneath the English Channel. Were these fears well justified [4, 5]? Let's take a look at the data. First of all, fire hazards were a well-known risk management issue for such a long tunnel. Consequently, its designers incorporated plenty of fire prevention and control. For example, there are smoke, flame, and carbon monoxide detectors throughout the tunnel and fire hydrants every 125 meters. A separate service tunnel parallels the main tunnel, with access every 375 meters.

This fire was the only one of any significance in the history of the Chunnel. Over 5000 freight trains had crossed without incident before it occurred. What this fire actually showed was that the active and passive fire control factors incorporated into the Chunnel worked! The fire was extinguished, the Chunnel, though slightly damaged, was still in good condition, and no lives were lost. But even though the systems worked, the headlines didn't mention this fact. Instead they placed the controversial tunnel project in the middle of another debate. The stock of Eurotunnel, the tunnel owner, dropped more than 17% after the incident. Public perception, fueled by an emotional press, didn't care about engineering. Their myopic focus was on the fear of being "caught in a tunnel of fire" and risk management could not make any difference.

Another fire occurred on September 11, 2008 [6]. Again the design and safety worked as intended. In total 32 people were evacuated safely and no one died. This time headlines were more precise, such as "Passengers Safe After Chunnel Fire," and others referring to the real problems associated with the time delays associated with the repairs. Because of the fire's particular date, the media spent more time speculating on terrorism connections than on the fear tactics they employed 12 years earlier. Once the cause was determined to be simply an accident, the subject was dropped for more newsworthy stories.

In the spring and summer of 2010, one of the biggest daily news topics was the oil spill in the Gulf of Mexico. The news media showed, and virtually anyone on the internet could view, the uncontrolled oil gushing into the ocean nearly one mile below the surface. News coverage aired stories of toxic oil dangers and showed oil cleanup crews in full protection suits with the heat index over 100°F (38°C). Journalists also showed the large amount of oil dispersant chemicals being used, the uncertainty in possibly adverse environmental effects to the affected ocean ecology, and the closing of fishing regions due to oil contamination. Then guess what hap-pened? Even though the affected coastline was small compared to the Gulf coast of the nearby states, tourism dropped dramatically. Even seafood restaurants in central Texas saw a large reduction in customers [7].

As these examples have shown, the power of the public's perception of risk is tremendous, volatile, and unpredictable. Logic and order do not prevail. When a battle develops between the emotions of public perception and the calculations of the technical or scientific community, you can guess which will dominate.

Now I need to take a minute to talk about the press, since I've just been much less than complimentary about their choices in headlines and reporting. Frankly, the journalists are just doing their jobs. And their jobs, referring back to Chapter 2, Measurement, are to sell papers, magazines, and get on-air ratings. And fear sells. In almost all cases, the details that follow the headlines are accurate, but the headlines that gets your attention may not be the true summary that best describes the whole story.

When I first began to research risk perception, I thought that the measurement process would be simple. I'd just write up a questionnaire and use good statistical methods to compile the responses. The more I learned about this type of risk measurement, the more I began to appreciate the tremendous difficulties in constructing good survey instruments. Measuring perception is a vastly different task than traditional scientific measurement. Let me explain. Risk perception is an opinion or a value judgment regarding a hazard to a person, to people, or to a valued object. Now how do you measure a value judgment? We don't have a device that provides a simple readout of a "risk perception value." To find out what a person is thinking, you have to ask questions, hence the questionnaire or survey becomes the measurement tool.

Now we have a new set of challenges. Referring once again to Chapter 2, one of the basic laws of measurement tells us that the process of measurement can affect the result. In this case, a survey's wording and construction can greatly bias responses and produce misleading results. How would you answer this question?

"Do you advocate a lower speed limit to save lives? Yes or No"

Yet suppose we asked it like this:

"Does traffic safety require a lower speed limit? Yes or No"

Your responses may be quite different, even though the same issue is being examined.

Another law of measurement tells us to use the right tool for the job. A simple, direct, multiple-choice question results in a simple answer with little room for interpretation. But this type of approach doesn't work well when you're assessing attitudes. Opinions are reflected in the answers to sets of questions. The interactions between questions can yield important information about attitudes on complex issues. For example, you could put together a series of questions to measure a group's risk aversion attitudes. Such a set might include questions about car seat belts, types of recreational activities, and eating habits. Together the answers present a tapestry of facts and a picture of the group's risk aversion characteristics. Every time you complete a survey or even fill out the warranty card on new products, you are providing profile information describing yourself, your family, and, most importantly in this case, your buying habits. In a sense, your buying risk–benefit perception is measured.

Survey use gets further confused when we look at survey objectives, funding sources, and who is collecting the data. Just put yourself in the shoes of a small town resident in the following situation. You are stopped in the local supermarket by a pollster conducting a survey for the Environmental Protection Agency. The pollster tells you he or she is gathering residents' views about the risks related to chemical plants in the area. The first thing you would probably think is that there is problem with one of these plants, otherwise the EPA wouldn't be spending the time and money required to conduct the survey. You've just formed an opinion—your perception has changed—and you haven't answered a single question! This is a prime example of measurement affecting the process. The researcher assures you that nothing is wrong, but you also know that people are not always straightforward with bad news, and your skepticism grows. This is just one brief example showing that it can be difficult, if not impossible, to collect unbiased data about risk perceptions.

Developing good questionnaires is a genuine art and takes a lot of work. It includes studying the effect of every word and every punctuation mark on each question's meaning. There are numerous books and courses on the subject that will provide you with more detailed information than is available in this book, so we'll move on.

I've just painted a picture of risk perception as a huge, vague monster that is difficult to measure and understand. The good news is that it has been the subject of intense research over the past three decades and much has been learned. We now have several structures and guidelines to understand the many variables of this pervasive phenomenon. Risk communication experts have developed a framework to assess the components of public risk perceptions. The criteria can help identify the key indicators of the public's hot button issues, and you can use the following lists as a general framework to assess risk perception. The lists' characteristics are related, and you'll probably be able to add your own experiences to some of them. I'm going to present both a long and short list. Here's the long list first [7].

WHAT INFLUENCES OUR PERCEPTIONS OF RISK? THE LONG LIST

1. *Voluntary or Involuntary Risk*: People don't like to be pushed. You can often get a person to do something by asking, but if you attempt to force the behavior they'll probably resist. It's a natural, knee-jerk reaction. This is where corporations and governments have gotten into trouble. For example: Company A buys a parcel of land to build a new plant. They move through all of the zoning and environmental obstacles required for them to build on it legally, yet still encounter strong resistance from the local community. The people in the community are not exposing them-

selves voluntarily to the hazards associated with the company and therefore resist exposure to any possibility of risks connected to it. The actual risks that the people in the community already have through their work and home activities may be many times greater than those of the new plant, but those risks are undertaken at their choice. This is the difference. Chauncey Starr, from the Electric Power Research Institute (EPRI), performed some interesting work on this subject. He found that people will voluntarily accept risks that are about 1,000 times higher than those risks associated with involuntary activities [8].

2. *Controlled or Uncontrolled Risks*: When you are sitting in the back of a large jet airliner, have you ever felt a little nervous as the plane takes off, transforming itself from a giant, very fast tricycle to an aircraft? Be honest! Now compare this feeling to those you experience when you drive your car to work, or the grocery store, or out to dinner on a weekend. What's the difference? In the airplane, you have virtually no control over your destiny. You are literally along for the ride. But in the car, since you are at the controls, you perceive that your risks are lower. Of course, you are much safer in that aircraft than in your car, but somehow logic is not usually the prevailing emotion in unfamiliar settings. Safety statistics aside, I would bet that most people who are afraid to fly have no fear of riding in or driving automobiles. Some feel so safe in cars that they don't even wear seat belts. This certainly isn't logical from a statistical perspective, but it makes sense to them; they're in control! (One caveat: This perception of safety in an automobile dissolves when your 16-year-old is behind the wheel!)

3. *Fair or Unfair Risks*: There is a term that accurately describes this characteristic that is often used in insurance circles. It is "distributional justice." This term simply means that those who share the risk should share the associated costs. If someone wants to skydive or bungee jump off of bridges, these people should uniquely share the risks. In this frame of reference, it would be "unfair" to someone of the general public who doesn't share the risk to have to pay for it or be injured as a result of the risky activity. Looking back at the chemical plant example in our last section, distributional justice means that even though the local community shares the hazards, it also shares in the economic benefits. This is fair as long as the benefits are as least as great as the risk. However, it could be judged unfair to people in towns outside of the economically influenced zone which have to bear the costs and effects of any risks associated with the plant.

4. *Familiar or Unfamiliar Risks*: Some of the most familiar risks, at least in the United States, are those involving home-use chemicals. One look beneath your kitchen sink probably will show you several toxic chemicals you have purchased for specific purposes, including disinfectants, insect repellents, pesticides, cleansers, waxes, varnishes, and spray paints. We come into contact with and breathe these chemicals without any concerns

about the exposures. After all, we need these products to clean, keep the bugs out, to keep metals painted and shiny. You can name you own "need to" reason. But depending on the product and how you using it, your risk exposure could be very high for latent illnesses.

Household cleaners are familiar risks. On the other hand, electromagnetic fields, radiation, and nuclear power are some areas in which the general public has a limited frame of reference or experience. Hurricanes, tornadoes, and earthquakes fall into both categories, depending where you live. People who have experience with a phenomenon will have a different risk perception than those who don't. A snowstorm that would be considered uneventful in Burlington, Vermont would bring the city of Atlanta to its knees. There is enough documentation about what happens in these situations that everyone could, at least in principle, understand their potential effects. However, I doubt if people in Phoenix, Arizona will ever have the same risk perception of hurricanes as people along the Gulf Coast.

5. *Memorable or Not Memorable Risks*: Events that stand out in our memories clearly influence the way we view hazards. The example I always find curiously interesting is when a small plane lands in a field, road, or shopping center parking lot. Notice that I didn't say "crash." Engines do fail, small planes run out of gas, and weather sometimes forces a pilot to land away from airports instead of fly into a problem. Whatever the reason, the landing is under control. Such controlled events usually don't involve deaths or serious injuries, but they are guaranteed to be in the local evening news. They are rare, odd events and, as a result, memorable. Car accidents involving deaths or severe injuries are just too common to be memorable on a continuous basis. There are just too many of them. Airliner crashes, plant explosions, nuclear plant accidents, large oil spills—these are memorable events.

6. *Trustworthy or Untrustworthy Information Sources*: The obvious meaning here is that people are going to believe those in whom they trust. This is one reason that the field of risk communication has become a discipline almost unto itself. Industrial organizations have become more sensitive to public opinion and have worked hard to establish credibility and trust with their public constituents. They've found that consistent communications regarding normal plant operations and frank discussions about problems have helped them develop healthy relationships with the general public.

7. *Ethically Objectionable Risks*: Pollution is a good example. Everyone knows pollution is bad, but all pollution is not created equal. The *Exxon Valdez* oil spill in the pristine Gulf of Alaska and the Gulf oil spill in 2010 had more ethical potential (or should that be unethical potential?) than a spill in the Houston shipping channel. The latter is an industrial area, where some pollution from shipping is expected and unfortunately somewhat

routine. The former two, although still heavily trafficked, are surrounded by the unspoiled and unique beauty of nature. Oil spills in the Gulf of Alaska and the Gulf of Mexico violate our unspoken rules of ethics far more than one in Houston.

Here's another case of ethics shaping perception of risk. It, too, involves pollution. The EPA is charged with task of cleaning up toxic waste sites around the country. In some situations, the practical solution is to remove the major percentage of the wastes but leave the rest. The reasons are logical and make economic sense. Generally, removing the last traces of wastes generally costs more than removing the bulk of the contamination. Waste dilution and transport can contaminate a very large soil volume compared to the smaller volume that contains the higher contamination levels. The low-level residual waste is unlikely to cause problems unless, for example, a child eats pounds of the contaminated dirt every day for many years. However, just imagine the parents of young children listening to a government administrator say that a company is going to leave some toxic wastes in the family's backyard. Do you think the parents would care about its level of toxicity? A former EPA official said it very precisely. He said that trying to explain the optimal (cost–benefit) level of pollution is like trying to talk about an acceptable level of child molesters [9]. Any perceived risk exposure to children, born and unborn, is viewed an ethically objectionable.

8. *Natural or Unnatural Risks*: When an earthquake occurs, a volcano explodes, or a tornado strikes, people really can't blame the government, a negligent company, or another human being. These natural acts certainly cause their share of tragedies, but the public perception of their risk does not generate any unusual response or action. Peoples' experiences with natural perils play a definitive part. To some extent, this is obvious. Practical experience with a Category V hurricane (sustained winds >155 mph) would certainly be more compelling than a classroom lecture and slide show. Another factor that I think explains why the public's perception of natural risks is nonvolatile is due to the regional culture. Take, for example, California's web of fault lines and widespread areas of high seismic activity. Why would anyone live in an area where seismologists are warning that severe earthquakes are imminent? One reason may economic. People just don't want to incur the very real and immediate hardships and economic costs related to moving compared to the chance of earthquake losses sometime in the future. Another reason could be that people like where they live and the benefits of the location are perceived to outweigh the earthquake risks. What we call the "California lifestyle" is a regional person, cultural identity mechanism. You might argue that people could practice the California lifestyle in a much more earthquake-stable zone such as Nebraska, but the interaction of the weather, topography, recreational activities, and people represent a very strong bond. The benefits

of this bond exceed the perceived risks of earthquake. While no one has done a study to indicate that the same mechanisms are working for hurricane-prone locations, I'd bet that they are. (By the way, I loved living in Santa Barbara.)

As I mentioned earlier in this discussion, part of the reason for the public's acceptance of natural risks is because there's no one to blame. In my former home, we had very acidic well water. We lived in the woods, and the well water absorbed some of the organic acids produced by decaying vegetation. To reduce the acidity we had special pH-balancing equipment in the water system to keep the hot water heater, boiler, and the piping from corroding away (not to mention the people). We didn't perceive this issue as a problem nor did we complain about the associated costs. Now just imagine if the acidic water was from the waste of a local plant. This would be pollution and my willingness to pay for treating the high-acidity water would be very low. I would try to make a case for the polluter to pay. Again, if the condition is natural, people have no one to blame. They decide whether or not the benefits of living with the risk exposure are greater or less than the risks. Sometimes people do change their minds. After a severe earthquake or devastating hurricane, some people do relocate to reduce or, depending where they move, eliminate this risk exposure.

Here's one last point on this issue of natural versus unnatural risks. In the United States, tort litigation is a common way of assigning blame when a person is injured or killed due to negligence. In the world of general liability law, there is an interesting precedent that is called the Attractive Nuisance Doctrine [10], or sometimes the Doctrine of Attractive Allurement. It mainly applies to children, and says that a possessor of land is liable if a child trespasses and suffers bodily injury due to "an artificial condition or a tangible and visible object that is reasonably known to be both attractive to children and unreasonably dangerous to them." The doctrine has been applied in situations where small children have drowned in privately owned swimming pools or artificial lakes, injured by dangerous machinery, or bitten by aggressive pets. The interesting point is that it is seldom applied to situations involving a natural land condition.

9. *Dreaded or Less Dreaded Risks*: One of the most dreaded risks of the general population is cancer. Any risk exposure that carries a correlation with higher cancer incidence certainly will be perceived by the public as serious.

It is really true that not all risks are created equal. A friend of mine was riding a personal watercraft two miles off shore from Key West, Florida. A severe thunderstorm came up while he was sitting on top of this floating lightning rod in the ocean. He said that lightning was all around him but he didn't have time to be scared. It was happening too fast. In addition, he said something that made sense in a morbid kind of way. He said, "Even if I was killed by lightning, it would be quick and painless—and what a way to go—riding a jet ski at sixty miles an hour, two miles off Key West. I'd rather go that way than waste away in a cancer ward." Although I'm not sure about the painless part of a lightning strike, I think this is one way of looking at degrees of dread; anticipation breeds fear.

10. *Seen or Unseen Risks*: Just look at the furor generated about possible effects of unseen electromagnetic fields from power lines and appliances. It's hard to comprehend and defend yourself against things you can't see. In contrast, the high risks associated with cars, boating, swimming, and the like provide visible cues about what is going on. You might or might not see the bump in the road or the rock in the lake, but the general perception is that risks for activities like these are visible.

11. *Level of Scientific Understanding of the Risk*: Radiation is well understood by scientists as far as what it is and how we can apply it to our advantage. Medical diagnostics would be very different without technologies such as X-rays, computer assisted tomography (CAT) scans, and magnetic resonance imaging (MRI). However, while radiation is well understood technically, its long-term dose-response effects are not. A radiation application that has been the subject of public controversy is irradiating food [11]. This process is designed to kill hazardous bacteria without the addition of harmful (or potentially harmful) chemicals or other invasive methods. The food is exposed to radiation for a short period of time and that's it. Part, not all, of the public concern is about the possible long-term effects of such treatment and no one really knows the answer. The only real way to know is to try it for thirty or forty years and see if there is a correlation between the process and public health. The weight of scientific evidence is clearly on the side of tremendous benefit to the consumer, but again no one can say for sure that the process will not make someone sick some day in the future. Of course, no one can say for sure that you are going to live through the day anyway. However, fate is understood. We live it every day. The long-term effects of irradiating our food are just not known yet. And we are just beginning to think of questions about some of our new airport screening devices and the possibility of long-term effects from their widespread use.

This list seems long and I hope you took the time to go through it. I think the topics give you an idea of how risk perception can be gauged, as well as a brief discussion of some potentially major issues in assessing its potential. The items on the long list sometimes complement one another, giving the criteria an inherent system of checks and balances. It's also representative of the fact that there is seldom, if ever, just one dimension to risk perception. You need interconnected measurement criteria to reflect that kind of phenomenon.

Okay, here's the short list. If you were going to measure the public's risk perception (or lack thereof) and you need to do it quickly, here are the four big-ticket items. These terms are pretty much commonplace so I'm not going to spend much time on them.

WHAT INFLUENCES OUR PERCEPTIONS OF RISK? THE SHORT LIST

1. *The Media*: The media tries to do two things: present the news while keeping you sufficiently entertained so you don't change the channel or radio station before their commercials. This tightrope act requires journalists to script their news reports and make them dramatic, entertaining, and interesting. And, oh yes, factual. Newspaper journalism is the best place to go for real news, once you get past some of the headlines. Radio and TV reports are so abbreviated and filled with drama, entertainment props, or "interesting" sound bites that finding the facts behind the news is a hard job for anyone. I'm not even going into tweets . . .

2. *Technology*: The more exotic, esoteric, foreign, complex, new, and outside of peoples' experiences an activity, technology, or process is, the more risk it will appear to contain. If you want more discussion of this one, read the Long List.

3. *Control*: People perceive lower risks for activities where they have a degree of control; the higher the degree of control, the lower the perceived risks. Skiing, driving a car, drinking alcoholic beverages, working around your house or apartment, and even smoking are risks that people believe they can control at some level. Living in close proximity to a prison, nuclear power plant, chemical plant, or riding in a commercial airliner, are situations where people have less control and therefore perceive higher risks.

4. *Time*: When are the effects (both the benefits and costs) experienced? If you are driving a car and choose not to wear a seat belt, the effects are observed immediately when an accident occurs. However, you can experience the "benefits" of smoking for years before you experience any disease. Exercise, dietary discipline, and learning are examples of situations in which you pay now and are rewarded later. All three are definitely risk-reduction activities. Yes, risk is always in the future, but I differentiate between short-term and long-term future effects here. The ways people view both of these futures will influence how they view risk.

As you can see from these attributes of risk perception, the field is a mix of many disciplines. In fact, the whole subject is receiving serious attention from scientists worldwide as this planet's residents work to allocate and manage finite financial and natural resources.

PERCEPTION OF RISK: SOCIOLOGICAL INFLUENCES

There are two basic scientific approaches to measuring risk perception: psychometric and cultural [12]. The psychometric paradigm has its foundations in psychology. This approach attempts to develop a cognitive map or framework of how people make decisions about risk. For example, the two lists of risk perception attributes we just discussed could be viewed as frameworks for a theory about how people develop their risk perceptions. These attributes are then measured as a function of some basic variables, called observables. Some "observables" of risk perception are gender, age, and occupation. These are the things you can practically measure in a population. In addition, scientists recognize that experiments may involve factors that can't be measured. These are called (you guessed it) nonobservables. The knowledge of external sources of variation helps quantify differences in results and gives us direction on how to improve the overall, long-term scientific effort.

The other way scientists study and predict risk perception is by studying cultural theory. You might have been thinking about this as you went through the attributes of risk perception. This perspective is the method used by anthropologists and sociologists who say that risk perception is a function of sociological relationships and cultural values. I'm sure you can think of examples where general cultural factors have influenced the ways people view authority, government, health, and safety. Just to give you the general framework of the cultural theory approach to risk perception, here are the four major dimensions. Into which group(s) do you fall?

Individualistic: These people are generally stable and rarely reactionary in nature. They have a positive outlook on the world and require definitive proof for change.

Egalitarian: This philosophy advocates equal social, political, and economic rights and privileges. With regard to risk perception, nature is viewed as delicate and dangerous. Thus changes in policy regarding personal behavior are adopted or rejected based upon their effects on the environment.

Hierarchical: Hierarchists believe in order. They respect authority, both scientific and administrative, and are advocates of big business and big bureaucracy. Those at the top give the orders, those at the bottom accept them, and those in the middle do a little of both.

Fatalistic: Fate is the hunter. A fatalist believes there is no point in attempting to change behavior or the environment because there is nothing that will make a difference. History is irrelevant since knowing the past can't help change what will happen in the future.

Are you a little bit of each? I've discussed the psychometric and cultural approaches here to give you a basic measurement framework for assessing your own risk perceptions and to possibly help you understand why seemingly irrational views can be logical to those who have them. I admit this stuff is not party conversation material, but, as I said before, perception of risk is one of the most powerful forces on earth. We all need a basic understanding of its "physics" for our personal and professional risk management.

PERCEPTIONS OF RELATIVE RISK: AN ASSESSMENT EXAMPLE

Now that we've looked at many of the complications involved in understanding perceptions of risk, let's take a look at a simple problem. In some cases, such as the one I'm about to show you, a level of perception can be assessed fairly easily. It's especially simple to measure variations in perception of relative risk in a group of activities. Here's a "do it yourself" or "how to" method for you to measure this type of risk perception [13]. You can use this for any specific set of risks. When I performed this assessment I wanted to see if there were any major differences in risk perception among three very different groups of people. In order to do so, I made up a list of 30 risks and asked each member of the three groups to rank the risks from highest to lowest. Here is the form I used.

Risky Business! A survey of perceptions of risk

Directions: (This exercise must be done on your own.) Read through the following list. Number the entries in the list from 1 to 30. Number 1 will indicate the item in the list you feel results in the *most* deaths in a given year. Number 30 is the list item you feel results in the *fewest* deaths per year in the United States.

boating	nuclear power
choking	power mowers
drowning	railroads
getting hit by a falling object	riding a bicycle
firearms accidents	riding in a motor vehicle
flying in a small plane	setting off fireworks
flying on a commercial airliner	slipping or tripping
getting bumed in a fire	smoking
lightning	snow skiing
getting overheated	suicide
getting too cold	tomadoes
hunting	using electricity

I chose three groups that I thought would give some interesting perspectives on risk perceptions. The three groups were 18 state troopers from Troop F of the Connecticut State Police, 25 insurance professionals, and 26 eighth-grade students and their teacher.

When I asked people to complete the survey, the first question was, "Why are you doing this?" I explained my reasons and promised everyone that the results would be shared once all the data was compiled. For each group, the average rank was computed for each entry. The risks were then ranked from highest to lowest risks. To make the comparison even more interesting, I located the actual number of deaths for each activity in the National Safety Council reports and other several other sources.

The results are shown in Table 7.1.

Table 7.1   Risk Perception Survey Results

Activity	Actual	Students	Insurers	Police
smoking	1	5	1	2
riding in a motor vehicle	2	2	2	1
AIDS	3	1	3	3
suicide	4	4	7	5
murder	5	3	4	4
accidental poisoning	6	18	15	11
drowning	7	7	5	7
getting burned in a fire	8	10	9	9
choking	9	8	6	8
flying in a small plane	10	9	25	14
firearms accidents	11	6	8	6
railroads	12	14	18	23
using stairs	13	30	11	22
boating	14	16	13	12
getting hit by a falling object	15	21	29	25
riding a bicycle	16	22	14	19
getting too cold	17	17	21	24
using electricity	18	19	10	10
slipping or tripping	19	28	12	17
using your bathtub	20	29	16	20
getting overheated	21	20	20	13
hunting	22	25	23	21
power mowers	23	23	27	29
lightning	24	24	17	26
tornadoes	25	13	24	15
snow skiing	26	27	26	27
flying on a commercial airliner	27	11	22	18
hurricanes	28	12	19	16
setting off fireworks	29	26	28	28
nuclear power	30	15	30	30

Let's take a look at some of the differences in risk perceptions and think about the reasons for the variations. For teenagers entering young adulthood, it's not surprising that AIDS is perceived as the number one risk. The 8th grade class's overall perception of the top five risks was accurate. Notice, however, that the students ranked nuclear power as 15 when its actual rank was 30. These students lived in a state with four operating nuclear reactors at the time of the survey and the students were just old enough to remember Chernobyl. The nuclear utility in Connecticut was also in the news routinely due to difficulties with the Nuclear Regulatory

Commission. The press attention may have raised the children's anxieties regarding nuclear power.

The students' perception of dying while using their bathtub was the corresponding perceived lowest risk, when in essence about 345 people a year die in this way. Young people's balance, agility, and strength are taken for granted. My own teenager, like many others, viewed the bathtub more as a health spa than a potentially hazardous area. Notice, however, the two adult groups recognized the hazard.

There are several additional interesting observations in the other two groups. State Police troopers are all too familiar with many of the risks in this list. I think their views reflect their duty experiences. They're on the scene when automobile accidents occur. To them, traffic fatalities are very real. It's no wonder they chose riding in a motor vehicle as the most risky activity of the list. Their almost routine experiences with tragedy must have had an effect on how they would view the risks. Notice that the State Police were very close to actual numbers when it came to other activities that they see in their profession. Some of these include murder, suicide, drowning, choking, fires, fireworks, and nuclear power risks. By the way, they got the bathtub ranking exactly!

Insurance professionals are, as a group, a rather conservative bunch. Since their jobs all involve risk management in some way, it's not surprising that their ranking is pretty close to the actual. There is one curious deviation from the actual rank that deserves some discussion. The largest difference between actual risks and the insurers' ranking was for the risk associated with flying in a small plane. I had difficulty understanding how they missed the mark so broadly for this category. My wife then made an interesting observation. Even though these participants came from many different professional areas of insurance, each of them knows me personally and had been in my office numerous times. I have had the privilege of owning a small aircraft, a Super Cub on floats to be exact, and my office had several pictures of the airplane. My wife speculated that my personal view of light airplane safety was probably implicitly transferred to the participants. After all, if I'm the risk "expert" and I fly my small plane, how risky can it be? And this is the only explanation that makes sense to me.

The point of this example is to show you that even though risk perception may seem constantly changing and elusive, it can be measured and you can do it yourself. Equally important is putting yourself in the respondents' shoes before you use the survey and evaluating the answers you receive. I asked only for a relative ranking rather than any direct measure of the risk. This information is the easiest exercise for people to do. An actual risk score could be obtained either by a multiple choice format like (very low, low, medium, high, very high) or by asking respondents to give a score within a prescribed range; for example 1 (low) to 100 (high). These semi-quantitative responses enable analysts to statistically measure differences risk perception and also to provide relative risk values in addition to risk-ranking results. The design you use needs to reflect your objectives and address the practical reality of having respondents reliably complete the survey.

Experts have made an interesting observation regarding the public's perception of overall risks of different activities. It appears that the people overestimate the

Table 7.2   Bias in Judged Frequency of Death

Most *overestimated*	Most *underestimated*
All accidents	Smallpox vaccination
Motor vehicle accidents	Diabetes
Pregnancy, childbirth, abortion	Stomach cancer
Tornadoes	Lightning
Flood	Stroke
Botulism	Tuberculosis
All cancer	Asthma
Fire	Emphysema
Venomous bite or sting	
Homicide	

actual risks for events which are highly sensational or spectacular and very low frequency and underestimate events that are familiar and occur at a relatively high frequency. There is good evidence to show this is true. Table 7.2 shows a list of some examples, comparing actual risks to the general public's perceptions [14, 15].

Since public opinion is an important aspect of managing everything from a technology to a global company, or even the electorate's voting preference, the risk equation, discussed in Chapter 6, can be customized to encompass the influence of risk perception [15]. For example, one way the U.S. nuclear industry considered modeling actual risk plus perceived risk is to modify the risk equation as [16]:

$$\text{Risk} = \text{Frequency} \times [\text{Severity}]^k$$

where "k" is a factor someone makes up to indicate the perceived risk. If $k = 1$ then the standard risk equation is obtained. For values where $k > 1$, risk will increase more for high severity–low frequency than for high frequency–low severity events. This is a purely mathematical model, but it does incorporate perception or event "importance" in the risk equation by changing the effective event severity. This behavior roughly describes the reality of a group's reaction to risk.

Figure 7.1 shows how the standard risk equation is altered by including public risk perception by using $k = 1.2$ [17]. As you can see, the risk curve is shifted above the standard version when perception is added. This increases the effective risk of low frequency–high severity or uncommon events (Point A) more than the high frequency–low severity or common events (Point B). This idea is one of several models that have tried to incorporate public risk aversion into mathematical equations. While it is not suitable, in many cases, for actual application to real life or death situations, it does illustrate graphically how perception can influence mathematical risk.

To bring this lofty discussion down to earth, consider the following situation that was used as example in Chapter 1. You are an emergency response leader in

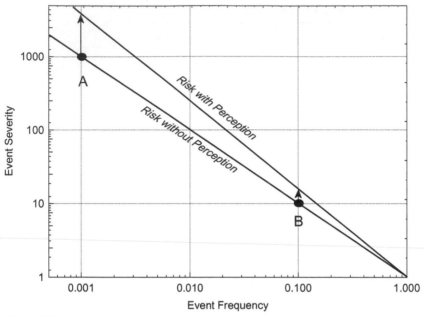

Figure 7.1   Effects of perception on the risk equation.

your town of 10,000 people. A sudden emergency arises, such as a dam break, tornado, or terrorist attack. You have two alternatives to save lives.

Alternative #1: You are absolutely certain that you can save 5000 people.

Alternative #2: There's a 50% chance you save everyone. There's also a 50% chance everyone will be killed.

Now alternative #1 really has no risk in terms of uncertainty. You know that you can save 5,000 people. However, there is uncertainty with alternative #2. You might be able to save the whole town, but you also might kill everyone. Let's look at these alternatives through the lens of the risk equation:

Alternative #1: Risk = 1.0 × 5,000 deaths = 5,000 deaths

Alternative #2: Risk = 0.5 × 0 deaths + 0.5 × 10,000 deaths = 5,000 deaths

The two alternatives have the same risk (deaths) value. But this result probably feels very inadequate to you. You need to make a decision now. Will you definitely kill 5000 people and save 5000 people—among them your friends, neighbors, and relatives, or will you gamble on the group as a whole? How lucky do you feel?

This hypothetical, and somewhat melodramatic example clearly shows the emotional and practical inadequacy of the risk equation. The mathematics tells you that the two alternatives are equivalent. But the mixture of certain loss with the possibility of "winner takes all" throws mathematics aside. Now, all else being equal, which would you choose?

In my experience in asking groups to choose between alternatives #1 and #2, fewer than 25% of the people choose alternative #1. More than 75% would try to save the entire group. By the way, in this study, people were not allowed to choose the 5,000 people they'd save if they had picked alternative #1. I've got a good hunch the results would be different had they been permitted to save 5,000 select people. I am not implying personal favoritism here. People may choose for example to save children over the elderly. If selection was allowed, then the ethical issues related to "lives" versus "life-years" saved adds an additional perspective to the problem.

When I respond to the people who choose alternative #2 by saying that you just killed everyone when you had a chance to save 50% of them, the most common reply is that people feel better trying to save everyone than to guarantee killing half of them. In this way no one can say that they didn't at least try to save everyone. Do you agree?

NOT ALL RISKS ARE PERCEIVED AS EQUAL

According to the Centers for Disease Control (CDC), more than 30,000 people in the United States die as a result of the common influenza each year, but do you ever hear about it? And by the way, do you get a flu shot each year? More than 30,000 people are killed in motor vehicle accidents each year. But do you always wear your seat belt? Whatever your answers are to these questions, the point is that these events are just not news that sells newspapers or keeps you glued to your TV so you'll stay around and see the commercials. They are, unfortunately, the mortality of the commonplace.

Another interesting phenomenon that seems to be a growing trend is the increase in social acceptance of cigar smoking [18]. In some social circles, cigar smoking is becoming an enjoyable part of the gustatory experience. First you have a good meal, and then you retire to another room to savor the aroma and tastes of a cigar, balanced with a snifter of cognac or brandy. Just as wine connoisseurs use an unusual vocabulary to describe the subtle dimensions of a wine's attributes, a cigar smoker now must learn the "subtle veins," earthy aromas," and "spicy aftertaste" characteristics of cigars. Cigar stores are becoming popular in malls and even in some hotel lobbies. You can spend well over $1,000 for a hand-made, wooden humidor that will keep your stogies at just the right humidity. In magazines, instead of cigarette ads you now see advertisements for cigar paraphernalia such as silver-plated cigar cases, khaki cigar hats, stainless steel cigar scissors and books like *The Art of the Cigar Label*, *Playboy: The Book of Cigars*, and, of course, *The Complete Idiot's Guide to Cigars*, 2nd edition.

Hollywood has also connected cigar smoking to control and success. For instance, the movie *First Wives' Club* portrayed powerful, successful women in control as cigar smokers and in *Independence Day*, the storyline identified cigar smoking with success. There are also the some well-known Hollywood icons who don't mind being connected to cigars. Arnold Schwartzenegger, Bruce Willis, Demi Moore, and Pierce Brosnan, all have posed for magazine photos with stogies.

Yet the actual risks associated with cigar smoking are significant. Here are a few statistics, just to give you a sense of how bad cigar smoking actually is. Cigars contain many of the same ingredients as cigarettes, and give off the same gases, such as carbon monoxide, hydrogen cyanide, and ammonia. Since cigar smoke is usually not inhaled as deeply as cigarettes, the lung cancer risk is only three times that of a nonsmoker, compared to the factor of nine for smokers. Lung cancer is only one possible fate of a cigar smoker. Smoking the "earthy aroma" and "spicy aftertaste" stogies increases the risk of lip, mouth, larynx, and esophagus cancers by a factor of five. True, there are some role models who have beaten the odds with good genes, luck, or a little of both. George Burns and Winston Churchill did beat the odds, but just remember, there are many, many more people who have either died or have become disfigured from this behavior. The rise of the social trend implies that if society accepts the behavior it must not be harmful. Perhaps neophyte cigar connoisseurs are unaware of the medical facts or maybe they are just being ignored for the perceived social benefit.

Cigar smoking risk is significant, but it isn't the kind of risk exposure that jolts you to another level of consciousness. These types of memorable events can cause a dramatic change in risk perception. Spectacular events can stimulate people into making conscious, rational decisions that are illogical from a risk perspective. Here are some classic examples:

- Between 1991 and 1994, U.S. Air had four unrelated fatal crashes. In the days immediately following the events, some business people actually chose to take limousines from Washington to New York rather than fly. The risk statistics didn't matter. This is "risk denial" at its best. Even though they were exposing themselves to much higher risk of injury, they "felt" safer.
- The entire grape industry almost stopped overnight due to the discovery of one or two poison-tainted grapes on a single bunch.
- The pharmaceutical industry changed its pill designs in response to the public's refusal to purchase certain products after a small number of innocent people died from cyanide poisoning after taking tainted medication for minor body pain.
- The food industry also redesigned its packaging to guard against tampering. The redesigns were taken to proactively guard against the possibility of such events occurring with their products.

Maintaining the confidence of customers and keeping a low perception of risk together are like walking a tight rope across a never-ending gorge. It doesn't matter for how many miles or how many years you have kept your balance. If you fall off, you're dead. Trust that has taken years to build can be lost overnight. Even though the actual risk to the public from these types of hazards may be small compared to the risks from routine, familiar hazards, the public doesn't always follow the numbers.

We have discussed factors that influence perception, how a group's collective risk perception can be measured, and some examples of how risk perceptions can be formed. This is just the beginning. Perception is a process that helps shape our attitudes and behaviors. It is just as dedicate and just as strong as the human behaviors related to trust, integrity, and faith. All of these attributes or qualities are part of the intangible components of our character and behavior. You can't see them, yet they essentially form who we are. There is no clear, deterministic scientific basis for many intangible qualities of human behavior. You don't take a pill, drug, or dietary supplement to obtain an attribute like integrity.

However, recent advances in our understanding of the neurological processes relating to the recognition and interpretation of stimuli or information (perception) are providing valuable insights about how this seemingly intangible human characteristic is marked in the human brain. Let's explore this interesting aspect of understanding risk perception through this case study illustration.

A well-educated woman in her mid-twenties got married and had two children [19]. After seven years of marriage, the relationship ended in a bitter divorce. She lost possession of her home and, perhaps from the financial and emotional stress, possession of her children. Dating a new man brought companionship but also drinking, and then drugs. The drug use grew to heroin use and eventually she contracted HIV from a shared needle, forcing her to leave her job and lose visiting rights to her children.

From this state of health and mind she developed HIV complications, including shingles with painful sores across her scalp and forehead. The medical treatments brought her HIV under control and drug rehabilitation helped defeat the heroin addiction. Over the next years she began to rebuild her life and things were good. And then, right after a recurrence of the shingles, she got an itch on her scalp in the area of the blisters. The blisters and pain subsided with medication but the itch persisted. In fact, the area of her head that had both the pain and the itch sensation now became numb to the touch but the localized itch was still there. The sensation to scratch was overwhelming. Even in her sleep she scratched, as evidenced by the blood on her pillowcase. Doctors initially treated the symptoms as a dermatological condition but no evidence of any disease was found. Having ruled out disease, doctors suspected a psychiatric cause. Her life history had plenty of stress and itching can be one possible symptom from the body's release of endorphins, nervousness, or increased sweating.

She tried antidepressants and other medications, but with no success. The itching persisted. To reduce the injury at night, she would bandage her head as protection from her fingers when she slept. Still the trauma from her self-inflicted, albeit unintentional, injury was getting worse. The last straw of voluntary care occurred when her doctor realized that a bacterial infection had eroded the skull to a point at which her brain was becoming exposed.

Neurosurgeons replaced the scalp tissue two times with skin grafts. Each time, her itching tore them away. At this point, she met a neurologist who tested her scalp for nerve function. The test results showed that 96% of the nerve fibers were gone. Also, when a local anesthetic was applied to the area, the itch subsided temporarily. The neurosurgeons convinced the woman to undergo surgery to cut the main sensory nerve to the scalp area. Their theory was that the remaining 4% of the nerve fibers were causing the itching, so cutting all the fibers would stop the sensation. There was a competing theory. This one was full of questions but it speculated that the origin of the itching sensation, the neurons in her brain, were stuck in a biological loop that could not be corrected since the large majority of the nerve fibers were destroyed. There was no way for these receptor cells to be turned off.

She had the operation and, after a short respite, the itch returned. The woman eventually learned to live with this sensation and her story is far from concluded. However, her case study highlights some new insights on the nature of some aspects of perception. How can the brain perceive a sensation in a part of the body with no functional nerves? The answer appears to support the biological loop theory. The nerve receptor cells are transmitting the last signals they received. With the local nerve endings destroyed, there appears to be no way for the body to reset the receptors, at least not in the standard neurological manner.

The most clear-cut and dramatic examples of these physical sensations are seen in amputee patients with a persistent sensation and pain in the limbs that have been removed. The medical community has been aware of phantom pains since the 1550s when the French surgeon Ambroise Pare [20] noted patients' discomfort and sensation in limbs that were amputated. Today over 90% of amputee patients experience these phantom pains to some degree [21]. Yet a plethora of medications, as well as invasive and noninvasive treatments, over the past four centuries have not been able to solve this complex neurological mystery.

There is a relatively new type of noninvasive, actually very simple therapy that appears promising in helping to reset some brain receptor cells that are stuck in a biological loop. Mirrors are set up so the patient sees the mirror image of the good limb in the space normally occupied by the amputated limb. The visual images sent to the brain appear to help reset the receptors and over multiple sessions, patients have experienced real relief. Research using this practical therapy performed with simple furniture store mirrors is providing a more robust understanding of the biological mechanisms between brain sensory reception and perception.

As humans we perceive situations that are analogous to brain receptors receiving signals of sensation and receptors just like those of the woman with the itch and most amputees, we maintain the same interpretation of stimuli or information until new data is received. If no new data are received, there is no change in perception. For the mirror therapy patients, the receptors appear to be reset by visual-related information. For us to change our perceptions, new data can come from a wide range of sources.

The physical sensation examples here are only a small facet of human perception. Nevertheless, perception is the awareness and interpretation of sensory stimuli that are not going to change without new data. And even with new data there is no guarantee. The risk perception factors discussed earlier in the chapter also have a major role in the formation and modification of how we perceive our world. For amputees and other patients, the new information created by mirrors helps and this new therapy is providing valuable insights on how the brain receives and perceives. I think researchers in this field would agree that we are just beginning to understand these extremely complex mechanisms. Understanding the even more complex nature of human risk perception is a larger challenge. And the better we are at this art, the better we will be in understanding one another and ourselves. Using an analogy from lawyer, American Civil War general, and *Ben Hur* author General Lewis Wallace: "Beauty is altogether in the eye of the beholder."

Without a doubt, risk is in the mind of the perceiver.

DISCUSSION QUESTIONS

1. Describe specific examples from your experience for each the eleven factors that influence risk perception.

2. Develop a risk perception survey that asks respondents to rank 10 common activities. Distribute the survey to different groups of people and compile the results. Using actual data for the number of deaths, injuries, or other risk measures, compare the different group results. What can you infer from your ranking results?

3. Identify five examples in the journalistic press where you think the headline or content suggests a higher risk perception than the actual risk related to the event being described.

4. In this chapter a problem was discussed that described two equivalent risk management alternatives: saving 50% of the people with complete certainty or having a 50% chance of saving or killing everyone. Construct two examples of how this problem could be stated in terms of life-years instead of lives saved. The two examples should apply different age distributions for the exposed population.

5. In the case of amputees, mirror therapy is showing some promise in reducing and eliminating phantom limb discomfort. This is medical-physical example of how the brain can be fooled into providing false information and how the visual stimulus helps reset the receptors with the correct "data." However, changing the risk perception of a group is not that simple. Based on the information in this chapter, choose a risk perception and construct a plan to accomplish this objective for a specific group. How would you measure your results?

6. The March 11, 2011, earthquake and tsunami in Japan caused enormous human and economic losses. These facts are indisputable. However, contrast the facts with the media descriptions of the events that occurred at the Fukushima Daiichi nuclear plant. How do they compare?

ENDNOTES

1 Gene Weingarten, "Pearls Before Breakfast," *Washington Post*, April 8, 2007.
2 Fred Mayernick, private communication.
3 Roderick Allison, Edward Ryder, Sandra Caldwell, Jeremy Beech, Peter Moss, Victor Coleman, Roger Lejuez, François Barthelemy, et al., "Report and recommendations: Inquiry into the Fire on Heavy Goods Vehicle Shuttle 7539 on 18 November 1996," Channel Tunnel Safety Authority, May 1997.
4 E. Comeau and A. Wolf, "Fire in the Chunnel!" *NFPA Journal*, March/April 1997.
5 "Chunnel Fire," *Chicago Tribune*, September 11, 2008.
6 V. Covello and P. M. Sandman, "Risk Communication: Evolution and Revolution," in *Solutions to an Environment in Peril*, Anthony Wolbarst, ed. Baltimore, MD: John Hopkins University Press, 2001, pp. 164–178.

7 Andrew D. Brosig, "Seafood Slump," *Killeen Daily Herald*, August 6, 2010.

8 Chauncey Starr, "Social Benefit versus Technological Risk," *Science*, Vol. 165, No. 3899, 1969, pp. 1232–1238.

9 Barbara Sattler and Jane Lipscomb, *Environmental Health and Nursing Practice*. Springer Verlag, 2003, p. 110.

10 Brien Roche et al., *Law 101: An Essential Reference for Your Everyday Legal Questions*. Sourcebooks, 2009, p. 235.

11 J. Adlon, "The Food Irradiation Debate," January 8, 2010, available at http://www.associatedcontent.com/article/2566881/the_food_irradiation_debate.html?cat=4 (accessed April 23, 2010).

12 Edward J. Burger, ed., *Risk*. Ann Arbor: University of Michigan Press, 2003.

13 P. Slovic, "Perception of Risk," *Science*, Vol. 236, Issue 4799, 1987, pp. 280–285.

14 S. Lichtenstein et al., "Judged Frequency of Lethal Events," *Journal of Experimental Psychology: Human Learning and Memory*, Vol. 4, 1978, pp. 551–578.

15 Paul Slovic, B. Fischhoff, and S. Lichtenstein, "Facts and Fears: Understanding Perceived Risk," in *Societal Risk Assessment: How Safe is Safe Enough?* ed. Richard C Schwing and Walter A. Albers, Jr. New York: Plenum Press, 1980.

16 Ralph R. Fullwood, *Probabilistic Safety Assessment in the Chemical and Nuclear Industries*. Butterworth-Heinemann, 2000, p. 7.

17 "An Approach to Quantitative Safety Goals for Nuclear Power Plants," NUREG-0739, October 1980.

18 C. D. Delnevo, "Smokers' Choice: What Explains the Steady Growth of Cigar Use in the U.S.?" *Public Health Report*, Vol. 121, 2006, pp. 116–119.

19 Atul Gawande, "The Itch," *The New Yorker*, June 30, 2008.

20 G. Keil, "So-Called Initial Description of Phantom Pain by Ambroise Paré ["Chose digne d'admiration et quasi incredible": The "douleur ès parties mortes et amputées"]." *Fortschr Med*, Vol. 108, No. 4, Feb. 10, 1990, pp. 62–66. In German.

21 S. R. Hanling, S. C. Wallace, K. J. Hollenbeck, B. D. Belnap, and M. R. Tulis, "Preamputation Mirror Therapy May Prevent Development of Phantom Limb Pain: A Case Series," *Anesthesia and Analgesia*, Vol. 110, 2010, pp. 611–614.

Chapter 8

Risk, Regulation, and Reason

The Lord's Prayer is 66 words, the Gettysburg Address is 286 words, there are 1,322 words in the Declaration of Independence, but government regulations on the sale of cabbage total 26,911 words.

—National Review

You're probably thinking that a chapter starting with this title will be another dose of government bashing. True, it's easy to find fault and inefficiencies with our government. Its size and visibility makes our government an easy target for those who want to poke holes in the oldest democracy in the world or, for that matter, the government of any country. Remember that we have little practice as a global and national society in dealing with government in the age of instant communication and information transfer. Legacy bureaucracies designed by previous generations are hard and slow to change, compared to the modern practice of downloading new "apps" when improvements are identified. Nevertheless, this is our first try, and the only one we'll get at governance. No one has ever had a democracy or another similar political system last so long. Perhaps in 500 to 1000 years, people will be able to look back and learn from all of our successes and mistakes, but for now, we're collectively doing the best we can. For simplicity, let's review the U.S. government here, although I suspect every government has its own risk inefficiencies.

Policy decision stories like giving the Russians $8 million [1] to "expose local officials to the best practices of community administration" or financing $30 million for a spring training baseball complex for the Arizona Diamondbacks and the Colorado Rockies [2] are commonplace. While these are true and seem a little strange, to say the least, there are always two sides of an issue. It's unfair but often tempting to cast judgment from a unilateral viewpoint. As you will see in this chapter, while contradictions in risk policies exist, the seemingly black and white issues are actually complex shades of gray.

Suppose you could redesign or, to use today's popular term, "reinvent" government. How would you do it? First, in your mission statement and business objectives,

20% Chance of Rain: Exploring the Concept of Risk, First Edition. Richard B. Jones.
© 2012 John Wiley & Sons, Inc. Published 2012 by John Wiley & Sons, Inc.

you would have phrases like "customer (i.e., taxpayer) focus," "protection or improvement of health, safety, and the environment," and maybe something about "efficiency," "reasonable costs," or "quality."

Let's skip over the obvious departments of Justice (Legal Department), Treasury (Financial), and Defense (Security) and look at those departments that deal with the basic aspects of your customers' businesses.

So how would you start? Your business falls into several rather specialized categories. To focus on the delivery of a quality product at a reasonable cost, you might "organize around the needs of the customer." Here are some possibilities.

Your customers want affordable, safe transportation, so you would have a Department of Transportation. This department could handle the usual cars and buses, but the other modes of transportation are so different they might require separate administrations. Going down the list we could have:

Aviation:	Federal Aviation Administration
Railroad:	Federal Railroad Administration
Ships:	Coast Guard

All customers eat and get sick at various intervals, so you'd have to manage food and medicine risks. The large challenge of managing these activities suggests that separate departments are required. Let's call it exactly what it is: the Food and Drug Administration.

In an active and growing technological society, protecting the environment is essential. This specialized activity has its own concerns, so let's call it the Environmental Protection Agency. I could go on, but you get the idea. The basic design of the U.S. government is consistent with the needs of its constituents. The problem is not with its structure; instead it is with how the different departments, administrations, and agencies operate.

While large differences and inconsistencies exist in how various governmental entities manage risk, there are also examples where risk management coordination is underway. Case in point: to better coordinate human health hazards relating to food safety risk management across the government, an agency acting as a central risk analysis clearinghouse has been proposed [3]. Its role would be to standardize (provide consistency), coordinate, and communicate risk-related programs and policies for all of the 18 identified governmental organizations with food safety responsibilities or interests.

Risk management segmentation is necessary sometimes due to unique differences between the types of risk-producing activities, public perception drivers, and associated costs. This type of efficiency produces a two-edged sword. Risk makes the parties unique, yet risk is also what makes them the same. All of us, as individuals, share risk as equal partners. Transportation risk, food risk, cancer risk, environmental risk—risk is risk. We all experience it as a common denominator of life.

I'm not implying that all government agencies should do things the same way. This wouldn't work simply because the problems in each area or category are too different. However, the same basic risk management principles can be used. To me, "reinventing government" is pure political rhetoric. We already have a good design. We only need to improve the existing structure, which is built on a good foundation.

Are any of these inconsistencies in risk management visible to you? Some are and some are not. Among the most visible evidence of inconsistent risk management are product warning labels. When you buy normal consumer products, look at the warning labels—if one exists. Here are some examples you should recognize: First, we look at the back of a common cough, cold, and flu medicine box. Keep in mind this product is a medical drug designed to be ingested.

Warnings

Do not take this product for persistent or chronic cough such as occurs with smoking, asthma, emphysema, or if cough accompanied by excessive phlegm (mucus) unless directed by a doctor. Do not exceed recommended dosage because at higher doses, nervousness, dizziness, or sleeplessness may occur. Do not take this product if you have heart disease, high blood pressure, thyroid disease, diabetes, glaucoma, or difficulty in urination due to enlargement of the prostate gland unless directed by a doctor. *Drug Interaction Precaution:* Do not use this product if you are now taking a prescription monoamine oxidase inhibitor (MAOI) (certain drugs for depression, psychiatric or emotional conditions, or Parkinson's Disease), or for 2 weeks after stopping the MAOI drug. If you are uncertain whether your prescription drug contains an MAOI, consult a health professional before taking this product. May cause excitability especially in children. Do not take this product, unless directed by a doctor, if you have a breathing problem such as emphysema or chronic bronchitis. May cause marked drowsiness; alcohol, sedatives, and tranquilizers may increase the drowsiness effect. Avoid alcoholic beverages while taking this product. Do not take this product if you are taking sedatives or tranquilizers without first consulting a doctor. Use caution when driving a motor vehicle or operating machinery. Do not take this product for more than 7 days (for adults) or 5 days (for children). A persistent cough may be a sign of a serious condition. If cough persists for more than 7 days, tends to recur, or is accompanied by rash, persistent headache, fever that lasts for more than 3 days, or if new symptoms occur, consult a doctor. If sore throat is severe, persists for more than 2 days, is accompanied or followed by a fever, headache, rash, nausea or vomiting, consult a doctor promptly. **Keep this and all drugs out of the reach of children**. In case of accidental overdose, seek professional assistance or contact a poison control center immediately. Prompt medical attention is critical for adults as well as for children even if you do not notice any signs of symptoms. As with any drug, if you are pregnant or nursing a baby, seek the advice of a health professional before using this product.

This information takes up almost the entire back of the box. Cold medicines are not singled out for ominous warnings. Check your aspirin bottle. Now let's look at the warning label for a common household kitchen cleaner, a product not designed for human ingestion:

Precautionary Statements: Hazards to Humans and Domestic Animals

Warning: Eye and Skin Irritant

Causes substantial but temporary eye injury. Do not get in eyes, on skin, or on clothing. Avoid contact with skin. Wear protective eyewear. Wash thoroughly with soap and water after handling and before eating, drinking, chewing gum, or using tobacco. Remove and wash contaminated clothing before reuse. Harmful if swallowed. For sensitive skin or prolonged use wear gloves. Vapors may irritate. Avoid prolonged breathing of vapors. Use only in well-ventilated areas.

Not recommended for use by persons with heart conditions or chronic respiratory problems such as asthma, emphysema, or obstructive lung disease.

This manufacturer also stated this product "Kills germs on hard, nonporous surfaces: Salmonella enteric, Cold Virus (Rhinovirus Type 37), and Flu Virus (Influenza A2, Hong Kong). In addition, you might also be relieved to read, at the bottom of the back label, "CONTAINS NO PHOSPATES." Aside from addressing health and safety risks, this company advertises some of its environmental risk management policies. That's not a bad idea.

It's hard to get precise figures on the actual risk associated with these products. National Safety Council figures indicate that in 2006, approximately 27,531 people died from accidental poisoning [4]. This includes medicines, drugs, mushrooms, and other recognized poisons. The number of accidental deaths specifically attributable to cold medicines and household cleaners are not known. My research indicates that in 2008, there were approximately 91 deaths from acetaminophen, 24 deaths from aspirin products, and at least 9 deaths from household cleansing cleaners and chemicals [5]. Keep these numbers and the length of the warning labels in mind as you read on.

Now let's consider another drug, very much designed for human ingestion: alcoholic beverages. Here's what's on the back of the bottles and cans.

Government Warning

(1) According to the Surgeon General, women should not drink alcoholic beverages during pregnancy because of the risk of birth defects.
(2) Consumption of alcoholic beverages impairs your ability to drive a car or operate machinery and may cause health problems.

The number of alcohol-related deaths is also very difficult to measure since its effects are so widespread. In 2007, a total of 23,199 persons died of alcohol-induced causes in the United States, including alcohol poisoning [6]. Annual data compiled inter-

Table 8.1   Alcohol-Attributable Death Fractions United States, 2006

5% of all deaths from diseases of the circulatory system
15% of all deaths from diseases of the respiratory system
30% of all deaths from accidents caused by fire and flames
30% of all accidental drownings
30% of all suicides
40% of all deaths due to accidental falls
45% of all deaths in automobile accidents
60% of all homicides

nationally reports 5 specific alcohol poisoning deaths in France, 166 in the UK, and an astonishing 10,637 in the Ukraine [7]. The long-term, chronic effects on health are not accurately measurable even though we all know intuitively that the casualty rate is high. According to the National Council on Alcohol Dependency, over 100,000 Americans die each year from alcohol-related causes. These people did not die from alcohol poisoning directly but from the behaviors or medical implications of alcohol abuse and misuse. Unintentional death forensics speak very clearly about the dangers and extent of alcohol abuse. Table 8.1 shows the percentage of unintentional deaths where alcohol was a contributing causal factor [8]. The U.S. statistics represent only a small fraction of alcohol's global effects. The World Health Organization estimates that alcohol is a cause for 20–30% of all instances of esophageal cancer, liver cancer, liver cirrhosis, homicide, epilepsy, and motor vehicle accidents [9].

To protect and warn the general public about the dangers of alcohol consumption, in 1988 Congress passed the Alcoholic Beverage Labeling Act. The legislation required the addition of the aforementioned 42 word warning label to all alcoholic beverages with greater than or equal 0.5% of alcohol by volume, intended for domestic human consumption. (For comparison, the cold medicine warning had 370 words.) Exported products do not require this label. I think you would agree, the so-called warning labels are weak compared to the magnitude of alcohol's consumption risk. Some experts argue today that the warning has done little to change drinking behavior and might have the unintended consequence of providing liability protection for the alcoholic beverage industry [10]. However, it is apparent that Congress did recognize the real and pervasive danger of alcohol in the Act's Declaration of Policy and Purpose. In Chapter 11, Section 231, Congress states

> that the American public should be informed about the health hazards that may result
> from the consumption or abuse of alcoholic beverages, and has determined that it
> would be beneficial to provide a clear, non-confusing reminder of such hazards, and
> that there is a need for national uniformity in such reminders in order to avoid the
> promulgation of incorrect or misleading information and to minimize burdens on
> interstate commerce. The Congress finds that requiring such reminders on all
> containers of alcoholic beverages is appropriate and necessary in view of the

substantial role of the Federal Government in promoting the health and safety of the Nation's population. [11]

Alcohol abuse has been ranked as the number 3 preventable cause of death in the United States [12]. Regardless of any national metric or statistic, I suspect that everyone reading this book knows at least one person on a personal level whose life has been affected by this powerful drug. Do you think Congress has achieved its purpose?

The United States is not alone in its attempt to caution people about alcohol's dangers by either mandating or suggesting warning labels [13]. Table 8.2 presents a listing of government required warning labels that shows an interesting variety of risk communication examples. Some examples of voluntary country warning labeling are shown in Table 8.3. Label wording, size, and the legal requirements to include such statements are evolving globally. However, these tables clearly show that the United States, usually thought of as a haven for litigation activities, is not alone in its legal requirements for communicating alcohol risk to consumers. In my opinion, some of the foreign examples better communicate the real hazards of alcohol than does the U.S. version.

If you think alcohol warnings are inadequate, consider an even more dramatic illustration of inconsistent government risk management. It's the lackluster warning for the single most preventable cause of disease and death in the United States: cigarette smoking. As of 2010, there are four warning labels that are rotated on a quarterly basis, to keep the warning information "fresh" to consumers. They all are prefaced by "SURGEON GENERAL'S WARNING," followed by one of the following sentences:

- Smoking causes lung cancer, heart disease, emphysema, and may complicate pregnancy.
- Quitting Smoking Now Greatly Reduces Serious Risks to Your Health.
- Smoking by pregnant women may result in fetal injury, premature birth, and low birth weight.
- Cigarette smoke contains carbon monoxide.

I can only speculate why the Surgeon General decided to fluctuate from direct hazard facts to almost a friendly warning for such a deadly behavior. Cigarettes contain carbon monoxide—so what? Perhaps the label writers were trying to placate the manufacturers by not scaring the consumer public too much.

Cigar packages show another interesting characteristic of how the government treats the tobacco industry. In 2000, the Federal Trade Commission and the Surgeon General [14] mandated the inclusion of the following warning labels (also rotated quarterly) on all cigar packages. Before 2000, no labeling was required. Notice the directness in these warnings and that one of the labels actually communicates the hazards associated with second-hand smoke. This is the first formal warning of the health hazards associated with passive smoking or environmental smoke for any tobacco product. The labels were also prefaced with "SURGEON GENERAL'S WARNING," followed by one the five following sentences:

Table 8.2   Government-Mandated Alcoholic Producer's Health Warning Labels

Country	Warning label text/guidance
Argentina	"Drink in moderation" "Sale prohibited to persons under 18 years of age"
Brazil	"Avoid the excessive consumption of alcohol"
Colombia	"An excess of alcohol is harmful to your health"
Costa Rica	"Drinking alcohol is harmful to your health" "Alcohol abuse is harmful to your health"
Ecuador	"Warning: The excessive consumption of alcohol limits your capacity to operate machinery and can cause harm to your health and family" "The sale of this product is prohibited for those younger than 18 years old"
El Salvador	"The excessive consumption of this product is harmful to health and creates addiction. Its sale is banned to those under 18 years of age"
France	"Drinking alcoholic beverages during pregnancy even in small quantities can have grave/serious consequences for the health of the baby" OR use the government-issued symbol showing a diagonal line being superimposed on an image of a pregnant woman holding a glass
Germany	"Not for supply to persons under 18, clause 9 Protection of Minors Act"
Guatemala	"The excess consumption of this product is harmful to the consumer's health"
Mexico	"Abuse of this product is hazardous to your health"
Russian Federation	"Alcohol is not for children and teenagers up to age 18, pregnant & nursing women, or for persons with diseases of the central nervous system, kidneys, liver, and other digestive organs"
South Africa	"Alcohol reduces driving ability, don't drink and drive" "Don't drink and walk on the road, you may be killed" "Alcohol increases your risk to personal injuries" "Alcohol is a major cause of violence and crime" "Alcohol abuse is dangerous to your health" "Alcohol is addictive" "Drinking during pregnancy can be harmful to your unborn baby"
Taiwan	"Excessive drinking endangers health"
Thailand	"Liquor drinking may cause cirrhosis and sexual impotency" "Drunk driving may cause disability or death" "Liquor drinking may cause less consciousness and death" "Liquor drinking is dangerous to health and causes less consciousness" "Liquor drinking is harmful to you and destroys your family"
United States	GOVERNMENT WARNING: (1) According to the Surgeon General, women should not drink alcoholic beverages during pregnancy because of the risk of birth defects. (2) Consumption of alcoholic beverages impairs your ability to drive a car or operate machinery, and may cause health problems"

Table 8.3 Voluntary Alcoholic Producer's Health Warning Labels

Country	Warning label text/guidance
Brazil	On packages and labels, it is reiterated that sale and consumption of the product are only for persons older than 18 years
Chile	"CCU asks you to drink responsibly"
	"Product for those 18 and older"
China	"Overdrinking is harmful to health"
	"Pregnant women and children shall not drink"
Japan	"Drinking alcohol during pregnancy or nursing may adversely affect the development of your fetus or child"
	"Be careful not to drink in excess"
	"Drink in moderation"
United Kingdom	"The Chief Medical Officer recommend men do not regularly exceed 3–4 units daily and women, 2–3 units daily"

- Cigar Smoking Can Cause Cancers Of The Mouth And Throat, Even If You Do Not Inhale.
- Cigar Smoking Can Cause Lung Cancer And Heart Disease.
- Tobacco Use Increases The Risk Of Infertility, Stillbirth, And Low Birth Weight.
- Cigars Are Not A Safe Alternative To Cigarettes.
- Tobacco Smoke Increases The Risk Of Lung Cancer And Heart Disease, Even In Nonsmokers.

Smokeless tobacco products have similar mandated warnings.

How many people succumb to tobacco products each year? The Centers for Disease Control and Prevention 2010 data show 443,000 deaths from tobacco use and the effects of second-hand smoke [15]. Oh, and by the way, the warning labels vary between 5 and 16 words. These deaths include lip, esophagus, pancreas, larynx, lung, and other cancers, cardiovascular diseases, respiratory diseases, and pediatric diseases of the innocent children who are victims of smoking caregivers.

It's clear that the government is trying to please two masters here: the taxpayers and the tobacco political machine. After reading these numbers, it's obvious that the warning on cigarette cartons packs the punch of a down pillow. Aside from the pain, suffering, disfigurement, and death, there are tremendous healthcare costs associated with the tobacco industry. The healthcare companies constantly are fighting a losing battle to contain and reduce these expenses due to the increasing need for long-term medical services. It's appropriate for people who choose to smoke to personally suffer and pay for the consequences, but that's not what happens. Everyone pays the increased healthcare and workers' lost productivity costs as the affected businesses transfer the added costs to consumers.

This is an excellent place to demonstrate the concept of "distributional justice" discussed in Chapter 3. In doing so, we'll also compute the average annual dollar loss associated with these healthcare costs. As of 2010, in the United States, approximately 46 million people, 24.8 million men and 21.1 million women, smoke [16]. The associated healthcare and lost productivity costs are estimated to be $193 billion each year [17], split about equally between direct healthcare expenses and lost productivity costs. Add in another $10 billion for second-hand smoke healthcare costs [18] and you get a total annual cost estimate of $203 billion.

If we were to apply distributional justice, we would require smokers to pay all of the associated medical and lost productivity costs. Since they share in the "benefits" of smoking, they must also likewise share in the subsequent costs, according to the distributional justice precept. Suppose, on the average, each of the 46 million smokers daily consumes one pack of cigarettes. This means that 920 million (almost a billion!) cigarettes (46 million packs) are consumed each day. On a yearly basis this comes to (46 million × 365) or 16.79 billion packs per year. To determine what the cost per pack should be, we divide the estimated total annual cost of $203 billion by the 16.79 billion packs smoked each year. The result tells us that smokers should pay a distributional justice tax of $12.09 per pack to account for the added costs. Even though this estimate is clearly dependent on several assumptions, it may not be too far off. The Centers for Disease Control's Office of Technology Assessment in 2006 calculated that each pack of cigarettes sold costs Americans $10.47 in excess health costs and lost economic productivity [19]. In addition, they have recommended that states invest $9.23 to $18.02 per capita on comprehensive tobacco control programs [20], which have been shown to decrease cigarette smoking.

I made assumptions in this example, but the logic is sound and every analysis has its limitations. However, in this case the real data are available. The government knows the healthcare costs, it knows the number of smokers in the United States, and it knows the average daily cigarette consumption rate. And the idea of additional taxes on tobacco is not a far-fetched notion. A poll, conducted by International Communications Research, found that 60% of voters in the wake of the 2008 recession would support an increase in the cigarette tax to help struggling states and would prefer it over other tax increases or budget cuts [21]. The chief executive of the American Cancer Society Cancer Action Network states:

An increase in tobacco tax rates is not only sound public health policy but a smart and predictable way to help boost the economy and generate long-term health savings for states facing deepening budget deficits.

He also stated:

We have irrefutable evidence that raising the tobacco tax lowers smoking rates among adults and deters millions of children from picking up their first cigarette.

In fact, January 2010 poll data, reported in February 2010, indicates that 67% of Americans favor a $1 increase in their state's cigarette tax [22]. If every state raised their cigarette tax rate by just $1 per pack, the financial and human effects would:

- Raise $9.1 billion in new annual state revenues;
- Save $52.8 billion in immediate and long-term health care costs;
- Prevent more than 2.3 million kids from becoming smokers;
- Prompt more than 1.2 million adult smokers to quit; and
- Prevent more than 1 million premature deaths from smoking.

Just imagine what a $10 or $12 tax per pack would do. Independent of the Washington, DC, tobacco lobby, a tax of this high relative magnitude compared to the basic cost of the product would most certainly stimulate a robust black market cigarette sales industry and cause more problems. And this means that state and federal governments could lose out on major sources of income and some people would still smoke. Under the taxation environment of 2009, the federal government collected about $8.7 billion with the states pulling in $15.7 billion [23].

Where this money goes is another story but let's see what the taxes on cigarettes really are. As of 2010, the federal tax per pack is $1.01. Now let's look at the state taxes. Rather than list all 50 states, here in Table 8.4 are the 5 states with the largest and the 5 states with the lowest per pack tax [24].

Over all 50 states, the average state tax is $1.41 cents per pack. In the major tobacco-producing states, the average tax is $0.42 per pack. In the other states it is $1.54. Where's the reason in this regulation? The political power of the industry in the major tobacco-producing states is clearly stronger than the risk implications. In states where the tobacco industry is a major economic factor, when tobacco lobbyists talk, politicians who want to be re-elected listen.

If the distributional justice tax is not paid by the smokers, who do you think is paying? Mr. or Ms. Taxpayer? You might think that a product that does so much harm should be banned from production. Remember, however, the tobacco industry is a multibillion dollar contributor to the economy, and cigarettes are legal. For the tobacco-producing states, replacing tobacco farming and manufacturing with that of other products is probably harder than replacing heroin-producing poppy farms in Afghanistan.

The U.S. government's hands-off legal policies toward the tobacco industry are changing. Since the 1950s, lawsuits have been filed against the tobacco companies alleging that they knew of their products' health and addiction hazards and failed to adequately warn smokers. In about 2000, they started to succeed. In

Table 8.4   The Five Highest and Lowest State Cigarette Taxes

Top five states		Bottom five states	
Rhode Island	$3.46	South Carolina	$.07
Washington	$3.025	Missouri	$.17
Connecticut	$3.00	Virginia	$.30
Hawaii	$3.00	Louisiana	$.36
New York	$2.75	Georgia	$.37

December, 2008, the U.S. Supreme Court even decided that plaintiffs can now begin to sue tobacco companies over the advertisement of the so-called light cigarette products [25].

The sheer number and size of the lawsuits against the tobacco companies today represents a litigation landscape that is too complex to adequately describe here. Let me just say that it appears there is a growing awareness that smoking is bad for you and your neighbors. People are becoming actively (rather than passively) resistant to smoking. There is a growing societal acceptance and, in some cases, demand, for "smoke-free" environments that hopefully can continue the decreasing addiction trend to this devastating but legal drug.

However, the risk remains. The good news: U.S. per capita cigarette consumption is at its lowest since the Great Depression. The bad news: at least worldwide, the tobacco industry is not going away. In fact, it's growing in terms of the number of cigarettes consumed. The long-term financial and human toll of tobacco use may not be paid for a generation and sad to say, unless there is some miraculous medical discovery, history will repeat itself.

Well, a miracle, albeit of a different kind, occurred. The government decided to kick the be-nice-to-the-tobacco-industry habit. On June 22, 2009, Congress passed the Tobacco Control Act [26] giving the Food and Drug Administration authority to regulate tobacco products. Among the several provisions that went in effect on June 22, 2010, are:

- Rules (with enforcement provisions) to limit the sale, distribution, and marketing to young people
- Provisions that prohibit the words "light," "mild," or "low" to be used in advertisements
- Requirements for new, larger warning labels on tobacco products.

Until 2010, in spite of the obvious, well-documented carnage, tobacco was one of the least regulated consumer products. The tobacco industry's political and economic power has kept its products virtually exempt from every health and safety law enacted by Congress including:

The Consumer Product Safety Act,

The Food, Drug, and Cosmetic Act,

The Fair Labeling and Packaging Act,

The Toxic Substances Act,

The Federal Hazardous Substances Act, and

The Controlled Substances Act.

Health risks from tobacco use seem like pretty clear-cut issues, don't they? In reality however, people are not logical machines. From smokers' perspectives, the perceived risks effectively are lower than what the actual risks indicate. How could so many people be so wrong? Aside from the addiction issue which certainly plays a major role, there are two primary perception factors that distort reality:

1. Smokers control the risk: it's a voluntary behavior, and

2. the mortality of smoking is not immediate but delayed.

Smoking shortens your life; it doesn't make you drop dead as you inhale. Unlike cyanide, it takes 20–30 years or more to kill, rather than 2–3 minutes.

The tobacco companies have large, state-of-the art research facilities. Like any other industry, the tobacco industry continuously researches new products to increase its shareholders' value. I asked a tobacco executive once why they didn't make a safe or safer cigarette. He replied, "We can and have made safe cigarettes. The trouble was nobody bought them." He was right that nobody bought them, but wrong about the safety of his "safe" products.

Researchers have found that tobacco-free cigarettes may be more carcinogenic than tobacco products. Using the same technique developed to document the harmful effects of tobacco products, the research team found that cigarettes made without tobacco or nicotine may be more carcinogenic because they induce more extensive DNA damage than tobacco products [27]. The novel technique used in this research was awarded a U.S. patent: 7,662,565.

Policies regarding drugs, alcohol, and tobacco use risks have a long way to go and it is clear that the length and content of warning labels, especially text labels, may not accurately communicate the risks. Under the new Tobacco Control Act, future warning labels are expected to include explicit health images. Researchers have observed that graphic health images communicate the hazards better than text only labels [28]. I guess one picture is worth more than 16 words!

There is also another conclusion that you could make from the label warning research: there may be other products you consume or use that possess formidable risks and have no warning labels. The basic lesson here is that in spite of the apparent disparity between the risk and the risk warnings, each of us needs to be the final, accountable judge of our actions. No government, however perfect, can or should play that role.

There are several feedback mechanisms to measure the effects of policies and laws. Re-election perhaps is the ultimate means, but on a more detailed level, cost–benefit studies and risk reduction statistics supply evidence of government performance. At the highest level, common metrics include the unemployment rate, inflation rate, currency valuation, and stock market indicators. At the policy or program level, the improvement or effectiveness metrics can include estimates of lives, life- years, or quality-adjusted-life-years saved. In others, the costs and benefits are measured in terms dollars. From a risk perspective, the two methods can be connected through the value of a statistical life relevant to the situation involved in the regulations.

Reporting benefits in the same units regardless of the activity may appear on the surface as the most efficient mechanism of communication, but unless you are a statistician or an account, this really doesn't make any sense. Using uniform statistics across risk exposures can hide the powerful effects of what is being measured and its effects. People need to hear benefits put into a frame of reference to which they can relate. Just using, for example, the ratio of dollars saved to dollars spent may not communicate the true value of the policy or activity. For example: from the Centers of Disease Control's statistics and health promotion information, here are some illustrations of the return on investment for certain health-related programs. Notice how the ROI is stated in each case.

- Every $1 invested in water fluoridation yields approximately $38 savings in dental treatment costs.

- Each dollar spent on diabetes outpatient education saves $2 to $3 in hospitalization costs.

- Each $1 invested in school-based tobacco prevention, drug and alcohol education, and family life education saves $14 in avoided health costs.

As you can see, it's not practical for all parties to use the same methods since risk management and the communication of risk-related information is too diverse. What should be common are the guidelines on how to and how not to compute these figures and how to state the conditions, assumptions, and approximations used in each case. To some extent, science has implicitly set some standards on the best (or most accurate) way to develop these values, but no formalized, uniform criteria have been established.

There are inconsistencies in how different agencies measure risk and compute policy costs and benefits. As I've said, some of these differences are justified due to the unique aspects of the characteristics being measured and communicated. In other cases, differences exist simply because method standardization was never historically required. There has been real progress in risk quantification and financial method standardization in government, but this topic most likely will be debated and discussed for a long time.

But who do you call for an unbiased, third party, and trusted opinion on financial matters? Your accountant, of course! And so it is with Congress and governmental agencies when it comes to assessing the real value associated with its changes and policy additions. The government's CPA is the Office of Management and Budget (OMB). The OMB is charged with the responsibility of being the third party accountant, or financial umpire that "tells 'em like they see 'em."

One of its responsibilities is to measure the quantitative benefits of Federal Budget programs. A sample of how the OMB shows its results is displayed in Table 8.5 [29] taken from a 2010 White House Draft Report. The officials writing this report consider a 10-year window to accumulate sufficient data to improve the confidence in the estimates and the intervals reflect discount rate and other assumptions. When the intervals between the costs and the benefits do not overlap, there is a good chance they are different. When they do overlap, then it is unclear if the program's benefits are truly different from the costs. In Table 8.5 you'll see several programs where the intervals overlie. The greater the crossover, the greater the chance that the program is not cost-beneficial. I suspect you won't find many lawmakers touting these achievements in their pre-election speeches.

But how many people actually read this stuff? Actually, more than you might think. There appears to be a growing trend of scrutiny by TV pundits, Congress, and

Table 8.5   Office of Management and Budget Program Cost and Benefit Assessment

Rule	Agency	Benefit $M	Cost $M
Energy Efficiency Standards for Commercial Refrigeration Equipment	DOE/EE	186–224	69–81
Energy Efficiency Standards for General Service Fluorescent Lamps and Incandescent Lamps	DOE/EE	1,111–2,886	192–657
Patient Safety and Quality Improvement Act of 2005 Rules	HHS/AHRQ	69–136	87–121
Revisions to HIPAA Code Sets	HHS/CMS	77–261	44–238
Updates to Electronic Transactions (Version 5010)	HHS/CMS	1,114–3,194	661–1,449
Prevention of Salmonella Enteritidis in Shell Eggs	HHS/FDA	206–8,583	48–106
Real Estate Settlement Procedures Act (RESPA); To Simplify and Improve the Process of Obtaining Mortgages and Reduce Consumer Costs (FR-5180)	HUD/OH	2,303	884
Part 121 Pilot Age Limit	DOT/FAA	30–35	4
Washington, DC, Metropolitan Area Special Flight Rules Area	DOT/FAA	10–839	89–382
Hours of Service of Drivers24	DOT/FMCSA	0–1,760	0–105
New Entrant Safety Assurance Process	DOT/FMCSA	472–602	60–72
Passenger Car and Light Truck Corporate Average Fuel Economy Model Year 2011	DOT/NHTSA	857–1,905	650–1,910
Reduced Stopping Distance Requirements for Truck Tractors	DOT/NHTSA	1,250–1,520	23–164
Roof Crush Resistance	DOT/NHTSA	374–1,160	748–1,189

DOE/EE, Department of Energy / Energy Efficiency; HHS/AHRQ, Health and Human Services / Agency for Healthcare Research and Quality; HHS/CMS, Health and Human Services / Centers for Medicare and Medicaid Services; HUD/OH, Housing and Urban Development / Office of Housing; DOT/FAA, Department of Transportation / Federal Aviation Administration; DOT/FMCSA, Department of Transportation / Federal Motor Carrier and Safety Administration; DOT/NHTSA, Department of Transportation / National Highway Traffic Safety Administration.

government leaders in the effectiveness of government spending. The once back-room OMB accountants and analysts are now being placed in center political ring. Their methods provide the standardization and their results the judgment of program effectiveness. I suspect we will be hearing more about this office in the future. I should point out that the director of the OMB is appointed by the president and confirmed by the Senate.

Now with all of this talk about federal programs, policies, cost, and benefits, it makes sense to look at the previous volume of these activities to give you a sense

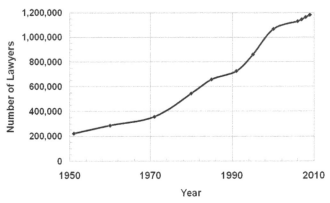

Figure 8.1   Number of lawyers, 1951–2009.

about the importance of policy risk, cost, and benefit calculations in the future. To be a law maker doesn't require you to be a lawyer, although it apparently helps. Washington, D.C., has the highest rate of lawyers per capita in the United States and most likely in the world [30]. Its ratio is over 10 times higher than that of any other state in the Union.

Let's start by looking at how our law-making performance has changed, starting with 1950. In Figure 8.1, the number of lawyers is plotted as a function of year from data obtained from the American Bar Foundation [31]. Figure 8.1 gives you a rough idea of the law-making proliferation potential by showing how the number of law professionals has changed since 1950. The growth shown in this plot by itself may not surprise you, since the U.S. population has also grown considerably over this time period [32]. Despite the growth in population, however, in 1951 there was one lawyer for every 700 Americans. In 1970 the growth rate in the number of lawyers exceeded the population showing a lawyer to population ratio of 1 in 509. In 1990 this ratio grew (or shrank depending on your point of view) to 1 in 299, and in 2009, the lawyer to population ratio was 1 in 259. So the proliferation of lawyers exceeded the proliferation of people. Between 1951 and 2009, the U.S. population increased by a factor of 2, but the lawyer population increased by over a factor of 5. There are more lawyers today than at any time in recorded history.

Now, to see how our regulatory system responded to all of these lawyers, refer to Figure 8.2 that shows how the length of the Code of Federal Regulations [33] has increased over roughly the same period of time.

By this point you probably have guessed where I am going with this. Figure 8.3 shows the correlation between the length of Federal Regulations and the number of law professionals. This is a classic chicken-and-egg paradox as to which came first, the increase in regulations or the increase in law professionals; my guess is the increase in the number of law professionals.

The next question that arises is which variable has been increasing the most, the lawyers or the laws? Table 8.6 answers this question. In 1951, there were approximately 14 lawyers per page of federal regulations. However, since about

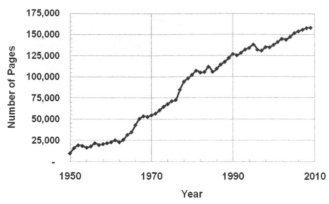

Figure 8.2   Proliferation of federal regulations, 1950–2009.

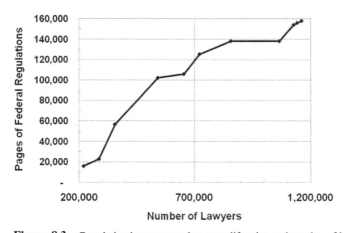

Figure 8.3   Correlation between regulatory proliferation and number of lawyers.

Table 8.6   Relationship Between Law Professionals and Regulation Growth

Year	Number of lawyers	Number of CFR pages	Lawyers per page
1951	285,933	15,932	13.9
1960	285,933	22,877	12.5
1971	355,242	56,720	6.3
1980	542,205	102,195	5.3
1991	755,694	125,331	5.8
2000	1,066,328	138,049	7.7
2009	1,180,386	158,369	7.5

2000, the number of lawyers and number of CFR pages have seemed to reach parity at about 7.5 lawyers per page of regulations. It will be interesting to see if this parity remains constant for the next decade.

I recognize that this correlation is an oversimplification of a complex set of relationships. Nevertheless, over the past 20 years, while both law professionals and the number of laws have increased at a substantial rate, the number of lawyers per page of CFRs has apparently reached a plateau.

The apparent invariance in this ratio does not imply that the regulation creation rate has been constant. Referring to Figure 8.2, Congress has created about 2,800 pages in new regulations each year since 1950. The actual number of regulations is difficult to count but the conclusion is clear regardless: the creation rate of new regulations is not decreasing and, as a result, regulatory compliance and its growing set of rules has become a major operating cost.

As businesses compete for market share, good margins, and rising stock prices, regulatory compliance represents a growing and significant expense. And with these costs, the safety mission to protect employees, the public, and the environment can become lost in the labyrinth of rule compliance. Thus the stage is set for a more logical, rational, and scientific approach to regulations: risk-based regulations.

In the 1970s and '80s, the nuclear power industry was swamped with regulations [34]. The regulatory arm of the industry, the Nuclear Regulatory Commission (NRC), developed its regulatory requirements from deterministic assessments primarily from experience, test results, and expert judgment. The strategy was called "Defense in Depth" and relied on the creation of redundant, independent layers of accident prevention. The philosophy was adequate to cover the identified failure modes but did little to identify the unintended, unknown accident sequences that were lurking in the complex nuclear and control systems.

Compliance was costly and an alternative method, consistent with a technology called probabilistic risk assessment (PRA), was suggested to ensure public and environmental safety. This logical and analytical description of reactor operations considers safety in a more comprehensive manner by examining a broad range of accident-initiating events. In addition to identifying the events, PRA computes the likelihood and the severity (or risk) of various accident scenarios.

So here is a case where the technology tools were available, the industry was ready, and the regulatory body was legally constrained to manage from the outdated defense-indepth philosophy. In 1993, perhaps too late for many nuclear plant investors, Congress finally did pass a law entitled the Government Performance and Results Act (GPRA). One objective of the law was to "improve federal program effectiveness and public accountability."

In response to this new mandate, the NRC adopted new policies designed to provide reactor operators with better regulations that utilized "risk-informed regulation" using the PRA methodology. In 1995, the NRC issued this statement as part of its new policy:

The use of PRA technology should be increased in all regulatory matters to the extent supported by the state of the art in PRA methods and data, and in a manner that

complements the NRC's deterministic approach and supports the NRC's traditional
defense-in-depth philosophy. [35]

This is a clear statement of the direct commitment to risk-based regulation. Aside from the financial savings achieved from reduced compliance documentation, an argument was made that the industry would actually be safer if regulations were risk-based" instead of prescriptive in nature. The argument went like this: rather than spending time on complying with the command and control doctrine of prescriptive regulations, a company could place its energies and financial resources in the areas of its highest risk. The change in activities would reduce risk more effectively than by just following a standard "one size fits all" set of rules.

The implicit message is not implying that the prescriptive stuff is bad. These rules, procedures, and subsequent documentation methods are adequate, on the average. However, in certain situations, if a company knows that the prescriptive actions pertain to areas of operation that are not posing the greatest risk, regulatory compliance is missing its fundamental objective. Risk-based, sometimes called performance-based, regulations enable a company to apply its resources most effectively while also meeting the objectives of government regulatory control.

I had the opportunity to see risk-based regulations actually work in the nuclear industry at a Connecticut nuclear utility. The NRC would identify a potential problem or hazard most likely from other operating nuclear power reactors and then require other operators of similar equipment to take preemptive action. In some cases, the utility used their system-wide PRA to show that the prescribed "fix" would actually increase risk and that either no action or another procedure was more appropriate in their case. The NRC accepted these results. The utility saved literally millions of dollars, risk was either reduced or not increased, and the NRC's regulatory role was satisfied. This is one way risk-based regulations can work.

Some states are actually changing their regulations from prescriptive to risk-based with some impressive results in the short time since the changes [36]. There are many reasons for this but in essence, the risk-based approach was taken because the states identified specific widespread problems affecting many taxpayers. They assessed risks, benefits, and alternatives and reached a rational decision that made sense to all constituents.

There are also other more practical reasons why risk-based regulations concepts are being considered around the globe. The reasons are size, cost, and importance. The costs to administrate prescriptive regulations for industries that have either grown in size or in importance are simply not sustainable. Take, for example, the safety and environmental issues associated with oil and gas pipelines.

There are 2.3 million miles of gas and oil pipelines in the United States [37]. Chances are, if you live in any reasonably dense population area, you are within a stone's throw of an oil or gas pipeline. Most gas is natural gas but the hazardous liquid pipelines carry everything from refinery feedstock crude oil, heating oil, diesel fuel, to jet fuel, and, of course, gasoline.

Pipeline safety and operations are managed in the Department of Transportation by the relatively small Office of Pipeline Safety (OPS). This office administers the department's national regulatory program to assure the safe transportation of natural gas, petroleum, and other hazardous materials by pipeline and develops regulations and other approaches to risk management to assure safety in design, construction, testing, operation, maintenance, and emergency response of pipeline facilities.

The application of risk-based regulations and inspections in the OPS represents a more common application in practice. Given the 2.3 million miles of pipelines, it's impractical to mandate effective prescriptive regulations or actively inspect all pipelines to ensure the safe and secure movement of hazardous materials. This office has been a leader in the development and implementation of risk-based regulations going back to the mid 1990s when risk-based regulations were first studied by the government in regards to pipeline safety [38].

There are now several applications of risk-based regulation and risk-based inspection across industries and governmental agencies. Prescriptive regulations have not been eliminated simply because they do have their place in risk management. Risk-based applications are viable alternatives for or additions to situations where the exposure domain is much greater than the available oversight resources. A relatively recent addition to the list is the Transportation Security Administration (TSA) which shares the issue of exposure versus resources with the Office of Pipeline Safety. Their risk-based applications are not in the public domain but are in active, real time operation. I think the best way to conclude this chapter is by quoting the former Administrator of the Pipeline and Hazardous Material Safety Administration, VADM Thomas J. Barrett USCG (Ret.) in a prepared statement before the 110th Congress on January 10, 2007:

> When it comes to improving transportation security, we follow a systems risk-based approach, recognizing that safety and security are connected, and that significant safety and economic consequences will flow from our decisions. The success of our efforts over time lies in our ability to mitigate overall risk, while avoiding undue burdens on transportation systems, operators, and the public. Effective coordination within the federal government is essential to addressing security concerns in the way that the American public deserves.

We have always had, and always will have, "risk" and "regulation." Maybe in the not-too-distant-future we can add a little more "reason."

DISCUSSION QUESTIONS

1. Using only 42 words, what warning label would you put on alcoholic beverages?
2. What warning labels would you put on cigarettes?
3. Explore the history of risk-based regulations in your state.
4. What three areas do you think will be the next to apply risk-based regulations?
5. This chapter used the number of pages for the Code of Federal Regulations as an indicator for the rate of regulation creation. What other types of measures could be used? Repeat the exercise given in Figures 8.1 through 8.3 and Table 8.6 for one of them.

Case Study: An Accident Waiting to Happen

Most, if not all, accidents involve a sequence of precursor events where the accident would have been prevented if any element of the sequence could have been changed to stop its subsequent progression. In this case study, you'll see how the regulations themselves and several other factors together caused a tragic outcome.

A chemical manufacturing startup company [39–41] considered constructing its first production-level plant in a good location near their first customer. However, the township's zoning ordinance did not permit the building of a manufacturing facility "whose primary uses involve chemical manufacturing or hazardous chemicals or materials" without conditional approval. The company was subsequently denied a building permit. The officials then looked at a secondary site in another township and were granted a certificate of occupancy in an industrial park where the zoning law permitted chemical manufacturing facilities. The approved site was in a multiple-tenant industrial park close to other businesses, across the street from a children's daycare center, and relatively close to private residences.

The company supplied the local township with material safety data sheets (MSDSs) for raw materials and finished products, but did not alert officials about the process hazards and uncertainties associated with the chemical production process. The product of the facility was hydroxylamine (HA), an oxygenated derivative of ammonia, represented by the chemical formula NH_2OH. Hydroxylamine is a chemical which can be unstable and explosive at concentration levels above 70 weight-percent. (Weight percent is the weight of HA in a solution divided by the total weight of the solution.)

The company's business objective was to manufacture HA at the commercial sales requirement of 50 weight-percent. Over the previous decade, semiconductor manufacturers have used HA solutions in cleaning formulations to strip process residues from integrated circuit devices. HA and its derivatives are now also used in the manufacture of nylon, inks, paints, pharmaceuticals, agrochemicals, and photographic developers.

The company began development of its own HA production process through laboratory-scale experimentation that included the construction of a 10-gallon HA production pilot plant, which was successful in showing proof of concept. Now with the new facility, 20,000 square feet in a multiple-tenant building, the officials began to set up shop to manufacture the first large-scale commercial grade HA batch-processing plant in the world. The 10 gallon model plant was going to be scaled up to 2500 gallon HA production batches.

The known explosive nature of HA when it was concentrated above 70 or 80 weight-percent was the general concern. Yet the company management was confident they could accomplish the 250× scale up safely and the facility was completed on time. The accident occurred during the first (and last) production attempt. The accident timeline goes as follows:

Monday afternoon:
   The company began its first distillation to produce 50 weight-percent HA in the new facility. The charge tank contained approximately 9000 pounds of 30 weight-percent HA.

Tuesday evening:
   The concentration of liquid solution in the charge tank was approximately 48 weight-percent. The process was shut down for maintenance to fix a water leak.

Thursday afternoon:
   The manufacturing process was restarted.

Thursday evening:
   The manufacturing process was stopped due to other equipment failures.

Friday morning:
   A heater line was replaced and the process startup was reinitiated (charge tank concentration approximately 57 weight-percent)

Friday night: 7:00–7:15 P.M.
   Liquid solution concentration in charge tank was 86 weight-percent.

Friday night: 7:45 P.M.
   Process shut down and diluted with 30-weight-percent HA to reduce the unanticipated high concentration.

Friday night: 8:00 P.M.
   The manufacturing and engineering supervisor was called and arrived at the site.

Friday night: 8:14 P.M.
   Explosion resulting in:
   • Five fatalities: four company employees and a manager of an adjacent business
   • Two company employees were buried in the building rubble until rescued by emergency responders; their injuries were moderate to serious.
   • Four people in a nearby building were injured.
   • The explosion damaged 10 buildings within the industrial park and several local residences.
   • Six firefighters and two security guards received minor injuries
   • A toy vending machine business adjacent to and in the same building as the company and a package delivery service facility across the street received significant structural damage.
   • A nearby daycare center had minor structural damage. No one was inside at the time of the explosion.
   • Most of the residential damage was limited to broken windows.
   • A crater approximately 6 feet wide, 26 feet long, and 16 inches deep was found in the concrete floor directly below where the charge tank had been located. Based on observed crater dimensions, the explosive force was estimated to be equivalent to 667 pounds of TNT.

Federal authorities charged the company's president with violations of the Occupational Safety and Health Act (OSHA). Specifically, 12 counts were filed under 29 USC §666. This section imposes criminal liability on an employer when he "willfully violates" regulations under the act and the violation results in the death of an employee. The Government's position was that he "willfully violated the regulation governing Process Safety Management of Highly Hazardous Chemicals, 29 C.F.R. §1910.119 (the 'PSM regulation')." The government argued that he failed to:

"(1) perform an initial process hazard analysis appropriate to the complexity of the process and failed to identify, evaluate, and control the hazards involved in the process,

"(2) develop and implement written operating procedures and provide clear instructions for safely conducting activities consistent with the process safety information, and

"(3) train each employee before becoming involved in operating a newly assigned process, train each employee in an overview of the process and in operating procedures, and provide each employee with effective information and training on the physical and health hazards of the work area chemicals."

The indictment contained facts claiming to prove he was an employer under OSHA, a Ph.D. chemist, the president of the company, a majority shareholder, and the highest-ranking officer. Furthermore, the indictment included the duties and responsibilities of:

(a) directing the overall operations of the corporation, including its two chemical processing facilities,

(b) actively supervising work in which OSHA regulated activities where conducted,

(c) having final decision making authority on all matters and

(d) ultimately assuming the proper training and safety of his employees, and

(e) complying with all OSHA standards and regulations.

Because the indictment had the proper allegations, the judge refused to dismiss the criminal indictment. Following the denial of the defendant's motion to dismiss, the president filed an additional motion to dismiss, claiming that the PSM regulation did not apply to their hydroxylamine production process because:

"1. the PSM regulation does not cover solutions in which the concentration of hydroxylamine is less than 100% pure, and

"2. the amount of hydroxylamine in the production process, when calculated without the weight of water in the hydroxylamine solution, was far less than the threshold quantity at which the PSM Regulation would apply."

The court, in analyzing this motion, ultimately determined that the case should be dismissed because of the ambiguities in the regulation. The court held that in upholding the rule of lenity in criminal cases, if there is uncertainty as to the meaning of what is forbidden, the indictment must be dismissed. Furthermore, a regulation is void for vagueness if it does not have sufficient definiteness that ordinary people can tell what conduct is prohibited and in a manner that does not encourage arbitrary and discriminatory enforcement.

The PSM regulation applies to any process which involves an identified chemical at or above the specified threshold quantities. Hydroxylamine is listed with a threshold of 2500 pounds. It was not listed with a requisite concentration level. OSHA interpreted the lack of concentration level to suggest PSM coverage at "(the chemical's) highest commercially available concentration." In the case of hydroxylamine, the commercial concentration was 50%. The president argued that the lack of concentration level suggests coverage only when pure and thus the regulation, on its face, does not cover hydroxylamine in 50% aqueous solution. Furthermore, he claimed that the 2500-pound limitation for noncoverage related only to the weight of the chemical and not the chemical plus the aqueous solution. The government claimed that the 2500-pound quantity related to the total solution weight, not the weight of the chemical alone. Under the president's interpretation, the quantity of hydroxylamine would be well below the amount required to support coverage. Under the government's interpretation, the limitation was exceeded such that coverage was supported.

The government argued that the court should defer to the agency's interpretation of the ambiguous regulation, citing civil case law. The court disagreed with this view, finding that when criminal sanctions are involved, "a regulation cannot be construed to mean what an agency intended but did not adequately express." Furthermore, the court decided that the interpretation would not affect the case, because to satisfy the due process requirements of a criminal conviction, the defendant must have had fair warning of what was required of the regulated community.

The judge held that the PSM regulation is ambiguous with regards to the company's production of hydroxylamine.

> It is clear that the PSM regulation is ambiguous with respect to the following two issues: 1) whether it applies to liquid hydroxylamine at a concentration level substantially less than pure, and 2) whether the threshold quantity . . . considers the weight of the entire mixture, including the weight of the solvent. As a result of these ambiguities, the President could not reasonably have known whether the company's hydroxylamine production process was covered by the PSM regulation. The President, therefore, cannot be guilty of willfully violating OSHA's regulation.

Because the regulation lacked clarity, causing not only confusion but also misleading information to those being regulated, the court granted a motion to dismiss the case. The judge stated that a person

> may not be prosecuted for violation of a regulation if the administering agency affirmatively misled the person into believing his conduct was not a violation. It would be grossly unjust to impose criminal sanctions on an individual for the violation of a regulation that even the administering agency had difficulty applying and which had been interpreted in a misleading manner.

The government did not appeal the district court decision.

QUESTIONS

1. Who or what entities are responsible in part for the fatalities and other costs of this "accident"?

2. How would you assign comparative negligence or fault percentages for these parties or entities?

ENDNOTES

1 Russia: USAID Mission Business Forecast Competitive FY 2010, http://russia.usaid.gov/opportunities/business_opportunities/financialyear2010/ (accessed April 29, 2010).

2 http://www.azcentral.com/12news/news/articles/2009/09/17/20090917stadiummoney917-CP.html (accessed May 1, 2010).

3 Marianne Miliotis et al., "The Interagency Risk Assessment Consortium: Improving Coordination Among Federal Agencies," *Food Safety Magazine*, June/July 2006.

4 National Safety Council, *Injury Facts®*, 2010.

5 Alvin C. Bronstein et al., "2008 Annual Report of the American Association of Poison Control Centers' National Poison Data System (NPDS): 26th Annual Report," *Clinical Toxicology*, 47, 2009, pp. 911–1084.

6 Jiaquan Xu, Kenneth D. KochanekSherry, L. Murphy, and Betzaida Tejada-Vera, "Deaths: Final Data for 2007," Division of Vital Statistics, *National Vital Statistics Reports*, Tables 12 and 23, Vol. 58, No. 19, May 20, 2010.

7 http://www.medicine.ox.ac.uk/bandolier/booth/Risk/accidents.pdf (accessed May 1, 2010).

8 NIDA Report, the Scientific American and Addiction Research Foundation of Ontario. Available at http://www.come-over.to/FAS/alcdeath.htm. (Also see CDC report, "Alcohol Consumption and Mortality, Alcohol Poisoning Deaths.")

9 http://www.who.int/substance_abuse/facts/alcohol/en/print.html (accessed May 28, 2010).

10 E. L. Deaver, "History and Implications of the Alcoholic Beverage Labeling Act of 1988," *Journal of Substance Use*, Vol. 2, Issue 4, October 1997, pp. 234–237.

11 Laws and Regulations under the Federal Alcohol Administration Act and Other Related Provisions of Title 27, United States Code, Laws and Regulations under the Federal Alcohol, Administration Act and Other Related Provisions of Title 27, United States Code and Title 27, TTB P 5100.8 (5/2007).

12 Ali H. Mokdad, James S. Marks, Donna F. Stroup, et al., "Actual Causes of Death in the United States, 2000," *JAMA*, Vol. 291, 2004, pp. 1238–1245.

13 International Center for Alcohol Policies, February 2010, http://www.icap.org/PolicyIssues/DrinkingGuidelines/StandardDrinks/tabid/126/Default.aspx (accessed May 5, 2010).

14 "FTC Orders Cigar Warning," *Los Angeles Times*, June 27, 2000.

15 "Smoking and Tobacco Use: Tobacco Control State Highlights 2010," available at www.cdc.gov.

16 National Center for Health Statistics, National Health Interview Survey (NHIS), 2008.

17 Centers for Disease Control and Prevention, "Smoking-Attributable Mortality, Years of Potential Life Lost, and Productivity Losses—United States, 2000–2004." *Morbidity and Mortality Weekly Report [serial online]*, Vol. 57. No. 45, 2008, pp. 1226–1228 (accessed March 31, 2009).

18 D. F. Behan, M. P. Eriksen, and Y. Lin, "Economic Effects of Environmental Tobacco Smoke Report." Schaumburg, IL: Society of Actuaries, 2005. Accessed March 31, 2009.

19 Centers for Disease Control and Prevention. "Sustaining State Programs for Tobacco Control: Data Highlights 2006." Atlanta: U.S. Department of Health and Human Services, Centers for Disease Control and Prevention, National Center for Chronic Disease Prevention and Health Promotion, Office on Smoking and Health, 2006. Accessed November 24, 2008.

20 Centers for Disease Control, "Best Practices for Comprehensive Tobacco Control programs—2007." Atlanta, GA: U.S. Department of Health and Human Services, 2007.

21 http://www.reuters.com/article/idUSTRE6194SD20100210 (accessed May 9, 2010).

22 "Tobacco Taxes: A Win-Win-Win for Cash-Strapped States," February 10, 2010. Available at http://www.tobaccofreekids.org/what_we_do/state_local/taxes/state_tax_report (accessed June 15, 2011).

23 "The Tax Burden on Tobacco," 2009. Available at http://www.nocigtax.com/upload/file/1/Tax_Burden_on_Tobacco_vol._44_FY2009.pdf (accessed June 15, 2011).

24 http://www.tobaccofreekids.org/research/factsheets/pdf/0097.pdf (accessed May 9, 2010).

25 "FDA Marks First Anniversary of Tobacco Control Act," FDA News Release-877-CTP-1373, June 21, 2010.

26 Adam Liptak, "Top Court Lets Smokers Sue for Fraud," *New York Times*, December 15, 2008.

27 Ellen D. Jorgensen et al., "DNA Damage Response Induced by Exposure of Human Lung Adenocarcinoma Cells to Smoke from Tobacco- and Nicotine-Free Cigarettes," *Cell Cycle*, Vol. 9, Issue 11, 2010, pp. 2170–2176.

28 V. White et al., "Do Graphic Health Warning Labels Have an Impact on Adolescents' Smoking-Related Beliefs and Behaviors?" *Addiction*, Vol. 103, 2008, pp. 1562–1571.

29 http://www.whitehouse.gov/omb/assets/inforeg/draft_2010_bc_report.pdf (accessed May 14, 2010).

30 http://www.averyindex.com/lawyers_per_capita.php.

31 Clara Carson, *The Lawyer Statistical Report*. Chicago, IL: American Bar Foundation.

32 http://factfinder.census.gov/servlet/SAFFPopulation?_submenuId=population_0&_sse=on.

33 National Archives and Records Administration, Code of Federal Regulations Page Breakdown: 1975 to 2008, May 26, 2010 (on file with the Office of the Federal Register).

34 http://www.nrc.gov/about-nrc/regulatory/risk-informed/history.html (accessed May 19, 2010).

35 "Use of Probabilistic Risk Assessment Methods in Nuclear Regulatory Activities; Final Policy Statement," 60 FR 42622, August 16, 1995,

36 "Risk-Based Site Evaluation Process Guidance Documents," Minnesota Pollution Control Agency, May 10, 2010.

37 http://www.phmsa.dot.gov/portal/site/PHMSA (accessed May 25, 2010).

38 Senate Bill 1505, "Accountable Pipeline Safety and Partnership Act of 1996," Sponsor, Sen. Trent Lott (introduced December 22, 1995).

39 U.S. Chemical Safety and Hazardous Investigation Board: The Explosion at Concept Sciences, Case Study No. 1999–13-C-PA, March 2002.

40 *United States v. Ward*, No. CRIM.00-681, 2001 WL 1160168 (E.D. Pa. 2001).

41 R. M. Howard et al., "Prosecution of Clean Air Act," *San Diego Law Review*, Vol. 44, No. 173, 2007, pp. 220–222.

Chapter 9

Earth, Air, and Water:
The Not-So-Bare Essentials

The dose alone makes the poison.

—Paracelsus, Swiss physician (1495–1541)

At the beginning of the 20th century, our life expectancy was about 45 years. Children born at the end of the millennium can expect to live into their 70s. We must be doing some things right. Medical advances, better lifestyles, and changes in diet are among the many things that have helped the human body sustain itself longer. This is the first time in history that we have a population consisting of such a large number of old people.

In 1900, the elderly (those over 65 years of age) comprised 3.1 million people, or only 1 in every 25 of Americans. At the beginning of the 21st century's second decade, the proportion grew to 1 in 8: 33.2 million; and the trend continues. According to the Census Bureau's projections [1], the elderly population will more than double by the year 2050, to 80 million, the number over 84 will more than triple, and the number at or over 100 years old will grow by a factor of over 7.

We've never before managed a civilization like this. We're writing the geriatric handbook as we go. On one hand, the increasing precision of science has helped tremendously in extending our lives. Yet, it seems, the more we know and the more problems we solve, the more problems we seem to discover. We are frustrated by the inability of science to answer all of our questions in this growing world of new hazards, new chemicals, and changing safety policies. Life expectancy statistics alone clearly show us that we've done a pretty good job. Yet the same great science that has given us longevity has created the chemicals, food additives, and pollution sources whose long-term effects are not known. When it comes to our health and our family's health, risk and risk management are no longer just numbers. Risk requires an up close and personal examination.

We hear and read phrases like "the risk of heart disease," "risk of cancer," and "risk of injury" so often that I doubt if most of us have thought a great deal about

20% Chance of Rain: Exploring the Concept of Risk, First Edition. Richard B. Jones.
© 2012 John Wiley & Sons, Inc. Published 2012 by John Wiley & Sons, Inc.

what they actually mean. Since this is the way most of us are informed about risks, it makes sense to spend a little time examining the meaning of these phrases. We'll use the fundamental components of risk, probability and consequence (or frequency and severity, take your choice) to examine this type of risk communication.

Notice in all of the phrases above, severity strictly means whether or not you experience a specific fate, such as heart disease, cancer, or injury. Since risk can be described as the product of frequency and severity, when severity is constant, risk is directly related to the probability of experiencing the disease or injury. This is how risk information generally is communicated in public forums. Usually, when you can replace the word risk with "the chance of" or "the likelihood of," the same rule will apply.

You might ask why this style is used. I believe it's because statistical hazards are not real unless you understand and believe statistics. I don't know about you, but if someone told me my "expected dollar loss" was x per year from engaging in a certain activity, I probably wouldn't care too much. On the other hand, if you told me the risk of dying from the activity was y, you would have my attention regardless of the number. In the majority of discussions regarding public health, safety, and environmental contamination, people use this approach. It is a sound, straightforward way to communicate risk information. As you will see, the science of risk analysis and the art of risk management are still very much in their infancies. In many ways, our individual intuition can be just as good, if not better, than laws, regulations, or instructions set.

It would be easy to take sides in the many controversies that are churning around the human and ecological effects of exposure to radiation, air pollution, food, and chemicals. I admit, it's tempting to voice personal opinions on these complex issues. My intent here, however, is to provide you a balanced perspective in plain English and to cast the technical details of all of this risk stuff into a frame of reference that illustrates how fundamental the risk concept really is. I believe that we are our own best risk management experts. We just never think of our actions in this way. All of the press about "frequency and severity," "risk analysis," "risk assessments," and "risk management" is, in many ways, the current spin on behaviors as old as human existence.

Before we go any further in this chapter, here is an example that illustrates some of the fundamental issues that make risk management a very difficult business. I call it "the water paradox."

Suppose you were a visitor from a distant planet, with a life span of about 5000 Earth years. You discover these so-called "human" life forms and a strange, clear liquid they call water. The humans tell you they need water to live. You, being a curious and scientifically oriented alien, decide to perform an experiment to measure the effects of this chemical, water, on the human life forms. You initially provide the test group of humans with no water. They all die within days. You systematically provide the next groups with more and more water. You note that the life span increases with the dose of water, so you extrapolate the dosage required for your new friends to live 5000 years. Upon delivering this dosage you notice that they all die immediately. Further experiments show that that there is a definite correlation between water and health, up to a certain dosage, that produces an average life span of around 76 years. Additional amounts of water produce no additional increase in life span and eventually life spans start to decrease as dosage increases. From all of these experiments you have shown that a certain amount of water is necessary for life. However, outside a certain dosage range, either too little or too much is fatal. The paradox is that water is necessary for life, but in excess or deficiency it can also be an instrument of death.

This very simplified problem gives you an idea of how the science of toxicology works. It's based on the principle that there's a relationship between a human biological reaction (the response) to the amount of chemical (dose) received by a person. The simple quotation from the father of modern toxicology, the Swiss physicians Paracelsus, that began this chapter stated the paradox in an elegant, simple way over 400 years ago. He is remembered for more than this one phrase. Paracelsus' original name was Theophrastus Philippus Aureolus Bombastus van Hohenheim [2, 3]. Like others ahead of their times, he was almost burned at the stake, and most of his works were not published until after his death. He was the first physician to recognize the use of herbal remedies and water in cures, and to understand the psychosomatic aspects of physical illnesses.

Scientific research efforts regarding the toxicity of chemicals are literally changing by the week. Never before in the history of the human species has so much time and money been devoted to learning about how the body reacts to chemicals. On one hand, we have our free enterprise industries developing new ways to improve our lives with products of all types. If industries were required to fully document the human effects of their products, the experimental trials could last for generations, and we would still be living in a chemical stone age. Consequently, we are forced to make compromises and tradeoffs, weighing the known scientific evidence with the benefits.

Assessing the health risks of chemicals and substances is a difficult, time-consuming, and complicated task for standard scientific procedures. For example, in discussing the exposure effects of water in the water paradox, there were several factors left out for simplicity. In real life, risk exposures' biological effects are dependent upon a large set of variables, too large to contain in a laboratory experiment. Look at Table 9.1. It discloses just a few variables that are used to measure exposure in risk assessments. Just one glance at this table and you probably could add some factors on your own. Also, to complicate matters, the listed factors cannot be taken independently, as interactions may exist between two or more of the exposures [4].

The scientific method inherent in this type of research requires a standardized, repeatable protocol to ensure the result's integrity. The EPA employs a risk assessment paradigm developed by the National Academy of Sciences. It divides the big problem of risk assessment into four smaller components: hazard identification, dose-response assessment, exposure assessment, and risk characterization. This paradigm is a useful tool for health and environmental risk management whether you work for a government, a private company, or yourself. So let's take a brief look at each of these smaller (but still big) problems.

Table 9.1   Some Common Exposure Factors in Health Risk
Assessments

Physiological	Behavioral
Body weight	Total tap water ingestion
Skin area (total)	Total water intake
Skin area (by body part)	Total fish consumption
Inhalation rates	Fish consumption (marine)
	Fish consumption (fresh)
	Shellfish consumption
	Soil consumption
	Soil adherence to skin
	Residence occupancy period
	By residence type
	By age
	Time since last job change
	Time in shower

Hazard identification is the process by which chemicals are recognized as harmful to humans or the environment. As you might expect, the toxicological data comes mostly from animal studies over varying time periods. Results from human toxicity observations are obtained mostly from occupational exposures to the substances and are derived from analyzing data after the fact. Ideally, the scientists want to perform what are called cohort studies. In cohort studies, scientists monitor a group of people exposed to a substance along with a control human group that has not been exposed. In medicine, doctors use this same method except that the "exposed" group receives the experimental drug or treatment and the control group gets a placebo. Studies like these may appear rather heartless. However, we only have two ways to get data in a timely fashion—either by animal or human cohort studies. The worst thing we could do is sit and wait, simply recording the data about how and why people die.

Dose-response assessment studies deal with the issue that got Paracelsus' attention: the biological response as a function of the dosage. Most of these studies are done on animals. The responses range from animal illness to death. Since in some cases death is certain, the response is measured in terms of reduced life span.

Exposure characterization studies the various ways specific chemicals can be transmitted to humans. In principle, there are only three pathways: ingestion, inhalation, and skin contact. In addition, there's the issue of how much of the chemical is retained in the body as a function of each pathway. Here, scientists use mathematical models of chemical transport across ecological systems along the various pathways to get to humans. This allows them to model the contaminant accumulation over a period of time.

Risk characterization combines the information from the three previous segments to compute the probability that the chemical will affect a member of the general public.

From these steps the overall challenge is how we can use our collective scientific knowledge and experience to *predict* the effects of a chemical or substance on humans. Models developed using the structured scientific approach, from hazard identification to risk characterization, are excellent tools when interpreted correctly, but, like any tool, their value relies on the expertise of those who use them. The EPA has the responsibility for guarding the public and environment from chemicals and other substances that can prove harmful. Their results are often controversial, and this kind of risk management often involves difficult controversial decisions. When it comes to this type of risk management, you need good science and a thick skin.

Public health risk management is one of the toughest jobs in the world. Not only is our scientific knowledge incomplete, but, in effect, right and wrong are based upon available resources, or, to be more direct, available money. When it comes to potentially harmful environmental exposures, the decisions we make determine who lives and who dies. This statement is not an exaggeration.

It would be easy to blame free enterprise for the environmental exposures caused by manufacturing emissions and product usage. However, not all environmental health risks are from manufacturing. Some are "all natural." One that has received and will continue to receive a great deal of attention is radon gas. Radon gas (actually ^{222}Rn) is a highly water soluble, radioactive substance created from the decay of uranium and thorium in the Earth's surface. It is common in the rocks and soils of most parts of North America and some other parts of the world. When ^{222}Rn decays into other radioactive elements, such as the polonium isotopes (^{219}Po and ^{214}Po), alpha particles are emitted. An alpha particle is the nucleus of a helium atom, consisting of two neutrons and two protons. From a radiation standpoint, alpha particles are huge, high-speed projectiles that destroy anything they hit. Fortunately, they don't travel very far. However, if you inhale ^{222}Rn or dust that has tiny particles of polonium on it, the alpha particles don't have to travel far to cause damage. In your lungs, the alpha particles can cause carcinogenic cell mutations.

In the outside environment, radon is diluted to such low concentrations that any dose you might get would be very small. The problem comes from indoor exposures where ventilation is restricted. Radon can enter homes from cracks in concrete floors, sump pump openings, and other areas that would allow gas from the Earth's surface to seep into your house.

Before going into a discussion of the radon issue, let's take a look at this from a public risk perception viewpoint. Here we have an invisible, involuntary public exposure to a highly feared radiation hazard for the most dreaded disease, cancer. The result? Americans spent over about $7.7 billion for indoor air quality (including radon measurement and abatement) in 2008 [5]. I must admit, a little bit of this money was spent by me. I have no irrational fear about radiation. I earned my Ph.D. working in the field of radiation transport theory. But when it came to involuntary exposure to the potential threat, I bought radon detection kits, and tested my home

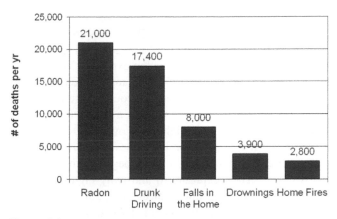

Figure 9.1   Radon health risk comparison.

for the perceived deadly gas over a period of time. So I am not surprised by the public's reaction. It was predictable and, in my opinion, justified.

But let's look a little deeper into this controversial issue. The scientific community identified radon gas in homes in the 1970s. It became a serious issue in the mid-1980s when high levels where observed in some homes in Pennsylvania. In 1986, the EPA published *A Citizen's Guide to Radon,* describing what they thought was a threshold for concern (the latest version referenced here was published in January 2009.) The guide indicates that if your household has a radon level above 4 picocuries per liter (4 pCi/L), you should take some action to reduce it. A picocurie represents the amount of radioactive material that produces 2.2 radioactive decays per minute. So a level of 4 pCi/L means that there are 8.8 radioactive decays per minute in a liter of volume. The guide also stated that radon gas in households was killing 14,000 Americans every year. I have seen other estimates with the number of deaths ranging from 7000 to 30,000 per year.

The graph in Figure 9.1 was taken from EPA document: EPA-402-K-93-008, dated March, 2010, entitled, "Radon—A Physician's Guide: The Health Threat with a Simple Solution." The EPA claims that radon is the second leading cause of lung cancer deaths and the primary cause for nonsmokers.

When I first wrote about the health effects of residential radon gas in 1995, I was not convinced of its connection to lung cancer. However, after reading the results of several scientific studies performed since then, I have changed my mind to believe the health risks of radon are real. And because of its pervasive nature (colorless, odorless gas) and the widespread geographic exposure, radon testing of your home, apartment, or condo is money well-spent, considering the alternative.

Still, people continue to discount radon's health effects for several reasons. For example, referring to Figure 9.1, the EPA claims 21,000 people die from radon-induced lung cancer every year. However, it is difficult to attribute lung cancer death directly to radon gas exposure. The victim could have smoked, been exposed to environmental smoke, or be a victim to any of several other risk factors. To measure and document high levels of radon residential exposures for several years prior to lung cancer diagnosis and subsequent death is difficult. The deaths from all other causes in Figure 9.1 are obvious, direct, and sudden, but for radon, all we know is that a death was caused directly from lung cancer, not radon exposure. The connection to radon takes some induction and detective work. With a half-life of only 3.8 days, radon-222 decays quickly. It is possible that the gas could cause the cancer from exposure over several years and then disappear without a radioactive trace.

However, there is no doubt that radon-222 in high doses over time is a definite killer. But no one knows for sure how radon kills at low doses in your house. The science behind radon-222's causal role with lung cancer began with studies of uranium underground mine workers. Their exposure to unusually high radon concentrations showed a definite connection to lung cancer for both smokers and nonsmokers [6]. Based on this evidence, the International Agency for Research on Cancer classified radon gas as a carcinogen in 1988. The issue then turned to how radon influences the occurrence of lung cancer for residential exposures. The so-called dose-response (or in this case exposure-response) knowledge would show the rate of lung cancer incidence as a function of exposure. This is what was needed to understand radon's lung cancer causality with relatively low residential exposures.

Radon exposure data for the underground mine worker studies were usually gathered posthumously from lung cancer victims by researching their work exposures, ages, and other pertinent characteristics. Analyzing this information, researchers were able to develop the first estimates of the exposure-response causality.

In 1999, a major review of 11 cohort studies was performed that included a total of 60,000 miners and 2,600 recorded lung cancer deaths in Europe, North America, Asia, and Australia [7]. All of the exposure-response curves showed an approximate linear increasing relationship between radon exposure and lung cancer incidence, but the results varied considerably. The variation was partly explained by the fact that eight of the studies involved uranium and the others tin, fluorspar, and iron mines. The mixed results could be from other contaminants and carcinogens that can occur in some of these underground environments.

The National Academy of Science's Biological Effects of Ionizing Radiation (BEIR) Committee VI follow-up studies and several other mining cohort studies were performed using radon-exposed miners in France [8], the Czech Republic [9], and Germany [10]. These research projects somewhat refined the exposure-response results but it was becoming clear that there was not much more we could learn from studying miners. The data was of great benefit in developing the scientific basis for the exposure-response epidemiology, but there also were weaknesses. The data was limited in its ability to identify miners' smoking and residential radon exposure history. Extrapolation of the miners' results to residential radon levels produces a large range of uncertainty. So while the data conclusively connected radon gas to lung cancer for underground miners' concentration levels, the health effects of radon in homes was still an open issue.

Earlier in this discussion I mentioned that it was difficult to connect radon exposure to lung cancer deaths because of its short half-life (3.8 days) and the fact that radon exposure can accumulate over many years prior to the onset of disease. New case-controlled studies have overcome these difficulties. These research projects identify people with lung cancer and also healthy control individuals from the same population. Detailed residential and lifestyle (e.g., smoking) data are collected on each individual. Radiation exposure generally has cumulative property so radon level measurements were taken at all residences where each individual lived for as many previous years as relevant. From these data sources, specialized statistical methods were applied to compute each study participant's effective radon exposure.

At least 40 case-control studies have been done but the results, when reviewed collectively, still contained a high level of statistical uncertainty. Scientists have attempted to pool or combine results [11] but with limited improvement, primarily due to the variability with which each study analyzed its own data.

New pooling studies have compiled the basic participant data from several case-controlled studies to produce the best and most conclusive quantitative evidence linking residential radon exposure to lung cancer. For example:

- A European pooling study [12, 13] considered 7000 lung cancer patients and over 14,000 controls. The results showed a linear relationship between radon exposure and lung cancer incidence with no evidence of a lower threshold below which cancer was not observed.

- A North American pooling [14, 15] studying involving 3662 lung cancer victims and 4996 controls obtained similar linear exposure-response results as the European study. In particular, they also observed no lower exposure threshold for radon-induced lung cancer.

- A Chinese pooling study [16] with 1050 cases and 1996 controls obtained similar results: a linear exposure-response relationship and no lower threshold.

These results have convinced me of the dangers of residential radon. In the mid-1990s, I viewed the EPA's radon warnings and how they were applied in the real estate and new construction industry as simply a way for contractors to generate more business. I did not question the high radon exposure results but I thought it was easy for realtors and home buyers to misuse the information the EPA published. However, the research done since then has provided a robust body of knowledge on residential radon risk that is difficult to refute.

Over the past decade, the EPA has done an excellent job of putting the risks in terms that are more easily understood than pure statistics. They have described residential radon risks using clear, simple and direct comparisons. For example, in their *Citizen's Guide to Radon* publication, the EPA communicates risk exposures in comparative terms as illustrated in Table 9.2.

Residential radon risk is not unique to the United States. In fact, it is a widespread hazard in many countries around the globe. Table 9.3 shows the estimated proportion of lung cancer attributable to radon and the corresponding number of deaths for selected countries [17].

Table 9.2 The Risks of Living with Radon

	Radon risk if you smoke	
Radon Level	If 1,000 people who smoked were exposed to this level over a lifetime* . . .	The risk of cancer from radon exposure compares to** . . .	WHAT TO DO: Stop smoking and . . .
20 pCi/L	About 260 people could get lung cancer	250 times the risk of drowning	Fix your home
10 pCi/L	About 150 people could get lung cancer	200 times the risk of dying in a home fire	Fix your home
8 pCi/L	About 120 people could get lung cancer	30 times the risk of dying in a fall	Fix your home
4 pCi/L	About 62 people could get lung cancer	5 times the risk of dying in a car crash	Fix your home
2 pCi/L	About 32 people could get lung cancer	6 times the risk of dying from poison	Consider fixing between 2 and 4 pCi/L
1.3 pCi/L	About 20 people could get lung cancer	(Average indoor radon level)	(Reducing radon levels below 2 pCi/L is difficult.)
0.4 pCi/L	About 3 people could get lung cancer	(Average outdoor radon level)	

	Radon risk if you've never smoked	
Radon Level	If 1,000 people who never smoked were exposed to this level over a lifetime* . . .	The risk of cancer from radon exposure compares to** . . .	WHAT TO DO:
20 pCi/L	About 36 people could get lung cancer	35 times the risk of drowning	Fix your home
10 pCi/L	About 18 people could get lung cancer	20 times the risk of dying in a home fire	Fix your home
8 pCi/L	About 15 people could get lung cancer	4 times the risk of dying in a fall	Fix your home
4 pCi/L	About 7 people could get lung cancer	The risk of dying in a car crash	Fix your home
2 pCi/L	About 4 person could get lung cancer	The risk of dying from poison	Consider fixing between 2 and 4 pCi/L
1.3 pCi/L	About 2 people could get lung cancer	(Average indoor radon level)	(Reducing radon levels below 2 pCi/L is difficult.)
0.4 pCi/L		(Average outdoor radon level)	

*Lifetime risk of lung cancer deaths from *EPA Assessment of Risks from Radon in Homes* (EPA 402-R-03-003).

**Comparison data calculated using the Centers for Disease Control and Prevention's 1999–2001 National Center for Injury Prevention and Control Reports.

Table 9.3   Proportion of Lung Cancer Estimates Attributable to Radon by Country

	Percentage of lung cancer attributed to radon exposure	Number of deaths
Canada	7.8	1,400
Germany	5.1	896
Switzerland	8.3	231
United Kingdom	3.3	1,089–2,005
France	5.1	1,234–2,913
United states	10–14	15,400–21,000

The wide range of percentage of lung cancers attributed to radon in Table 9.3 is not simply from the concentration of residential radon. Other factors are the amount of time spent indoors, the type of structure, the type of ventilation system, and the weather. There are additional factors but you can see the difficulty in statistically modeling radon effects when the exposure environments can be so different. Table 9.3 does not address the range of residential exposures that are common in the countries.

The combination of geology and building structures are important factors involved with understanding residential radon risk. The United Nations, World Health Organization, and several countries are beginning to measure radon exposures now that the scientific exposure-response relationships are becoming more detailed. Surveys have been carried out to determine the distribution of residential radon concentrations in most of the 30 member countries of the Organization for Economic Cooperation and Development (OECD). Most countries measure radon concentrations using the International System of Units (SI): Becquerels per cubic meter (Bq/m^3). The unit was named for Henri Becquerel, who shared a Nobel Prize with Pierre and Marie Curie for their work in discovering radioactivity. One Bq/m^3 is defined as one radioactive disintegration in a cubic meter. For comparison to the measurement scale used in the United States: $37\,Bq/m^3 = 1\,pCi/L$, $100\,Bq/m^3 = 2.7\,pCi/L$, or $4\,pCi/L = 148\,Bq/m^3$. The country results are shown in Table 9.4.

The worldwide average indoor radon concentration has been estimated at $39\,Bq/m3$ ($1.05\,pCi/L$).

Table 9.4 shows that the U.S. population is not alone in its exposure to residential radon [18–21]. In fact, country averages for 21 out of the 28 countries listed show higher overall averages than does the United States. The exposure of individual country residents is a function of the time they spend indoors, and several other factors previously mentioned, but regardless of the technical issues, the potential exposure to residents worldwide is significant.

To address this risk, the World Health Organization member states completed a detailed survey regarding each country's radon risk communication and mitigation policies. It is of particular interest that some countries do have different standards

Table 9.4   Indoor radon concentrations in OECD countries

	[Bq/m3]		[Bq/m3]
Czech Republic	140	Poland	62
Mexico	140	Denmark	59
Finland	120	Greece	55
Luxembourg	110	Republic of Korea	53
Sweden	108	Germany	49
Austria	99	Norway	49
Spain	90	Belgium	48
France	89	U.S.	46
Ireland	89	Portugal	45
New Zealand	89	Canada	28
Slovakia	87	Netherlands	23
Hungary	82	United Kingdom	20
Switzerland	78	Japan	16
Italy	70	Iceland	10

for new construction as compared to existing structures. Some of the survey results are shown in Table 9.5 [22].

Notice that the United States has the lowest level for action at $148\,Bq/m^3$ (4.0 pCi/L). This may change, however. In January of 2009, the World Health Organization reduced its suggested action or acceptable threshold to $100\,Bq/m^3$ or 2.7 pCi/L [23]. There is an expectation that the United States and perhaps other countries will follow their leadership and change their action threshold values to $100\,Bq/m^3$ or 2.7 pCi/L. This is a significant and noteworthy event. The World Health Organization's research involved over 100 scientists in 30 countries and collectively they have developed scientifically solid evidence of residential radon health effects.

This news is particularly important for Americans since about 87% of their time is spent indoors [24] and, of course, radon gas does not discriminate between residential or other indoor exposures. The potential silent, tragic irony could be that you think you are not a lung cancer risk because you don't smoke or are not exposed to environmental smoke, yet if you are living in a high radon residence, you still could possess a significant lung cancer risk.

Okay, so much for the bad news. The good news is radon measurement and abatement is really simple. The United States and other countries shown in Table 9.5 have radon programs and there are contractors who can design and implement radon reduction systems. The EPA even gives you good advice on how to choose and manage contractors performing this work. The cost of radon mitigation in the United States for a home usually ranges between $800 and $2500, with an average of $1200 [25]. There are plenty of sources of information for acquiring the measurement devices and what to do in case you want to reduce your home's radon levels. Here are some numbers you can call for help.

Table 9.5   International Survey of World Health Organization Member States' Radon Programs

	Do you have a radon program?	Do you have an action level?	Level for existing dwellings (Bq/m^3)	Level for new construction (Bq/m^3)
Argentina	Y	Y	400	200
Austria	Y	Y	400	200
Brazil	Y	N	–	–
Belgium	Y	Y	400	200
Bulgaria	Y	Y	500	200
Canada	Y	Y	200	200
China	Y	Y	400	200
Czech Republic	Y	Y	400	200
Denmark	Y	Y	200	200
Ecuador	N	–	100	–
Finland	Y	Y	400	200
France	Y	(N)	400	
Georgia	Y	Y	200	100
Germany	Y	Y	100	100
Greece	Y	Y	400	200
Ireland	Y	Y	200	200
Italy	Y	N	–	–
Japan	Y	N	–	–
Korea	N	–	–	–
Kyrgyzstan	Y	Y	<200	200
Latvia	Y	Y	200	200
Lithuania	Y	Y	400	200
Morocco	Y	Y	No specific number	–
Netherlands	Y	Y	30	30
Norway	Y	Y	200	200
Paraguay	N	–	–	–
Peru	N	Y	200–600	200
Romania	Y	Y	400	200
Russia	Y	Y	400	200
Slovenia	Y	Y	400	400
Spain	Y	N	–	–
Sweden	Y	Y	200	200
Switzerland	Y	Y	1000	400
UK	Y	Y	200	200
USA	Y	Y	148	148

1-800-SOS-RADON (1-800-767-7236)*	**National Radon Hotline** Purchase radon test kits by phone.
1-800-55RADON (1-800-557-2366)	**National Radon Helpline** Get live help for your radon questions.
1-800-644-6999	**National Radon Fix-It Line** For general information on fixing or reducing the radon level in your home.

The science is now clear and irrefutable. Residential radon-222 exposure is a very real and important risk for all of us. And the current state of knowledge of no lower exposure safety threshold makes it even more important for you to know the radon levels of your living and sleeping environments. This fact may be a transient point in our continual learning about radon or it may in fact be true. All we can say for sure now is that our knowledge appears to contradict Paracelsus. In most situations you can say that "the dose alone makes the poison." For radon, evidently, "any dose alone makes the poison."

Radiation triggers our inherent fear of the unknown and the invisible, but not all radiation is bad for you. This would make sense since we evolved in a natural environment of radiation exposure. We have some radioactive substances in our bodies. Tritium (^{3}H), carbon-14 (^{14}C), and potassium-40 (^{40}K) are the most notable. We receive about 11% of our overall radiation exposure from these sources [26]. And by the way, if you sleep next to someone you're probably doubling your radiation exposure. Potassium-40 in the body varies with the amount of muscle [27]. This is an unintended consequence of body building but it is also true for more common differences in body types. For example, potassium-40 levels are twice as high in younger men than in older women.

It is easy to make statements about loosely connected facts and draw inappropriate conclusions. You have to be careful when interpreting the basis for some of the reports you read. For example, in New York City there is a definitive correlation between the amount of ice cream consumed and the number of heat stroke victims. Does this mean that eating ice cream can cause heat stroke? I think you can see that not all so-called statistical correlations are really causal in nature. You can make your own conclusions about ice cream, but when similar statements are made regarding cancer incidence and nuclear power plant proximity, for example, it makes the national press. You are left to conclude on your own that the two are causal, even if this conclusion is unsubstantiated by the facts.

We cannot rely on science to save us from all harm. Science, by its nature, is ill-equipped to solve the types of problems we now have. Let me explain. Scientific investigations are designed to understand how nature works. I divide science into two parts: basic and applied research. Basic researchers learn about, for example, the universe, the oceans, and cellular mechanism. Applied researchers, on the other hand, figure out what to do with this knowledge. In both cases, scientific endeavors work with what can be measured. Given observable (that is, measurable) inputs and outputs of a process, scientists try to figure out what is happening in between.

The problem with trying to manage risk via scientific methods has more to do with decisions than the knowledge about the risks themselves. Someone has to decide how much of a toxin is safe over what period of time. The numbers alone can't make that kind of a judgment. We are living in an era where our exposure to new substances is increasing faster than our ability to fully understand their biological effects. Scientists can give decisionmakers the facts as they know them today, but then someone or some group makes the call. Moreover, as time progresses, additional health effects surface, and sometimes in indirect ways. Here are just two examples:

- Three versions of Ex-Lax, the leading laxative for nearly a century, were removed from the market because an ingredient that has been in the product since its 1906 origin has been linked to cancer in rats and mice [28].

- The delayed health effects of the PCB (polychlorinated biphenyl) chemical exposures have been documented to show that children born to mothers who ate large amounts tainted fish have relatively low IQs [29]. Additional studies more recently have shown adverse health effects on children of other ages [30].

Now with these (and other) well documented examples of unintended chemical health effects in our collective knowledge, scientific and governmental organizations are now watching our backs the best they can for new problems. For example, in 2010 we had a situation where a ubiquitous chemical, bisphenol-A (BPA), was suspected of adverse health effects, but the research is still being performed.

BPA has a wide range of uses but simply put, if you eat or drink anything out of a can or plastic bottle you might be exposed to this suspected carcinogen. The chemical prevents the container's contents from interacting with the container thereby influencing their taste. It has also been found on paper receipts generated by supermarkets, automatic teller machines, gas stations, and chain stores. This may be one reason that, according to the Centers for Disease Control, BPA can be detected in an estimated 93% of Americans. We don't know how much BPA is absorbed by the skin or whether people ingest it by touching their mouths [31].

Research has also shown that BPA is linked in some studies of rats and mice to not only cancer but also to obesity, diabetes, and heart disease. As of 2010, consumer groups, food distributors, and manufacturers are arguing their side of this issue basically using the same research results to date. In the meantime, there are several states that have or will have various types of BPA. So far the FDA has not made any decisions. They have funded a $30 million study whose results are expected by late 2011 [32]. The U.S. government is preparing legislation but action isn't expected until at least 2012 pending the completion of the research.

Most of the chemical health effects case studies follow the same sequence of events. It takes time and a significant exposure base to first identify suspected chemical-dose response situations and then even more time to determine the true nature of the effects. In the meantime we all live life the best we can knowing that our society's chemical production far exceeds its ability to measure, let alone to understand the effects of our creations. This is one of the challenges for risk management in the future. Science can help in the risk analysis or risk assessment phases, but risk management implies that decisions are made by balancing losses on the scale of costs and benefits. But take a deep breath and let's now consider an important aspect of our environment that everyone desires: clean air.

Air pollution is not usually a stealth hazard. Its presence is seen, smelled, and sometimes even felt in real time. It is produced by cars, power generation, oil refining, chemical production, industrial manufacturing, and even by your fireplace. It comes in many forms and can vary due to weather conditions, time of day, day of week, and many other factors. So to understand the risks associated with this complex, multifaceted issue, we'll discuss its major components [33]:

- ozone,
- particulate matter,
- carbon monoxide,
- nitrogen oxides,
- sulfur dioxide, and
- lead.

Each item in this list represents either a direct pollutant, a family of pollutants, or a chemical which serves as a catalyst enabling the formation of even more dangerous stuff to breathe. And speaking of breathing, about 10,000 liters ($10\,m^3$) of air a day passes through each pair of lungs. Now this is really not a huge volume of air, but that's not the point. This $10\,m^3$ of air every day is your personal $10\,m^3$ of air. Air pollution risks are personally important to all of us. With this short discussion of air pollution components, we begin to put a face on this global monster.

Ozone is a naturally occurring and artificially produced gas composed of three oxygen atoms with the chemical symbol O_3. It is a prophylactic or poison depending on where it is. The prophylactic ozone occurs in the troposphere between 10 km (6 miles) and 50 km (31 miles) above the earth. Ozone at these altitudes has the valuable property of shielding us from the high levels of ultraviolet radiation from space. You can think of it as the "earth's sunscreen protection." The bad ozone is produced near ground level by the same reaction of sunlight on O_2 molecules, but because ozone is so highly reactive, it combines with other hydrocarbon and nitrogen oxides in the air to form even more toxic pollutants. You can't see ozone but you can smell it. It is commonly produced by the interaction of lightning and air or when there are electrical sparks, for example, around electrical motors. In the stratosphere, ozone is beneficial as a radiation shield but at ground level, ozone is a respiratory irritant that has been linked to asthma [34, 35] and other pulmonary disorders.

Next in the list, particulate matter (PM) refers to suspended particles and liquid droplets from acids (such as nitrates and sulfates), organic chemicals, metals, and soil or dust particles. From a health effects perspective, aside from the type of particle or droplet, there are two additional characteristics that are important: particle size and concentration. For air pollution, size does matter but not in the conventional sense. The smaller the particulate size, the more dangerous! Smaller particles are more likely to pass through the throat and nose and enter the lungs where the tissue can be irritated and damaged.

Particle size is measured in terms of microns or micrometers. A micron is 1 millionth of a meter or about 1/250,000th of an inch. For comparison, a human hair is between 40 to 180 microns in diameter, depending on the type of hair, and the thickness of copy paper is 100 microns. Scientists use a special nomenclature to describe airborne particle sizes. For example, PM_{10} refers to particles that are less than or equal to 10 microns in diameter.

The EPA groups particle pollution into two categories:

- "Inhalable coarse particles," such as those found near roadways and dusty industries, are larger than 2.5 micrometers and smaller than 10 micrometers in diameter.

- "Fine particles," such as those that can form when gases emitted from power plants, industries, and automobiles react in the air are 2.5 micrometers in diameter and smaller, or $PM_{2.5}$.

Particle size is important but the amount, or particulate density, plays an equal role in determining how the human body will react to these types of foreign and airborne potential poisons. Particulate densities are measured in terms of micrograms per cubic meter or $\mu g/m^3$. These units are applied worldwide in the development of air pollution standards.

Another general air pollution characteristic is time. Air pollution is often a transient phenomenon caused by changes in weather, wind patterns, traffic density, industrial activity, and human actions. Have you ever worked or lived in a place where air quality was a function of the direction the wind was blowing? In some of areas the country, when weather conditions exist, for example, temperature inversions, cities place temporary bans on using fireplaces. (Wood stoves have become more popular recently due to the increase in heating fuel costs.) Weather variability suggests that one standard for air quality would not be a fair way to represent the true nature of the problem. The EPA and the World Health Organization agree. They publish pollution standards using annual and various short-term values to account for possible variations for transient situations. The short-term time intervals I've seen range from 1 hour, 8 hour, or 24 hour averages, to quarterly averages.

Carbon monoxide (CO) is colorless and odorless gas produced from incomplete combustion in internal combustion engines, some space heaters, industrial processes, and from cigarette smoke. When inhaled, CO has 200 times the affinity for hemoglobin than that of oxygen. Consequently, blood stream does not transport the necessary oxygen levels to the heart, brain, and other organs. This oxygen starvation, depending on the CO concentration level, can cause cardiovascular disease, nervous system problems, or death. CO also is a major element of the airborne cocktail called smog.

Nitrogen oxides (NOx) refer to nitrous oxide (NO) and nitrogen dioxide (NO_2). They are produced as emissions from internal combustion engines, power plants, and off-road equipment. The health effects from NOx are significant because they can react with ammonia, moisture, and other compounds to form small particles. The small particles, when inhaled, can penetrate deeply into the lungs and either cause or worsen respiratory diseases such as emphysema and bronchitis, and can aggravate existing heart disease, leading to increased hospital admissions and premature deaths.

Sulfur dioxide (SO_2) is produced primarily from fossil fuels such as coal, power generation. Automobiles are not major contributors to SO_2 emissions, due to the EPA's requirement for low-sulfur gasoline. The health effects and pollution inhalation pathways of SO_2 are similar to those of the nitrogen oxides family.

Automobile lead emissions have greatly declined over the past two decades due to country regulations requiring oil refiners to produce "unleaded" gasoline. About 90% of the world's countries use unleaded gas, with most of the leaded fuel being consumed in developing countries [36].

Lead's health effects are pervasive compared to the other pollutants in this list. It affects the blood's oxygen-carrying capacity, but that is just the beginning. Lead distributes itself throughout the body, accumulating in the bones. Depending on the exposure duration, lead can cause a wide range of nervous system, kidney function, immune system, and cardiovascular problems. Children and infants, due to their relatively low body weight, are especially sensitive to even low levels of lead which may contribute to behavioral and learning problems. In the United States, due in part to the elimination of leaded gasoline, the CDC has measured a consistent dramatic decrease in elevated lead blood levels in children [37] in testing done between 1997 and 2007. This is one type of air pollution where the mitigation efforts have dramatically improved our (and our children's) environment.

How are we doing in reducing the other types of pollution? A 2010 report developed by the American Lung Association shows that the air quality in many places has improved, but that over 175 million people—roughly 58% of the United States—still suffer pollution levels that are too often dangerous to breathe and the health and economic costs are significant. There is a large body of scientific evidence that ties fine particulate matter ($PM_{2.5}$) with shortened life spans, heart attacks, asthma attacks, missed work days, and other health impacts. California's Air Resources Board (CARB) has compiled the pertinent effect estimates from published, peer-reviewed studies that relate health effects to ambient concentrations of particulate matter. In addition, California has a system of ambient air monitoring stations to provide information on actual pollutant concentrations. By combining the dose- or exposure-response scientific results with the measured concentrations of ozone or PM in the air, California has estimated the health impacts associated with public exposures as shown in Table 9.6.

These numbers are just for one state, albeit for a state that is well-known for its air pollution. But if you consider that the national costs are summed over all states you can get an idea of the true magnitude of the economic costs associated with air pollution. There are research studies that estimate air pollution health and economic costs for cities, states, or regions, but no well-defined research has been done at a national level. The California estimates quoted above serve only as an indicator of the potential severity of this problem for the United States and other areas globally.

Table 9.6   Annual California PM2.5 and Ozone Health Impacts

Health outcome	Cases per year	Uncertainty range[1] (cases per year)	Valuation[2] (millions)	Uncertainty range[3] (valuation— millions)
Premature Death	19,000	6,000 to 33,000	$170,000	$55,000 to $300,000
Hospital Admissions (Respiratory & Cardiovascular)	9,400	5,600 to 14,000	$370	$220 to $540
Asthma and Other Lower Respiratory Symptoms	$220 to $540	110,000 to 430,000	$5.60	$2.3 to $8.6
Acute Bronchitis	22,000	0 to 44,000	$10	$0 to $20
Work Loss Days	1,900,000	1,600,000 to 2,100,000	$370	$310 to $420
Minor Restricted Activity Days	13,000,000	9,700,000 to 16,000,000	$830	$620 to $1,000
TOTAL VALUATION	NA	NA	$170,000	$56,000 to $300,000

[1]Range reflects uncertainty in health concentration-response functions, but not in exposure estimates.

[2]Economic valuations are for the year 2009 (in undiscounted 2007 dollars). These numbers will be updated as new valuations become available.

[3]Range reflects uncertainty in health concentration-response functions for morbidity endpoints and combined uncertainty in concentration-response functions and economic values for premature death, but not in exposure estimates.

If you have ever flown into the smog layer that forms over cities in stable air weather patterns, you would agree we have a long way to go.

So how are we doing? Table 9.7 shows the list of the "best" and the "worst" metropolitan areas for different types of air pollution [38]. The data on air quality for this research were obtained from the U.S. Environmental Protection Agency's Air Quality System and a private firm was contracted to transform the raw data into hourly averaged ozone and 24-hour averaged $PM_{2.5}$ concentration information for the 3-year period for 2006 through 2008 for each monitoring site. The methodology section of this report describes exactly how the ranking was accomplished. The results in Table 9.7 have uncertainty estimates but at least the results here are taken directly from documented measurements. In other words, there is very little wiggle room in these rankings. The results shouldn't be too surprising, although to me there were a few surprises. After reading this, depending on where you live, you may or may not breathe a little easier . . .

The science associated with quantifying the causality between health effects and air pollution will continue. Just like other aspects of environmental risk, the latent,

Table 9.7 2010 Worst and Best Cities for Air Pollution in U.S.

	Cities with the worst air pollution		
Rank	Ozone	Year Round Particle Pollution	Short-Term Particle Pollution
1	Los Angeles-Long Beach-Riverside, CA	Phoenix-Mesa-Scottsdale, AZ	Bakersfield, CA
2	Bakersfield, CA	Bakersfield, CA	Fresno-Madera, CA
3	Visalia-Porterville, CA	Los Angeles-Long Beach-Riverside, CA	Pittsburgh-New Castle, PA
4	Fresno-Madera, CA	Visalia-Porterville, CA	Los Angeles-Long Beach-Riverside, CA
5	Sacramento-Arden-Arcade-Yuba City, CA-NV	Pittsburgh-New Castle, PA	Birmingham-Hoover-Cullman, AL
6	Hanford-Corcoran, CA	Fresno-Madera, CA	Sacramento-Arden-Arcade-Yuba City, CA-NV
7	Houston-Baytown-Huntsville, TX	Birmingham-Hoover-Cullman, AL	Salt Lake City-Ogden-Clearfield, UT
8	San Diego-Carlsbad-San Marcos, CA	Hanford-Corcoran, CA	Visalia-Porterville, CA
9	San Luis Obispo-Paso Robles, CA	Cincinnati-Middletown-Wilmington, OH-KY-IN	Modesto, CA
10	Charlotte-Gastonia-Salisbury, NC-SC	St. Louis-St. Charles-Farmington, MO-IL	Hanford-Corcoran, CA

	Cities With the Cleanest Air		
Rank	Ozone	Year Round Particle Pollution	Short-Term Particle Pollution
1	Bismarck, ND	Cheyenne, WY	Alexandria, LA
2	Brownsville-Harlingen-Raymondville, TX	Santa Fe-Espanola, NM	Amarillo, TX
3	Coeur d'Alene, ID	Honolulu, HI	Athens-Clarke County, GA
4	Duluth, MN-WI	Anchorage, AK	Austin-Round Rock, TX
5	Fargo-Wahpeton, ND-MN	Great Falls, MT	Bangor, ME
6	Fayetteville-Springdale-Rogers, AR-MO	Tucson, AZ	Billings, MT
7	Honolulu, HI	Amarillo, TX	Bloomington-Normal, IL
8	Laredo, TX	Albuquerque, NM	Brownsville-Harlingen-Raymondville, TX
9	Lincoln, NE	Flagstaff, AZ	Cape Coral-Fort Myers, FL
10	Port St. Lucie-Sebastian-Vero Beach, FL	Bismarck, ND	Champaign-Urbana, IL

long-term exposure effects will continue to impede our risk perception when compared to direct risks such as drowning, choking, or automobile accidents. This part of the equation requires our acceptance of statistical or scientific causality before, tragically, enough data can be collected. True, in some cases our conclusions will be wrong, and we may have spent some time and money for little benefit. But what if we're right? Remember, if a statistician jumps out of a 100-story building, based on the data collected on the way down, he or she cannot predict the occurrence of hitting the pavement. Sometimes you have to consider the science and then use some plain common sense, even if it appears expensive, disruptive, and politically incorrect. Overcoming these barriers is the real environmental risk challenge.

The science is clear although the path to a solution is not. Here are some examples of the formidable research that shows the importance and pervasiveness of air pollution.

- *Air Pollution Linked To Early Death*: More than 5000 aged 30 and above were included in the study spanning 16 years in Britain [39].

- *Excess Pneumonia Deaths Linked To Engine Exhaust*: Data on atmospheric emissions, published causes of death, and expected causes of death in England were combined to calculate the impact of pollution on death rates between 1996 and 2004 [40].

- *High Hourly Air Pollution Levels More Than Double Stroke Risk*: The researchers assessed data on stroke deaths in people aged 65 years and older, occurring between January 1990 and December 1994 in 13 major urban areas in Japan and found that high hourly rates of particulate matter around two hours before death were associated with a more than doubling in the risk of death from a bleed into the brain [41].

- *Air Pollution Shrinks Fetus Size*: A Queensland University study compared the fetus sizes of more than 15,000 ultrasound scans in Brisbane to air pollution levels within a 14km radius of the city and observed that mothers with a higher exposure to air pollution had fetuses, on average, smaller in terms of abdominal circumference, head circumference, and femur length [42].

In reality, the science is the simplest part of the analysis. It is difficult to displace a chemical's application in practice. It may be embedded in a product or drug manufacturing process or another application that is either making money or serving an important purpose. Injury and disease evidence can accumulate relatively slowly in time and since there are no standards that warrant corrective action, human exposure and effects continue until some kind of critical mass is reached. Tragically, this delay time can take decades to manifest—even crossing generational boundaries if reproductive functions are affected. Here is an example spanning more than 40 years, from the Vietnam War and into the second decade of the 21st century. It illustrates the complexities, motivations, and unfairness present in these types of risk management decisions.

During the early stages of the Vietnam War, U.S. officials were confronted with increased casualties along rivers that had dense mangroves and other areas of intense jungle growth. The air war was also having difficulty in stopping the flow of soldiers and supplies into South Vietnam as the jungle canopy obscured the views of potential targets from bomber pilots. Several chemical manufacturers developed "deforestation products" intended to eliminate the dense vegetation. About 20 million gallons (75 million liters) of this stuff was dumped on 5 million acres (2 million hectares) of forest and another 500,000 acres (202,000 hectares) of crops in Vietnam over a period of 10 years. The Air Force pilots who flew the spraying missions (dubbed "Operation Ranch Hand") had a slogan: "Only you can prevent forests."

There were several products used for this type of mission. The most well-known herbicide used was nicknamed "Agent Orange" because of its color. About 70% of the total herbicide used was this product. Agent Orange contained the now infamous chemical dioxin. In fact, there were actually several defoliants used in Vietnam and not all of them contained dioxins. And since different manufacturers made the products, dioxin levels varied, even for the same agent. To further complicate matters, "dioxin" refers to a series of 35 chemically related compounds called furans. At the time, these chemicals were believed to be harmless to humans. It was not uncommon for U.S. and other troops to be sprayed during the aerial runs to the point where their uniforms were covered with the chemicals.

The spraying began in 1962. The first U.S. combat troops arrived in 1965. As time progressed and the number of troops exposed to high levels of defoliants increased, several health effects became apparent in this population. Rare forms of cancer, birth defects, and chronic symptoms like fatigue, chills, and weight loss were observed. After much debate, the root cause of the maladies was traced to the dioxin family of chemical compounds. In 1969, former President Nixon ordered the restriction of Agent Orange applications to unpopulated areas after the National Institutes of Health reported it could cause adverse health and reproductive effects in rats and mice. Within two years, by 1972, all Agent Orange and herbicide spraying ended, terminating Operation Ranch Hand.

This is the beginning of a story that is still evolving today. In 1978, about 500 Vietnam War veterans filed claims for illnesses they attributed to Agent Orange. The Air Force, the service branch responsible for the airborne dispersion, understood that these claims may be only the beginning and that a formal analysis of Agent Orange was justified. Subsequently, in 1979, the government funded a $200 million (37-year) study to determine what the exposure-response health effects. However, early in the work, questions relating to proper study protocols, conflicts of interests, and data validity raised doubts on the government's real intentions. In 1998, the San Diego Union Tribune reported that the government's systematic manipulation of the study protocols and results could render the results meaningless [43].

How could a $200 million, 37-year study on health effects from an obvious exposure of a well-defined population to a known cause of cancer and reproductive deformities in rats and mice even begin to be covered up? To understand this we need to go back in time to the political decision-making attitudes of the 1970s.

The Vietnam War was a military, political, and public struggle. Opposition to the United States' involvement was rapidly growing, with protests and demonstrations sometimes becoming tragically violent. In 1971, long before Wikileaks, the "Pentagon Papers" were published by the *New York Times* after the government fought the release all the way to the Supreme Court and lost. The Nixon administration was highly concerned about controlling information leaks since they knew that in many cases, the information was politically injurious to their strategies. The Pentagon Papers documented a detailed description of how the government systematically deceived the public regarding U.S. policies in Vietnam. Controlling information in the interest of national security is a delicate balance between political and public benefit. For many people, the government had gone, too.

Then, in 1972, the beginning of former President Nixon's ultimate demise occurred with the Watergate break-in followed by the many investigations, congressional hearings, and eventually trials. The apparent secrecy at the government's highest levels was still further exposed when the White House audiotapes were discovered. The tapes revealed an ethical vacuum in which information cover-ups were routinely justified by national security. In 1973, former President Nixon chose to resign rather than face impeachment on the charges of obstruction of justice, abuse of power, and contempt of Congress. Also in this year, a ceasefire with North Vietnam was signed and American combat troops left that country. However, the peace accord didn't last very long. North Vietnam's longstanding objective to unify the country came to fruition in 1975 when the South's resistance, represented by the fall of Saigon, was over. With the war over and no more need for defoliants, the Air Force incinerated 2.2 million gallons of Agent Orange.

In the 1970s, the Nixon administration was deep into a military war in Vietnam and a morale war back home with public opposition and mistrust. The leaders, political and military, believed that defoliating the jungle would save lives and hasten its end. History now shows this supposition was wrong. Regarding the chemical manufacturers' knowledge of the dangers of Agent Orange, what they knew and when they knew it is a matter of speculation and content for legal case studies.

Over time and after a long struggle for the truth, officials had little choice other than to state what many people already knew from tragic practical experience. Exposure to Agent Orange is linked to a wide range of serious health effects and birth defects. Updated reports, published every two years with new findings, have provided some basis for claim relief. Resolution, if not justice, came for many veterans and their families from a 2008 update report. The government approved Agent Orange claims for hairy cell and chronic B-cell leukemia, ischemic heart disease, and Parkinson's disease [44]. These illnesses now have been added to the long list of other diseases and birth defects known to be caused by exposure to Agent Orange and the other herbicides used in Operation Ranch Hand. Provisions are also being implemented for retroactive payments that will extend benefits to about 86,000 Vietnam beneficiaries, and, over the next two years, about 200,000 veterans are expected to file disability compensation claims. The 10-year cost for the new regulation is projected to be over $40 billion.

Another tragedy of this historical event is that the effects of Agent Orange are still with us. The genetic effects will remain for generations. In 1978, the Veterans Administration (VA) began a program to examine and record the names of veterans concerned about health problems related to their exposure to Agent Orange and other herbicides during their military service in Southeast Asia. Eligible veterans qualify for an Agent Orange Registry examination at the VA. Almost 50,000 veterans, or one out of every six who served there, are in the registry [45].

According to U.S. reports, these defoliants destroyed 14% of South Vietnam's forests. Yet this is just a small part of the tremendous damage still being seen in this country. Even though Vietnam joined the United Nations in 1977, the country struggled to recover from the war. At the conclusion of the fighting, little research was done to examine how the Vietnamese people who lived in the defoliated provinces fared from the exposure. In 1995, a study of Vietnamese people found the highest dioxin levels ever recorded in humans [46]. In this developing, war-ravaged country, medical records are rare, but a research group found sad evidence of cancers, birth defects, and other illnesses also found in U.S. soldiers. Even though Vietnamese officials can't prove it, they believe at least 500,000 people have died or contracted serious illness from their exposure to Agent Orange. In 2002, a joint U.S. and Vietnamese team started a program to measure the induced health effects [47], but regardless of their findings, it's clear that the Vietnamese people in the exposed provinces will be suffering from the effects of Operation Ranch Hand for generations.

Now that Vietnam's perception has changed to the point at which it's becoming a tourist destination, the U.S. government's attitude has changed from resistance to cooperation in cleaning up the resident dioxin-contaminated sites. Since 2007, the United States has supplied $9 million in cleanup assistance and in 2010, policymakers are reacting to a formal technical report indicating that $300 million is needed to adequately clean up the dioxin contamination [48].

To put these numbers in perspective, the 2010 Gulf of Mexico oil spill daily cleanup costs in the Gulf of Mexico tallied up to about $33 million/day with claims exceeding several billion dollars.

The world probably will never know the total human price tag of using defoliants in South Vietnam. In a country where just surviving is hard enough, the data will most likely never be collected to quantify scientifically the magnitude of this human devastation. And the effects on soldiers from many countries involved in the Vietnam conflict are camouflaged by life impairments from other causes.

Since the use of Agent Orange, the EPA and others have learned a great deal about dioxin dose-response and how dioxins interact inside human cells to cause dysfunction but there is more to accomplish. Dioxin "safe" level policies are still evolving. Although you would think that a highly toxic and carcinogenic substance would foster unity from all parties, outright elimination of dioxins is impossible. Here is just one illustration of contemporary considerations involved in the dioxin dilemma.

A review of research conducted by several independent groups and government agencies in 2010 has revealed that adults are exposed to 1,200 times levels the EPA calls safe, mostly from eating meat, dairy products, and shellfish [49]. Today dioxin is in our environment, not as a sprayed defoliant, but as a byproduct of combustion and various industrial processes. Dioxins are used in electric insulation materials and fire retardants and also occur naturally to some extent. Even forest fires inject dioxins in the air. And for you smokers, dioxins are in tobacco smoke, too. Everyone probably has a little accumulation in fatty tissues from this pathway, although infants are particularly sensitive. The current research involving nursing infants has observed that they can daily ingest up to 77 times higher dioxin levels than the level EPA proposes as safe limits from breast milk and formula. Infants' low body weight puts them at risk for health problems and further safe level limits are being evaluated.

Not all chemicals are exposure time bombs. For an example of a substance we have been using for decades without serious problems, consider chlorine (Cl). It's an "all natural" substance, but also a human health risk depending on the concentration. Chlorine is widely used in disinfectants, plastics, pharmaceutical products, agriculture, and, of course, water purification. On the surface, if a substance is known to cause disease, eliminating its use seems like a good idea, but what about all of the beneficial uses of chlorine?

First of all, the replacement products in some chemical industries might end up being more harmful than chlorine. If we just stopped using chlorine or even used it less, we would definitely see increased illness and death. Chlorinated water is a basic requirement for life in much of our overpopulated and polluted planet. Sure, there are health effects depending on its concentration, but the benefits exceed the risks. If you are really that concerned about your chlorine exposure from water, you can buy filters for your whole house for about $800 or just for a faucet or shower head for $50 each, with replacement filter elements starting at $15. The filters do remove odors and some taste attributes but these qualities are small prices to pay for the assurance of safety implied by their presence.

And when it comes to drinking water, the public's general perception of bottled water's superiority to tap water quality is hard to supplant. Over the years, many blind tastings done by the news media have systematically shown that people really can't tell the difference between most tap water sources and bottled water. True, bottled water suppliers have larger advertising budgets than public utilities, so I suspect this trend will continue. On the other hand, we're getting an idea of the effects of all the empty bottles.

But the water utility industry is not giving up its "tap water is better than (or as good as) bottled water" campaign. Each year at the American Water Works Association (AWWA) conference, a "best of the best" water taste test is held where, in 2010, 21 North American utilities battled for the supreme title [50]. The four judges were chosen from academia, industry, and the journalistic press. The winner was Stevens Point, Wisconsin, followed by New York City and Lincoln, Nebraska. Just in case you're interested in tasting *almost* the best of the best" quality tap water, here are the honorable mentions:

City of Blythe, GA	Kearns (UT) Improvement District
Aurora (CO) Water	Marshalltown (IA) Water Works
Ave Maria (FL) Utility Company	Moorhead (MN) Public Service
Massachusetts Water Resources Authority	San Patricio (TX) Municipal Water District
Village of Canajoharie (NY) Water Works	St. Charles (LA) Water District No. 1
Guaraguao Treatment Plant, Ponce, Puerto Rico	City of Stratford, Ontario
City of Hamilton, OH	Valley City (ND) Public Works
Hardin County (KY) Water District No. 2	Village of Park Forest, IL

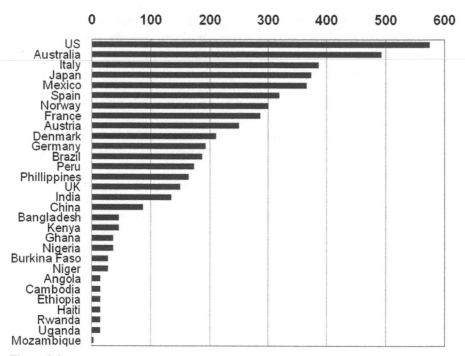

Figure 9.2   Average per capita daily water use (liters).

In case you're interested, the AWWA publishes guidelines on how to conduct a water taste test under the "Public Affairs" tab at their website: www.AWWA.org.

Tap water drinking is only one of many ways to consume this invaluable resource. Let's see who the biggest global consumers are. Figure 9.2 shows the average daily water use per capita by country developed by the United Nations. It's not a surprise that the United States leads the world with about 600 liters (150 gallons) of water use per person per day. The high water use for the countries at the top is primarily from manufacturing and agriculture activities. Also remember that a molecule of water can have more than one use, depending on how far upstream it is.

In the United States, about half of the water is used for power generation, about a third more for irrigation, 10% for public supply, and the remaining 10% for industrial, mining, and domestic applications [51]. Public supply is defined as water withdrawn by public and private suppliers that provides service to at least 25 people or has a minimum of 15 connections for at least 60 days a year. A community water system is a public supply that services customers year around. While this definition may sound trivial and mundane, it has very significant implications. About 268 million people get their residential water this way.

If you pump water from a well or other source, then you're on your own to check the water for problems. The EPA publishes guidelines but they're not coming into your backyard to do any testing. If you are in the large majority of people that are connected to a public supply, then your tax dollars are at work protecting you 24/7 from a plethora of contaminants. These supply systems serve as a common point for aggregation, decontamination, and distribution that makes them economical and safe for many people.

From a risk management perspective, these systems deserve (and get) a great deal of attention and investment. But do they get enough? Remember the proverb: "You never appreciate the water till the well runs dry?" Substitute "pipes" for "wells" in this sentence and you have a glimpse into the future for some folks. The water pipeline distribution system in the United States is aging, resulting in leakage losses and reduced flow rates due to mineral deposits buildups from decades of use. These conditions, plus the bacteria problems fostered by the mineral deposits, are chronic conditions that some public water companies deal with on a regular basis. It's common news to hear of a water main break under city streets causing local flooding and minor service disruptions, but these events are warnings of an ageing water distribution infrastructure.

Maintaining the physical integrity of these pipes is imperative for water conservation and health safety, but the maintenance costs—primarily the overall costs to replace worn out pipe systems—are high. I am not quoting any specific sources but the anticipated costs to replace old water pipelines could be in the range of $290 to $500 billion. I know this sounds unrealistically high, but consider the magnitude of the buried, quietly corroding infrastructure.

There are about 863,000 miles of water pipeline in the United States and about 40% of the network is at least 40 years old, with some parts over 100 years old. It's estimated that about 75% of the current network is older than 25 years and that about half of the current iron pipe network is older than 50 years [52]. Water pipes of this type are expected to last between about 50 and 100 years, depending on the wall thickness.

Of course, pipelines can and will be replaced over several years, most likely several decades, so the financial effects will be spread over time. However, one fact is certain in this environment of uncertainty: your water prices will go up as the supply availability changes, demand increases, and the infrastructure is eventually replaced. A cistern today seems like a good investment for tomorrow—a true liquid asset.

There are other benefits and risks of public supply systems that are important risk management concerns. The centralization of water decontamination is highly cost effective but at the same time it also places a huge burden of responsibility on utility leadership to install state-of-the-art technology equipment, perform proper maintenance, and monitor water purity. There are at least 33 microorganism, 4 disinfection byproduct, 3 disinfectant, 16 inorganic chemical, 53 organic chemical, and 4 radionuclide known water contaminants that are potential health risks. They make the function of water purification a very unforgiving and a continually important job. These systems are silently working everyday and you seldom, if ever, hear about water utilities except in reference to water rationing related to drought, or short-term quality issues related to major flooding. That is exactly how everyone likes it. For when they do get into the news, the damage usually has already been done. Here are some examples.

The largest waterborne epidemic scale outbreak in U.S. history occurred in Milwaukee, Wisconsin, in April, 1993. Over 400,000 people out of a population of about 1.6 million were stricken with cryptosporidiosis, a parasitic illness that causes stomach cramps, fever, diarrhea, and dehydration. Over 54 people, mainly elderly and immunocompromised people, died. The Milwaukee utility had state-of-the-art equipment and had maintained water quality to better than federal standards. The outbreak highlighted the need for cryptosporidiosis surveillance and prevention guidelines [53, 54].

This event reshaped water safety and health policies at virtually all levels of government and I think the results speak for themselves. There has not been a public water supply cryptosporidiosis outbreak since the Milwaukee incident. Notice I qualified this statement by stating "public water supply" outbreak. The reason is that cryptosporidiosis is still a major waterborne health risk in the United States, and, in fact, the number of cases is not decreasing.

Before we talk about our risk exposure from this parasite, let's get to know it a little better. Cryptosporidium is a microscopic parasite found in the feces of infected humans and animals. They exist initially in a protective shell called an oocyst that makes them resistant to bleach and most cleaners. Only prolonged exposure to ammonia or temperature can kill them. Once ingested into the body, multiple parasites emerge from the shell and mature in about 12 hours. They lodge and proliferate in the intestinal tract where the host has no symptoms for about 3 or 4 days. After that time period, the most common symptom is watery diarrhea that lasts for 6–12 days. There is no cure other than treating the dehydration and other secondary effects. The sickness is uncomfortable for healthy people but potentially fatal for children, the elderly, and immune system-compromised people.

From a public safety perspective, one danger from this parasite is that its resistance to normal levels of chlorine in the water so the parasite has to be caught before it can be killed. Special 1-micron polypropylene water filters are applied to screen for oocysts and if the filters test positive for the parasites, then additional purification actions are taken.

The testing process is the first line of defense. At this stage, officials can prevent any illness, but at the next level it's too late. People start to get sick. The danger here is the human exposure to the parasite has a 3- or 4-day head start before people start getting sick so health officials have some catching up to do in locating and eliminating the contamination source.

You probably won't pickup cryptosporidium from tap water but you might from outdoor water recreational activities. According to the Centers for Disease Control and Prevention there were 6479 reported cases of cryptosporidium in 2006; 11,657 in 2007; and 10,500 in 2008. These cases occurred from exposure to the bacteria in lakes, rivers, swimming pools, and even the ocean, that is, any place where animal or human feces can be present [55].

Now if you are swimming in a murky, animal-infested river, and accidently drink some of the water, or if you drink from a seemingly pristine stream on a mountain hike, ingesting this parasite is not too surprising. But what about your municipal pool? For example, three of the incidents in Arizona during the summer of 2008 occurred in public swimming pools and a children's splash park. Both swimming pool outbreaks were at large municipal pools. The first was reported by members of a swimming team where stool samples verified five cases with 52 cases labeled as probable cause. The second incident also involved a swim team at another pool. All 18 members reported being sick. At the children's splash park, there were four confirmed and five probable cases.

Once the parasite was identified the pools were shut down and decontaminated, but since the parasite takes about 3 days to start producing symptoms, those infected are unaware of the hazard during or shortly after exposure. Every time you go in a public pool, hotel hot tub, river, lake, or ocean you may be swimming with cryptosporidium. My advice is don't ingest even one drop. If you do, you are taking the risk and you'll know only later if the drink was worth it.

Of course, millions of people are in the water every day during the summer months without any problems, and I am not implying that parasites, bacteria, and other microorganisms are at dangerous levels everywhere. The lesson here is simple. Swimming, water sporting, tubing, skiing, boarding, and just plain wading around are great activities. Just don't drink the water unless you know it is safe. Cryptosporidium is just one of many contaminates in water that can change what appears to be a pure, clear liquid into a poison.

The world community knows the effects of contaminated water and complaints about taste, odors, and color are irrelevant compared to the 1.8 million children who die annually from diarrheal illnesses that are caused by unsafe water and poor sanitation. Because many people die in remote locations, no one knows the real toll. We only know that most waterborne diseases, illnesses, and deaths are never reported [56, 57]. This is one of many reasons why chlorinated water and state-of-the-art filtration systems are vital to support human life. As the world's population continues to grow to over 7 billion, water is the most valued commodity for the +1.1 billion people in developing countries that have inadequate access to safe water. The harsh facts are [58]:

- 2.6 billion people lack basic sanitation.
- Lack of water is closely related to poverty:

- ○ Almost two in three people lacking access to clean water survive on less than $2 a day, with one in three living on less than $1 a day.
- ○ More than 660 million people without sanitation live on less than $2 a day, and more than 385 million on less than $1 a day.
- 1.8 billion people who have access to a water source within 1 kilometer, but not in their house or yard, consume ~20 liters per day. (For comparison, the average daily water usage is about 150 liters a day in the United Kingdom and 600 liters a day in the United States).
- Close to half of all people in developing countries suffer at any given time from a health problem caused by water and sanitation deficits.

These facts show the magnitude of only one aspect of the world's environmental risk, but there is work being done to improve these statistics. The World Health Organization and other groups are helping developing areas improve their water supply and sanitation.

You would expect that the costs to improve the quality of life for so many people would be enormous. But as an investment banker once told me, it's not the cost or investment size that counts, it's the return on the investment. The benefits of clean, accessible water for people living in developing countries are staggering. The added productivity and reduced health costs show that for every $1 spent, the return is between $5 and $28. Even when pessimistic assumptions are applied, the benefits always exceed costs [59]. This is why the World Bank (http://www.worldbank.org/) is actively engaged in supporting water improvement projects. Work is being done, but as is the case with all problems of this size, change takes time.

When it comes to benchmarking all of the risk exposures we manage in life, earth, air, and water are not-so-bare essentials. After all, if we don't have unpolluted earth, clean air, and pure water, what else matters? But the real uncertainty with the world's environment and the bare essentials doesn't stem just from what we are doing now to manage risk. For better or worse, today's policies will leave a legacy. Read the following quote taken verbatim from the September 1946, issue of *Reader's Digest*:

> *The outlook for human beings with regard to DDT, is excellent. . . . The armed forces have proved that DDT is safe and effective. There is no reason why civilians cannot have the same benefits without danger to themselves or their community.*

What chemical exposures will we be writing about in 70 years? Are there other 2010 DDTs in the plethora of chemical compounds lurking our earth, air, and water?

DISCUSSION QUESTIONS

1. The European Union has a trading regulatory process in place to manage the gradual reduction in emissions for carbon dioxide. The program may expand its coverage eventually to include methane, nitrous oxide, hydrofluorocarbons, perfluorocarbons, and sulfur

hexafluoride. The U.S. emissions trading regulatory process only manages sulfur dioxide and nitrogen oxides, with no limits on carbon dioxide. Why?

2. The Agent Orange example illustrated the delay between exposure, disease, and regulatory changes. In the spring and summer of 2010, the oil spill in the Gulf of Mexico caused a large application of chemical dispersants to diffuse the millions of gallons of crude oil. Based on current knowledge, examine the long-term human and biological effects of this chemical cocktail.

3. Suppose you are the governor of a province or leader of a country and have the opportunity to determine public policy to manage air pollution. What actions would you require immediately and longer-term to improve air quality in your congested, polluted cities? Assume you want to be re-elected or otherwise remain in office.

4. Is there a difference between bottled and tap water? Perform a taste test using the American Water Works Association guidelines and discuss the results.

5. Examine how water quality has affected your area by researching the local pathogens present at different times of the year and the number of observed incidents of sickness, if any.

Case Study: *Time Will Tell the Truth*

In lawsuits where plaintiffs are seeking to recover damages for injuries related to chemical exposures, the outcome rests almost entirely on expert testimony. To prove their cases, plaintiff experts will attempt to connect the chemical to the injury with a reasonable degree of certainty while the defendant's team will opine that the injury is connected to other factors, such as family history, exposure to other chemicals, and pollution. The strength of an expert's logic, evidence, or scientific studies, and ultimately testimony, are perhaps the most significant factors relied upon by the jurors in deciding whether or not to award the injured party.

The trial judge is at the center of this process. If he or she decides not to admit one side's expert testimony for whatever reason, the other side will win as a matter of law, without the need to argue the case before a jury. The process by which judges assess an expert's qualification is critical to the involved parties. In 1993, the landmark Supreme Court decision *Daubert v. Merrell Dow Pharmaceuticals, Inc.* [60] established, for the first time, a uniform standard for expert qualifications.

In Daubert, the plaintiffs were two children and their parents who argued that the children's serious birth defects were caused by the mother's prenatal ingestion of Bendectin, a prescription drug marketed by the defendant. The District Court granted defendant's summary judgment (i.e., ruling that as a matter of law, and without a need for a jury's assessment of the facts), based on the defendant's well-credentialed expert's affidavit concluding that maternal use of Bendectin had not been shown to be a risk factor for human birth defects.

The plaintiffs had prepared testimony of eight well-credentialed experts whose conclusions were based on three sets of data:

1. "in vitro" (test tube) and "in vivo" (live) animal studies that found a link between Bendectin and malformations,

2. pharmacological studies of the chemical structure of Bendectin that purported to show similarities between the structure of the drug and that of other substances known to cause birth defects, and

3. the "reanalysis" of previously published epidemiological (human statistical) studies.

Yet the court determined that this evidence did not meet the applicable "general acceptance" standard for the admission of expert testimony. The District Court found the plaintiff's experts' epidemiological analyses were based on recalculations from previously published studies that showed no causal link between the drug and birth defects. The judge rejected these results because they had not been published or subjected to peer review. Because the plaintiff's evidence was found inadmissible, and the defendant had admissible evidence in support of its case, the court dismissed the case in the defendant's favor. The Appellate Court eventually upheld this decision.

But the Court of Appeals did provide more information. The Court explained that according to legal precedent started by *Frye v. United States* [61], expert opinion based on a scientific technique is inadmissible unless the technique is "generally accepted" as reliable in the relevant scientific community. Under the "generally accepted" rule, an injured plaintiff may be precluded from recovering damages for injuries caused by exposure to new chemicals, or a chemicals with latent effects, since there may not yet exist scientific techniques to establish causation that are "generally accepted" in the relevant scientific community. The flaws of this standard are especially highlighted in modern society's increased usage of, and reliance on, new chemical compounds with unknown side effects.

In the Daubert case, the Supreme Court did acknowledge, however, that strict adherence to the "general acceptance" standard would prevent the jury from hearing of authentic scientific breakthroughs. The court dismissed the "generally accepted" standard as a necessary precondition to the admissibility of scientific evidence, and prescribed a more flexible test to assess expert qualifications.

The Supreme Court took the assessment of scientific expert testimony for trial purposes away from what was generally acceptable (i.e., published) in the relevant scientific community, and assigned this responsibility directly to the judges. It is now the judge who decides whether an expert's testimony is to be admitted by considering several issues related to the pertinent theory or technique (method):

- General test results (if possible),
- Peer review and publication,
- Known or potential error rate,
- Operational standards (if in existence),
- Acceptance within the relevant scientific community.

The assessment process after the Daubert case is more flexible but also has significantly increased the role of trial judges. The new standard's flexibility reflects the growing diversity of chemical and technological risk exposures and also demonstrates how a legal system can adapt to our changing world. Nevertheless, the new standard's value is fully dependant on a single individual's (the judge's) ability to understand and assess scientific methodologies, new technologies, and their associated accuracies.

Questions
1. Make a list of what chemicals might be involved in health effect litigation in the next twenty years. Is there any evidence to date of potential effects?
2. Can you envision future programs like the Agent Orange Program for drug or chemical exposures in the future?

ENDNOTES

1 U.S. Census Bureau, Population Projections, Table 2: Projections of the Population by Selected Age Groups and Sex for the United States: 2010 to 2050, Released 2008.
2 Bruce N. Ames et al., "Paracelsus to Parascience: The Environmental Cancer Distraction," *Mutation Research/Fundamental and Molecular Mechanisms of Mutagenesis*, Vol. 447, Issue 1, January 17, 2000, pp. 3–13.
3 Larry R. Sherman and Scott Black, "Theophrastus Aureolus Bombastus Paracelsus von Hohenheim (1493–1541)," *Journal of Nutritional Immunology*, Vol. 2, Issue 2, January 1994, pp. 95–106.
4 B. Finley, D. Proctor, P. Scott, N. Harrington, D. Paustenbach, and P. Price, "Recommended Distributions for Exposure Factors Frequently Used in Health Risk Assessment," *Risk Analysis*, vol. 14, no. 4, 1994, pp. 533–553.
5 U.S. Indoor Air Quality Market, BCC Research, ENV003C, August 2009.
6 J. I. Fabrikant, "Radon and Lung Cancer: The BEIR IV Report," *Health Phys.*, Vol. 59, Issue 1, July 1990, pp. 89–97.
7 National Research Council, Board on Radiation Effects Research Commission on Life Sciences, Committee on Health Risks of Exposure to Radon (BEIR VI), Health Risks of Exposure to Radon—BEIR VI, 1999.
8 A. Rogel et al., "Lung Cancer Risk in the French Cohort of Uranium Miners," *Journal of Radiological Protection*, Vol. 22, 2002, pp. A101–106.
9 L. Tomasek and H. Zarska. "Lung Cancer Risk among Czech Tin and Uranium Miners: Comparison of Lifetime Detriment," *Neoplasma*, Vol. 51, 2004, pp. 255–260.
10 B. Grosche et al., "Lung Cancer Risk among German Male Uranium Miners: A Cohort Study, 1946–1998," *British Journal of Cancer*, Vol. 95, Issue 9, 2006, pp. 1280–1287.
11 M. Pavia et al., "Meta-analysis of Residential Exposure to Radon Gas and Lung Cancer," *Bulletin of the World Health Organization*, Vol. 81, 2003, pp. 732–738.
12 S. Darby et al., "Radon in Homes and Risk of Lung Cancer: Collaborative Analysis of Individual Data from 13 European Case-Control Studies," *BMJ*, Vol. 330, No. 7485, 2005, pp. 223–227.
13 S. Darby et al., "Residential Radon and Lung Cancer: Detailed Results of a Collaborative Analysis of Individual Data on 7148 Subjects with Lung Cancer and 14208 subjects Without Lung Cancer from 13 Epidemiologic Studies in Europe," *Scandinavian Journal of Work, Environment & Health*, Vol. 32, Suppl. 1, 2006, pp. 1–83.
14 D. Krewski et al., "Residential Radon and Risk of Lung Cancer: A Combined Analysis of 7 North American Case-Control Studies," *Epidemiology*, Vol. 16, 2005, pp. 137–145.
15 D. Krewski et al., "A Combined Analysis of North American Case-Control studies of Residential Radon and Lung Cancer," *Journal of Toxicology and Environmental Health, Part A*, Vol. 69, 2006, pp. 533–597.
16 J.H. Lubin et al., "Risk of Lung Cancer and Residential Radon in China: Pooled Results of Two Studies," *International Journal of Cancer*, Vol. 109, 2004, pp. 132–137.
17 World Health Organization, WHO Handbook on Indoor Radon, 2009, p 16. Table 9.1, Radon-222 Gas Lung Cancer Death Estimates by Country.

18 United Nations Scientific Committee on the Effects of Atomic Radiation (UNSCEAR), *Sources and Effects of Ionizing Radiation*. UNSCEAR 2000 Report to the General Assembly, with Scientific Annexes. New York: UNSCEAR, 2000.

19 World Health Organization, *International Radon Project Survey on Radon Guidelines, Programmes and Activities*. Geneva: WHO, 2007.

20 S. Billon et al., "French Population Exposure to Radon, Terrestrial Gamma and Cosmic Rays," *Radiation Protection Dosimetry*, Vol. 113, 2005, pp. 314–320.

21 S. Menzler et al., "Population Attributable Fraction for Lung Cancer Due to Residential Radon in Switzerland and Germany," *Health Physics*, Vol. 95, 2008, pp. 179–189.

22 World Health Organization, International Radon Project Survey on Radon Guidelines, Programs and Activities, Final Report, WHO/HSE/RAD/07.01, 2007.

23 *WHO Handbook on Indoor Radon: A Public Health Perspective*. World Health Organization, 2009.

24 "Green and Healthy Housing," *Journal of Architectural Engineering*, Vol. 14, Issue 4, December 2008, pp. 94–97.

25 http://www.epa.gov/radon/pubs/consguid.html (accessed July 7, 2010).

26 http://www.doh.wa.gov/ehp/rp/factsheets/factsheets-htm/fs10bkvsman.htm (accessed July 7, 2010).

27 Qing He et al., "Total Body Potassium Differs by Sex and Race across the Adult Age Span," *American Journal of Clinical Nutrition*, Vol. 78, No. 1, July 2003, pp. 72–77.

28 Sheryl G. Stolberg, "3 Versions of Ex-Lax Are Recalled After F.D.A. Proposes Ban on Ingredient," *New York Times*, August 30, 1997.

29 Joseph L. Jacobson and Sandra W. Jacobson, "Intellectual Impairment in Children Exposed to Polychlorinated Biphenyls in Utero," *New England Journal of Medicine*, Vol. 335, No. 11, September 12, 1996, pp. 783–789.

30 P. W. Stewart, E. Lonky, et al., "The Relationship between Prenatal PCB Exposure and Intelligence (IQ) in 9-Year-Old Children," *Environmental Health Perspectives*, Vol. 116, Issue 10, October 2008, pp. 1416–1422.

31 "Risk Debated as BPA Found on Paper Receipts," *Dallas Morning News*, July 28, 2010.

32 U.S. Food and Drug Administration, Update on Bisphenol A for Use in Food Contact Applications: January 2010. January 15, 2010. http://www.fda.gov/NewsEvents/PublicHealthFocus/ucm197739.htm (accessed July 10, 2010).

33 http://www.epa.gov/air/urbanair/, updated July 1, 2010 (accessed July 14, 2010).

34 W. Anderson et al. "Asthma Admissions and Thunderstorms: A Study of Pollen, Fungal Spores, Rainfall, and Ozone," *QJM*, Vol. 94, 2001, pp. 429–433.

35 Jonathan A. Bernstein, ed., "Health Effects of Air Pollution," Environmental and Occupational Respiratory Disorders, *Journal of Allergy and Clinical Immunology*, Vol. 114, Issue 5, 2004, pp. 1116–1123.

36 Marc Lacey, "Belatedly, Africa Is Converting to Lead-Free Gasoline," *New York Times*, October 31, 2004.

37 http://www.cdc.gov/nceh/lead/data/national.htm, updated May 5, 2010 (accessed July 15, 2010).

38 Quantified Health Impacts of Air Pollution Exposure, http://www.arb.ca.gov/research/health/qhe/qhe.htm (accessed July 18, 2010).

39 "Air Pollution Linked to Early Death." *Science Daily*, August 1, 2007. Available at http://www.sciencedaily.com /releases/2007/07/070731085554.htm (accessed July 18, 2010).

40 "Excess Pneumonia Deaths Linked to Engine Exhaust, Study Suggests," *Science Daily*, April 16, 2008. Available at http://www.sciencedaily.com/releases/2008/04/080414193025.htm (accessed July 18, 2010).

41 "High Hourly Air Pollution Levels More Than Double Stroke Risk," *Science Daily*. September 22, 2006. Available at http://www.sciencedaily.com/releases/2006/09/060921094534.htm (accessed July 18, 2010).

42 "Air Pollution Shrinks Fetus Size, Study Suggests," *Science Daily*, January 10, 2008. Available at http://www.sciencedaily.com/releases/2008/01/080107094944.htm (accessed July 18, 2010).

43 "How the Military Misled Vietnam Veterans and Their Families About the Health Risks of Agent Orange," *San Diego Union-Tribune*, November 1, 1998.

44 *Federal Register*, Vol. 75, No. 57. March 25, 2010, p. 14391.

45 "The VVA Self-Help Guide to Service-Connected Disability Compensation for Exposure to Agent Orange for Veterans and Their Families," *Vietnam Veterans of America*, February 2010.

46 J. H. Dwyer and D. Flesch-Janys, "Editorial: Agent Orange in Vietnam", *American Journal of Public Health*, Vol. 85, No. 4, April 1995, p. 476.

47 David Cyranoski, "U.S. and Vietnam Join Forces to Count Cost of Agent Orange," *Nature*, 416, March 21, 2002, p. 252.

48 Margie Mason. "$300 Million Needed for Agent Orange in Vietnam," *Associated Press*, Hanoi, Vietnam, June 16, 2010.

49 "Infants Ingest Dioxin at 77 Times EPA's Safe Threshold," press release, *Environment Working Group*, July 13, 2010.

50 "Taste Test," *American Water Works Association Streamlines*, Vol. 2, No. 15, June 23, 2010.

51 Joan F. Kenny et al., "Estimated Use of Water in the United States in 2005." Circular 1344, U.S. Department of the Interior, U.S. Geological Survey.

52 State of Technology Review Report on Condition Assessment of Ferrous Water Transmission and Distribution Systems EPA/600/R-09/055, June 2009.

53 W. R. MacKenzie, N. J. Hoxie, M. E. Proctor, M. S. Gradus, K. A. Blair, D. E. Peterson, et al., "A Massive Outbreak in Milwaukee of Cryptosporidium Infection Transmitted through the Public Water Supply," *New England Journal of Medicine*, Vol. 331, 1995, pp. 161–167.

54 Susan T. Goldstein et al., "Cryptosporidiosis: An Outbreak Associated with Drinking Water Despite State-of-the-Art Water Treatment," *Annals of Internal Medicine*, Vol. 124, No. 5, March 1, 1996, pp. 459–468.

55 "CDC Reports Increase in Cryptosporidiosis," American Water Works Association Streamlines, June 10, 2010.

56 World Health Organization and United Nations Children's Fund Joint Monitoring Programme for Water Supply and Sanitation, *Progress on Drinking Water and Sanitation: Special Focus on Sanitation*. New York and Geneva: UNICEF/WHO, 2008.

57 E. D. Mintz and R. L. Guerrant, "A Lion in Our Village: The Unconscionable Tragedy of Cholera in Africa," *New England Journal of Medicine*, Vol. 360, 2009, pp. 1060–1063.

58 United Nations Human Development Report, 2006, pp. 6, 7, 35.

59 Guy Hutton and Laurence Haller, "Evaluation of the Costs and Benefits of Water and Sanitation Improvements at the Global Level," In *Water, Sanitation and Health, Protection of the Human Environment*. Geneva: World Health Organization, 2004.

60 *Daubert v. Merrell Dow Pharmaceuticals*, 509 U.S. 579 (1993).

61 *Frye v. United States*, 293 F. 1013 (D.C. Cir. 1923).

Chapter 10

Food . . . for Thought

Food—the first enjoyment of life.

—Lin Yutang (1895–1976), Chinese writer and philologist

Ever since we left the security of the womb, food procurement has been high on our list of needs. It's one of those basics which, along with air and water, we can't live without. But food is special. Unlike air and water, food must be planted or born, nurtured, and then harvested. There are very few of us in the so-called civilized world who plant, grow, catch, or raise all of our own food. We don't grow the vegetables, fruits, and grains, nor do we raise the chickens, cattle, or pigs that we eat. Most of us trust others to do all of these things.

When it comes to food, everyone is an authority so it's not surprising that we have plenty of advice on the subject. While the content of this chapter is not intended to help you lose weight or change your eating habits, it may have that effect depending on your conclusions from the risk discussions to follow. The exposures are relevant today and, as you will see, they may be more important for future consumers. All of us may be food experts, but are we asking the right questions? Are we providing adequate regulations and safeguards for our and our children's generation?

Generally speaking, new scientific discoveries and, of course, health-related incidents have been the primary causes for changes in food safety regulations. But science is not always a part of the solution. Sometimes it's the cause, by creating new exposures. Other times technology helps us identify exposures that were always there, but which we just didn't have the ability to see. In both cases, regulations have been either created or modified to mitigate the identified risks. This is a continuous process.

There is one example from history of how U.S. food safety regulations have changed that is particularly noteworthy. It represents a milestone about how our world was viewed in the era before technology and how it changed once the risk horizon was extended.

20% Chance of Rain: Exploring the Concept of Risk, First Edition. Richard B. Jones.
© 2012 John Wiley & Sons, Inc. Published 2012 by John Wiley & Sons, Inc.

Paracelsus said, "The dose makes the poison." Unfortunately, we always don't know what dosage is "safe," over what period of time, and for what age groups. The time delay required to observe the effects of some chemicals, even in clinical studies of rats and mice, is sometimes years. There is no doubt that some of today's chemicals can cause problems. We know, for example, that our powerful, efficient pesticides can cause cancer in sufficient dosages. The question on our table is: How much residue is "acceptable" for us and for our children? To learn how U.S. policy started on this subject, let's return to the Eisenhower administration of the late 1950s where I'd like to introduce you to a key figure.

Does the name James J. Delaney ring a bell? Unless you're from New York City or a student of history, I doubt you know of him. Yet he probably has done more to shape the U.S. government's policy on food safety than anyone else.

James Delaney was born on March 19, 1901, in New York City and grew up attending public schools in Long Island City [1]. In 1944, he was elected to the U.S. House of Representatives where he started out as a liberal with close ties to the labor movement. However, as his constituency changed in New York from a blue-collar enclave into an economically healthy and vocal middle-class stronghold, he became increasingly conservative. In 1946, he lost re-election, but regained his House seat two years later. After this, he won 15 consecutive elections. Mr. Delaney retired from the House of Representatives in 1978 where he had served for 38 years and moved to Key Biscayne, Florida. Mr. Delaney died at his son's home in Tenafly, New Jersey, on May 24, 1987 [2].

Mr. Delaney's legacy is the amendment to the federal Food, Drug and Cosmetic Act that carries his name. The "Delaney Amendment," commonly called the "Delaney Clause," outlawed the use of any substance in processed foods that is known to cause cancer, without consideration for the degree of risk. It says if a substance, pesticide, preservative, nutrient, or coloring caused cancer in any dose, it could not be used. This zero-tolerance or zero-risk approach seemed manageable in 1958.

But technology presents a double-edged sword. In 1958, when the Boeing 707 airliner went into service and the first micro-chip was invented, we knew of only a relatively small number of carcinogenic chemicals. Since then there has been a continuous avalanche of new substances. Currently, the EPA has approximately 20,000 products registered as pesticides that are formulated from about 850 different active ingredient chemicals, manufactured or formulated by more than 2,300 different companies, and distributed by about 17,200 distributors [3]. A lot has changed since 1958.

Just to give you an indication of the difficulty in complying with the Delaney Clause today, there are over 60,000 industrial chemicals regulated under the Toxic Substance Control Act. No one knows for certain how many of these are in our food supply [4]. Free-enterprise science is creating about 10 new products a day, a rate much higher than our ability to understand their effects.

Obviously, the Delaney Clause is no longer a manageable law. In addition to the myriad chemicals we produce, we're now able to measure better than ever before. In 1958, scientists could measure concentrations down to the perceived minuscule level of 1 in a million. This seems pretty small. It's roughly equivalent to being able to identify one drop of a substance in twelve gallons. Now we can measure down to 1 in a trillion. In numerical terms this is 1 in 1,000,000,000,000. Identifying one part of a substance in a trillion others is like picking one drop of liquid out of 17 Olympic-sized swimming pools.

We can't perceive this level of detail. Most people have difficulty picturing a one-in-a-thousand or one-in-ten-thousand situation. Now that we can view our universe down to this level, we don't like what we see. In 1958, when we thought we were safe, it was only because we were blind to the world beyond the 1 in a million horizon. Many of the carcinogens that Delaney outlawed were present, but at levels below our ability to detect them. Now that we can see a little farther, reality has set in. As a nation or global community, clearly with more problems than resources, we just can't afford a zero-tolerance level.

In the first half of 1990s, the Delaney Clause was more of a curse than a blessing. Countermeasures were taken to work around the clause and its interpretation was liberalized. In other words, the once-rigid Delaney Clause was bent a little. People began to realize that it wasn't practical, but before the zero-risk level could be changed, Congress had to set other limits.

Delaney's zero-risk was replaced by a policy called "negligible risk." This is the risk management philosophy under which we are living today. The idea that risk can be eliminated is dated to an era where our technological looking glass had a narrow view of the universe and risk management was thought of solely as an insurance term.

Now we'll explore elements of our food chain and examine the systems that produce and transport these delicate commodities to our tables day after day after day. Each element integrates specific scientific and technological advances to enhance food quantity, quality, or safety. Just like technology caused the rejection of the Delaney Clause, I wonder how our "negligible risk" policy today will be viewed in fifty years. Think about this as you read the following food chain discussions.

Figure 10.1 presents the food chain components that will be discussed in terms of their risk exposures. There are many food chain characterizations and I am using this version just to acquaint you with the fundamental process components that take basic inputs and (most of the time) put sustenance in your body. Now let's take a look at each element of this life sustaining sequence for most people around the world.

Figure 10.1   Food transport chain.

FARMLAND

The issue here is simple. In 1980 there were about 2,428,000 farms on 1042 million acres. In 2009 there were 2,200,000 farms on 920 million acres. Also during this period, the U.S. population grew from 250 million to 300 million people. Figure 10.2 shows these relationships from 1980 to 2009 [5]. So as the number of farms decreased by 9% and the number of farming acres decreased by 12%, the number of people to feed increased by 35%. The reduction in number of farms could come from consolidation but the decrease in farming acreage is a serious food risk issue for our future. Once farmland is lost to a shopping mall, housing development, or industrial park, it's virtually gone. And since an inch of nutrient-rich top soil takes about 500 years to form [6], it should be treated with a great deal of respect.

Obviously, there has been a significant increase in yields or productivity per acre. Table 10.1 shows some examples to give you a sense of the tremendous improvements that have been achieved by farmers and those industries who support them [7]. Crop yields vary each year depending primarily on the weather but the

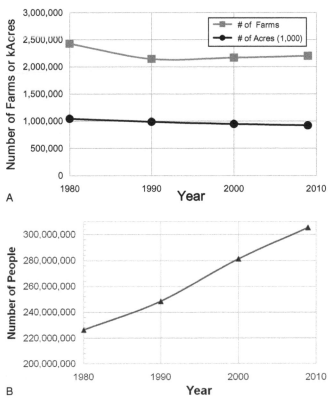

Figure 10.2   (A) Number of U.S. farms and acres. (B) U.S. population, 1980–2009.

Table 10.1   Crop Productivity Changes: 1980–2009

| | Yields (bushels/acre) | | |
	1980	2009	% Increase
Corn	91.0	164.7	81%
Soybean	26.5	44.0	66%
Wheat	33.5	44.4	33%

percentage changes shown in Table 10.1 are representative of actual productivity improvements over this time. If this wasn't the case we would have either a smaller population, much higher food prices, or both.

How can you get this much more food out of a reduced amount of land? Let's see what variables you can change. The length of a day and the weather control the amount of sunlight, temperature, and moisture so there isn't anything we can do about them except move farms to different latitudes. You can plant the crops closer together, but there are space limitations required by each seed or plant. (We'll talk more about this in the next section.) Or you can increase the amount of nutrition in the soil. This is something you can control. Put your soil on agricultural steroids, that is, fertilizers.

Fertilizers have been around for as long as people have been growing food. Advances in our understanding of plant growth mechanisms have changed fertilizer applications from a simple fish dropped in a corn hill to a scientifically precise recipe. And one size does not fit all. Soil nutritional requirements can vary on the same farm and with the crops being grown. Farmers today have added soil chemistry to their job skills as they periodically assess soil nutrients to maximize crop yields. Too little fertilizer, too much, or the wrong chemical mix all have detrimental effects on crop yields, the environment, and profits.

Here's a short primer on agriculture fertilizer that shows how precise fertilizer applications have become. There are 80 basic elements that occur in nature. Plants require only various amounts of 16 of them to support their lifecycle [8]. Crop production, first and foremost, requires carbon, oxygen, and hydrogen for photo-synthesis. Plants convert sunlight, CO_2 and H_2O into plant sugars. Fertilizers are not needed here but this is just the start. All living things are built from DNA, deoxyribonucleic acid, a double-helixed molecular structure that stores genetic information formed by the sequence of four chemical bases: adenine (A), guanine (G), cytosine (C), and thymine (T). Plants and humans alike also store energy in the form of adenosine-triphosphate or ATP. This molecule is unique in the sense that it is the primary mechanism used by cells to transfer energy to other chemical reactions, one of which is cell division or growth. DNA strands and ATP molecules include carbon, oxygen, and hydrogen, but they also contain nitrogen (N), phos-phorus (P), and potassium (K). To complete the list of the 16 elements, here are the remaining 10:

Calcium	Copper
Magnesium	Zinc
Sulfur	Iron
Boron	Molybdenum
Manganese	Chlorine

All of these chemicals together, in the right amounts, support healthy plant growth in a variety of ways including good leaf color, shape, resilient structures, size, and, of course, yields.

The main fertilizer ingredients nitrogen, phosphorus, and potassium are usually mixed together in different ratios to match soil needs. At feed stores and garden centers you'll see fertilizer bags with three numbers separated by dashes, like 5-10-15. The code is a sequence that indicates the percent weight of available nitrogen, phosphate (P_2O_5), and potash (K_2O), respectively. To convert the phosphate number to the percentage of phosphorus, multiply the second number by 0.44, and to get the actual percentage of potassium multiply the third number by 0.83. So a 100-lb bag of fertilizer with the code 5-10-15 has 5 lbs of available nitrogen, 10 lbs phosphate (4.4 lbs phosphorus), and 15 lbs of potash (about 12.5 lbs potassium). Today, farmers test their soil and then buy the right ratio and amount of nutrients to maximize crop yields. The days of qualitative judgment regarding fertilizer requirements are over for the high yield business farmer.

Nitrogen is a primary soil nutrient and it may sound odd that its scarcity could ever occur. After all, 78% of the air by volume is nitrogen so how can plants suffer from a lack of this element when it exists literally all around us? Nitrogen in the air is in the form of the stable molecule N_2. In order for nitrogen to be useful for plant growth, it needs to be in the chemically active molecular form of NH_3, or ammonia. The process of converting N_2 to NH_3 is called nitrogen fixation, wherein nitrogen is "fixed" into a form that chemically reacts with plant proteins to promote plant growth and health.

Some plants, like legumes (e.g., beans, soybeans, alfalfa, clover, and peanuts), perform nitrogen fixation as part of their natural reaction to common soil bacteria. Grain crops like wheat and corn lack this genetic advantage and they rely on the direct ammonia content of the soil or fertilizers. Nitrogen fixation does occur naturally to some extent in bacteria, algae, and lightning, but these sources are insufficient for soil nutrient enhancement.

Figure 10.2 clearly shows there is no shortage of fixed nitrogen and that farmers have been very successful in acquiring and using nitrogen-rich fertilizers. However, at the beginning of the 20th century, things were different.

In the early 1900s, most of the nitrogen-based fertilizers came from organic sources like concentrated bird dung called guano [9], imported from Peru, and from inorganic sodium nitrate mines in Chile. The demand for fertilizer was beginning to exceed supply and the political and business opportunities of fixing atmospheric nitrogen motivated entrepreneurs and governments alike to accelerate their research.

One reason why governments were interested is because ammonia and related compounds are also essential ingredients in explosives.

Several researchers worked on this problem, but only one succeeded with a cost effective, industrial scale, workable process. It was a German chemist, Fritz Haber [10, 11], who along with Karl Bosch solved this world-scale problem. The Haber-Bosch process is being used today still as the primary source of fixated nitrogen. In 1918, Fritz Haber was awarded the Nobel Prize in chemistry for this achievement [12]. But his scientific talents had other, controversial applications. As a loyal, patriotic German chemist during World War I, he applied his technical skills to help his country develop poison gas. The active role he played, not only in producing poison gas, but also in its promotion and application, caused him to be criticized and censored by some scientific groups.

The Haber-Bosch process made enough fertilizer to continually feed the world's growing population. The life-producing discovery made by Fritz Haber is demonstrated in Figure 10.3 [13]. The difference between the dotted line population forecast and the actual growth lines alludes to the significance of his achievement. To give Fritz Haber all of the credit for the marginal increase in population (~3.5 billion people) perhaps is an overstatement, but regardless, it is safe to say today most of us owe a debt of gratitude to this somewhat forgotten man. Even if you're not a science historian, you probably recognize names like Einstein, Maxwell, Planck, Mendel, or Freud, but the accomplishments of these and many more great scientists pale to the achievements of Fritz Haber. For this dedicated scientist created the process that enabled farmers to produce sufficient crop yields to support the population of this planet over the last century.

Even though farm acreage has decreased over the past several decades, unit productivity, thanks to science and engineering, has grown at record levels. The

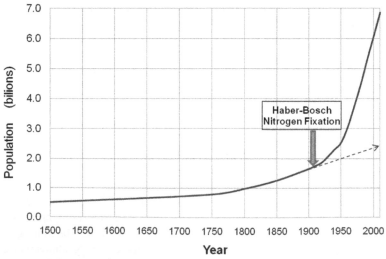

Figure 10.3   World population, 1500–2010.

question today is: How are we going to feed more and more people with the same amount of farmland in a safe, sustainable manner? Part of the answer lies in the next element of the food chain.

FARM PRODUCTION

Growing crops requires a mixture of science, art, and luck. The science relates to our understanding of the biological, chemical, and genetic drivers for robust plant growth, the art relates to our lack of knowledge of these drivers, and luck refers to the weather. But science, more than art or luck, is responsible for the tremendous gains in crop yields experienced over the past century.

Nature is dynamic. Everything living thing is constantly adapting to its environment. Survival demands this. This fact is common knowledge today but, as with all scientific knowledge, someone had to make the discovery. It was Charles Darwin [14] who first proposed the theory that plant as well as animal populations are constantly changing with variations or mutations that might improve productivity or proliferation. The animals and plants with such changes that make them stronger are more likely to survive and to pass these changes on to the next generation. This is a natural process with natural timing. Hence the term Darwin coined: "natural selection." Then humans got involved. Take, for example, corn or maize, which may well be the most intensively altered food source (from its original form) that we grow today.

Contemporary corn has its roots in a short, grass-like plant called teosinte found in Mexico over 5000 years ago. It has a small cob, about 4 cm in length [15], with kernels arranged in scattered patterns. This is far less than the corn cob lengths and straight line kernel patterns that produce over 570 kernels per ear commonly grown today [16].

Early corn farmers chose the best ears of teosinte and propagated these grains as seeds, accelerating natural selection. These early hybrid engineers saw the benefits in a higher number of kernels per ear and in larger cobs. This process continued for generations but it wasn't until Gregor Mendel's work was applied that the science of plant hybrids really began to grow [17]. The fruit of this research was realized in the 1930s as corn hybrid seeds gradually became the choice of most farmers. By the 1950s, almost all farmers were using corn produced from hybrid seeds.

This is what happened, but why did farmers change after generations of tradition? Improved yields in the future were not guaranteed. The uniformity of hybrid fields made harvesting easier but these benefits were often not convincing. The most likely reason for the rapid spread of hybrid plants was the Dust Bowl experience from 1934 to 1936. The hybrid strains were clearly more resistant to drought than were the open-pollinated varieties [18].

During this time of transition and improving crop yields, there wasn't any controversy about the ethics of the research or questions related to hybrid product safety. While scientists controlled the physical processes, the fundamental transformations were still regulated by natural forces. In a sense, nature had the final say as to what type of hybrid was produced. Nature protected itself. Then in 1953 James Watson and Francis Crick discovered the double-helix structure of DNA [19]. Francis Crick was quoted as saying, "We have discovered the secret of life" [20].

This statement was not completely accurate, but their discovery changed our view of genetics forever. DNA was actually first discovered in 1879 by the Swiss physician Friedrich Miescher [21]. Miescher recognized the unique biochemical properties but its connection to genetics slowly evolved over the 63 years before Watson and Crick's structure discovery. Knowing how the amino acids were combined provided scientists with the framework to map the genetic structure of animal and plant life. They were able to develop specific, detailed genetic maps of certain plants and eventually, at the conclusion of a 13-year project, even the human genome [22].

Armed with the knowledge of plant genetic structures, scientists took some of the guesswork out of developing hybrids and shortened the developmental time. How? By altering the genetic structure directly without going through the traditional lengthy, imprecise cross-fertilization process. The practice of improving crops by subjective generational improvement was quickly an antique, at least to some farmers.

As you might expect, even with the new genetic manipulation tools, the successful development of genetically modified products still required a considerable amount of financial investment and risk. To protect that investment, companies relied on patent licensing both in the United States and internationally to recover their development expenses. This makes sense from a business perspective. After all, what incentive would a company have to invest $500 million on a new product if it could be copied by a competitor with legal immunity? The protection issue is straightforward for business process, device, and inorganic manufacturing. But what about patenting new life forms?

In 1972, Ananda Chakrabarty, working for the General Electric Company, filed a patent application entitled "Microorganisms Having Multiple Compatible Degradative Energy-Generating Plasmids and Preparation Thereof." In other words, the patent assignee, General Electric, created and wanted to own a new life form. On March 31, 1981, the patent was approved [23] and legal precedent was set for the protection of genetically modified life forms, their precursors, and future generations. The companies that took the financial risk, created genetic innovations, and produced new products of value now had legal assurance that their investments were protected.

Armed with the genetic code and now favorable patent law, scientists developed methods to mask the expression of some genes and to insert new genes with desired traits that were not present in the host. An example of a genetic insertion from animal to plants occurred in the early 1990s by researchers at DNA Plant Technology Corporation [24]. They located a gene sequence in the Arctic flounder fish that coded for a protein that kept the fish's blood from freezing. Scientists produced DNA based on the gene sequence and, after some changes, they transferred it to tomato plants. It was hoped that the genetically modified (or transgenic) tomatoes would have better freezing properties. The result? Nature: 1, Science: 0; it didn't work. However, this was just the first inning of a game we are still playing.

What did work for tomatoes was a gene addition that prevented the production of an enzyme that produces rotting. The successful gene replacement enabled tomatoes to stay firm longer, basically increasing shelf life and resistance to physical bruising during harvesting. The U.S. Food and Drug Administration (FDA) approved it as safe for public distribution in 1994 [25]. In spite of it being the first genetically modified food source, the FDA did not warrant special labeling since it saw no obvious health risk and the nutritional content and characteristics were the same as those of conventional tomatoes. The business failure after a few years of sales was attributed more to the company's inexperience in food distribution than to the quality and viability of the transgenetic product [26]. Nevertheless, during the 1990s, the ability to change plant properties was firmly established and genetically engineered (GE) or genetically modified (GM) crops became used by a growing number of farmers.

In addition to the weather, weeds and insects are farmers' major impediments to high yields so, on the surface, these "new and improved" hybrid crop variations seem like good ideas. Farmers operate on tight budgets and the ability to improve or just make crop yields more reliable is important for their sustainability. Also, with the world's population growing to approximately 7 billion people, maximizing crop yields means more food for more people.

The plan was to have more reliable (and higher) crop yields, using fewer pesticides and herbicides. There are two major types of GM crops that are grown today: herbicide tolerant (HT), where plants are genetically modified to be resistant or immune to specific powerful herbicides, and "Bt", short for the *Bacillus thuringiensis*, where the GM plants "naturally" produce their own pesticides to protect them from insects. The GM plants were also engineered to be resistant to common viruses, fungi, and other bacteria to make them more resilient and to allow them to thrive in a wide variety of soils with less care [27, 28]. The expectations also included crops that would be more resistant to cold, drought, salinity, and even nutrition. There has been some success in these noble goals. I think the proof is reflected in Figure 10.4, which shows the fraction of total acres planted with GM soybeans, corn, and cotton, the three largest agriculture products in the U.S [29].

In 2010, over 90% of all soybeans, and over 60% of all corn, was produced from genetically modified seeds. Today farming is more of a science than ever before in history. So it makes sense as tractors today measure distances with onboard GPS equipment, farmers plant genetically modified seeds with pesticide and herbicide resistances that have taken the normal hybrid methods to new limits. It is also a good business model for the companies who supply the farmers. They perform the genetic research and periodically market new seed types with improved farming attributes. They additionally make the chemicals that the crops will require for treatment, and they own the second generation seeds since they, too, are covered in their patents. Farmers who plant GM crops are allowed only to sell their crops commercially. They also can't store the seeds (unless they pay the licensing fee). Each year, they are required to buy the newest seed products and start the cycle all over again.

It's almost like buying designer jeans. Every year, new styles are sold for basically the same type of clothes. Agribusiness has its own version of this model by

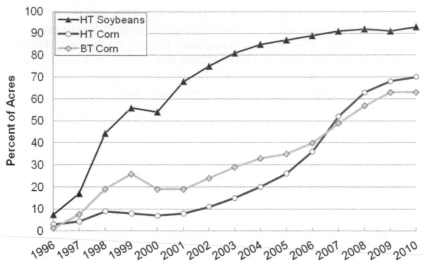

Figure 10.4   U.S. growth of genetically engineered crops.

marketing new seeds that promise better farming benefits. The difference here is that with clothes, once you pay the money, you own the jeans. With the GM seeds that isn't what happens. The current state of the patent law as of 2010 says companies claim new seed types, their crops and subsequent seeds as their "invention." Farmers can license use of the invention and pay the technology fees, but they don't own the plants or the seeds. Imagine going to a store and paying money to license a new pair of jeans with the restrictions that only you can wear them (no hand-me-downs) and only for certain types of situations. This is basically what farmers must do who grow GM crops. See this chapter's case study for further discussion.

Well, agribusiness' "business plan" appears to be have little risk from human threats but nature, apparently, is not accepting genetic modification as part of her plan. Figure 10.5 shows that pesticide use for GM corn and soybeans is now actually higher, on the average, than for the non-GM crops [30]. You might ask, "What has changed?" In simple terms, nature has adapted to the chemicals and developed its own "genetically modified weeds," through natural selection, to survive in the new herbicidal environment. Recent research has confirmed the weeds are developing a strong resistance to the current herbicides [31]. In fact, the new version (or perhaps a better term) is the more recent generation of these weeds, called "super weeds," which are fast growing and have thick stalks that can even damage harvesters [32]. So in the matter of a few years, weeds have evolved nearly just as quickly as the genetically modified plants and for now it appears that the weeds are gaining ground.

The United States has been the leader in developing and growing GM crops, although the Americans are not alone. The world's leading producers of GM crops in addition to the United States are Argentina, Brazil, Canada, India, and China. Other countries that are beginning to do this type of farming are Paraguay, South

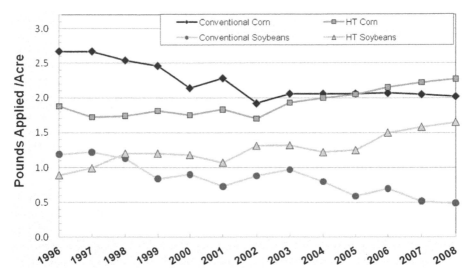

Figure 10.5   Average pesticide pounds applied per acre of conventional, herbicide-tolerant (HT), and *Bt* varieties, 1996–2008.

Africa, Uruguay, Australia, Iran, and the Czech Republic. The European Union has resisted GM crops as a major food source [33].

There are pros and cons to GM crops and foods, with arguments related to everything from reducing world hunger on the proponent side to "playing God without knowing all of the rules" on the opposition side. This is a classic risk management scenario where uncertainty fuels the arguments of both sides. Here is an outline of the dominant issues on this very important subject. Keep in mind that it's not an academic exercise—unless you don't eat.

The proponents tout the benefits of better taste, higher quality, improved crop yields, better stress tolerance, and "friendly" biochemical herbicides and bioinsecticides; the list goes on. Some of these benefits are also seen in normal hybrid research, except the process takes longer. Also the ability to genetically modify a plant to contain specific properties by design in a relatively short time period has its appeal to agribusiness.

The business environment for the seed and agrichemical producers may be favorable thanks to the current interpretation of patent law. However, regardless of the legal issues' morality or fairness, the intellectual property issues are trivial compared to the potential long-term health and safety questions that remain unanswered. To be fair, some proponents have answered these questions by saying that the science is in control and GM crops are good if not better than the natural varieties. Others are not convinced.

When genes are added, removed, or replaced in plants or animals for the purpose of developing new "products," we judge the initial success directly from the expressed characteristics of the result. Are there unintended consequences? Proponents claim that testing before release ensures safety and this is the best we can do if we are going to continue to produce and consume GM foods. However, our functional knowledge of the human and plant genome is incomplete. We simply don't know the potential human health effects from, for example, allergens, transfer of antibiotic resistance markers, unintended transfer of transgenes through cross-pollination, and unknown effects on other organisms such as soil microbes [34, 35]. Like most other areas of biology, the list of what we don't know is a lot longer than the list of what we do know. So here we are, genetically modifying plants and animals in self-declared confidence when we can't even predict the weather accurately 72 hours in the future. How are we then predicting food safety of genetically modified food elements for future generations? This is the risk. The benefits are a matter of continuous debate.

The issue of tampering with genes between species of plants and between animals and plants raises new questions that test our legacy standards. When you mix genes between species or between an animal and a plant, what do you call the result? Sure, it's a hybrid, but the precise name that takes into consideration its genetic structure may be complicated. This is the situation of labeling of GM foods. The regulatory requirements for GM food labeling vary widely between countries.

In the United States, as of 2010, there was no requirement that genetically modified foods be labeled. This is perhaps a little strange since apple juice requires a label to show if it is made from concentrate or juice, and milk is labeled as homogenized or not, but plant or animal foods whose genetic structure has been modified require nothing.

In 2008, regulations were proposed that would not require producers to label most genetically engineered meat, poultry, or seafood. The rules treated altered DNA inserted into livestock as "drugs." Companies are not required to alert consumers when antibiotics, hormones, or other drugs are used in animal development.

The FDA is making an exception in the case where genetic engineering alters the makeup of food. For example, companies are developing DNA that causes pigs to produce more omega-3 fatty acids in their muscles. In this case, the rules would require a label indicating the product is high in fatty acids, but not that it is genetically modified [36].

The U.S. government maintains this laissez faire, no labeling policy for genetically modified foods because:

1. Their analysis says there are no safety or health issues, and
2. The products are nutritionally equivalent to their nongenetically engineered counterparts.

Not everyone shares this view. In fact, there is a large grassroots effort to encourage food producers to voluntarily provide labeling. This sounds like a good idea, but voluntary labeling brings another set of problems. Every food producer can have its own terminology and definitions, so chances are that no two labels would have the same meaning. From the public's perspective, you would know that some type of genetic engineering was involved but without strict labeling standards, the information would be hard to interpret.

Not all countries have the same view on GM food labeling as does the United States. For example, since 1997, the European Union has required labeling for food products that consist of or are derived from genetically modified organisms. Even genetically modified seeds must be labeled. The current version of this regulation (EC 1830/2003) [37] covers the traceability through the food chain and labeling of genetically modified organisms. It is clear that the European Union regulators believe that people should know about not just the ingredients, but also the genetic modifications that are deeply embedded in the food they eat. The labeling provides them a choice that is not widely available in the United States.

Other countries are at various stages of requiring GM food and feed labeling. As the world's population continues to grow, the farming benefits of GM crops will continue to proliferate around the world. And let's not forget that new life forms, that is, "products," are being developed every year. We will learn either from science or time if genetically modified foods are as safe as the proponents claim, but in the meantime, in the spirit of personal risk management, everyone should at least have the right to know what they are eating.

ANIMAL PRODUCTION

Most animals destined for human consumption today are raised more in a factory than on a farm or ranch environment. One look at a typical supermarket meat section or fast food chain menu shows that there is no shortage of meat. And the prices are amazingly low considering the time, growing logistics, and transportation that are involved in getting the cows, chickens, turkeys, and ducks to restaurants and food stores. There is no magic here, only simple economics: the economics of scale.

Production of this much meat at reasonable prices requires large-scale operations. These "factory farms" are called concentrated animal feeding operations (CAFOs). These facilities have the layout of a factory where animals are kept and raised in confined spaces. Feed is brought to the animals, and feces, urine, and other wastes are removed by specialized systems or processes. To give you an idea of the number of animals in these concentrated areas, Table 10.2 shows the animal counts required for a CAFO to be classified as small, medium, and large [38]. The EPA designates 19,149 U.S. farms as CAFOs. They also estimate that hundreds of thousands more facilities that confine animals (AFOs), not large enough to be classified as CAFOs, exist [39]. These meat production factories produce the bulk of the meat, poultry, and dairy consumed in the United States.

From a risk management perspective there are benefits in the high animal concentration configuration to enable uniformity for applying growth hormones, regulating feed quality, and administering drugs. However, the same high animal concentrations require constant diligence to remove infected or otherwise diseased animals. Feed quality also becomes an important factor since all of the animals basically eat the same food at the same time. This was the root cause of the *Salmonella enteritidis* (SE) outbreak in the summer of 2010 [40].

Table 10.2 CAFO Type and Animal Count Categories

Animal Type	CAFO Designations		
	Small	Medium	Large
Cattle or cow/calf pairs	<300	300–999	1,000 +
Mature dairy cattle	<200	200–699	700 +
Veal calves	<300	300–999	1,000 +
Swine (weighing over 55 lbs)	<750	750–2,499	2,500 +
Swine (weighing less than 55 lbs)	<3,000	3,000–9,999	10,000 +
Horses	<150	150–499	500 +
Sheep or lambs	<3,000	3,000–9,999	10,000 +
Turkeys	<16,500	16,500–54,999	55,000 +
Laying hens or broilers (liquid manure handling systems)	<9,000	9,000–29,999	30,000 +
Chickens other than laying hens (other than a liquid manure handling systems	<37,500	37,500–124,999	125,000 +
Laying hens (other than a liquid manure handling systems)	<25,000	25,000–81,999	82,000 +
Ducks (other than a liquid manure handling systems)	<10,000	10,000–29,999	30,000 +
Ducks (liquid manure handling systems)	<1,500	1,500–4,999	5,000 +

A major practical issue associated with CAFOs is waste removal and processing. National and state governmental agencies monitor water quality and other variables to ensure the "emissions" are within the law [41]. With waste that is clearly identified chemically and by odor, violations do occur. Some are direct environmental infringements from unauthorized waste discharges with financial penalties. Others are violations of "Generally Accepted Agricultural Management Practices," for example, spraying liquid manure on the 4th of July weekend is not a good neighbor relations tactic [42].

Given the current volume of meat production required to satisfy consumer demands, CAFOs and AFOs are technically efficient at providing our steaks, chicken, and hamburgers at reasonable prices. But there is a price. The Union of Concerned Scientists estimates that CAFOs leave staggering bills behind for taxpayers, including:

- $26 billion in reduced property values from odor and water contamination
- between $1.5 billion and $3 billion annually in drug-resistant illnesses attributed to the overuse of antibiotics in livestock production
- $4.1 billion in soil and groundwater contamination from animal manure leakage [43].

CAFOs and AFOs operate not as farms in the traditional sense but as low-tech chemical processing facilities whose product is meat. The literature is full of descriptions, video documentaries, and discussions about CAFOs and what they include in terms of the intense corn diet and antibiotic and other drug combinations that produce large animals in a short period of time. Obviously, this is good for business but at what price to our health? The food risk exposures are clear. Corn-fed, antibiotic-drugged, and growth hormone-laden livestock represent the bulk of our meat products today. The state and national agencies that inspect this intense industry do a good job most of the time with very few disease or animal product contamination outbreaks. The real risk is that no one really knows the long-term effects on humans of consuming such artificially treated food products and, just like with genetically modified seeds, only time will tell.

Rather than repeat what others have identified in terms of the risk exposures associated with these operations, I am taking you to a sustainable cattle farm where quality trumps quantity in an elegantly simple manner. I like to introduce you Betsy Ross.

Sometimes it takes an event completely out of our control to build something renewed. In the case of Betsy Ross, the birth of a premature grandson provided the incentive and drive to address issues in our food supply.

Ross believes she's among the "last generation whose mamas knew where every bite they ate came from." And while she herself moved out of that group during her career as a civil servant, at least when it comes to beef, she is back on the front lines.

Seeking the healthiest food for her new little one, Betsy Ross came back to the farm as a retiree to run cattle. Studying under the likes of Sally Fallon, Malcolm Beck, and Elaine Ingham, Ross came to realize that the soil itself was fundamental to the nutritional value of the animals raised on it.

"We were losing our shirts," Ross said. "We were losing our grass to more and more weeds. Everything we were doing [to control the weeds] was unsuccessful."

And so Ross continued her studies in the soil, taking classes within a national network of folks who sought to re-establish the healthy biology of the soil itself. Purchasing a $3,500 "tea brewer," Ross soon began to recognize the benefits of reintroducing the natural organisms to the soil through composting and spraying the resulting organic "tea" on the pastures: "All of a sudden, the weights [of the cattle] were outstanding."

At the time, Ross was one of just three women in her area of central Texas driving a tractor. "They called me the crazy lady with the green pastures." And her results, in cattle weight, got the attention of her former "gurus." The quality of her grass-fed cattle spoke for itself, and the student became the teacher. Ross began instructing prospective soil biologists, including those at Texas A&M University.

Ross runs her grass-fed beef farm as an organic farm. It's labeled "grass-fed" rather than the even more restrictive term "organic," as there is always the possibility that in a drought, the cattle will require a load of hay and organic hay may not be available. Better to forego the label than risk the cattle starve.

And Ross' agenda remains with assuring the healthiest starting place for all she grows—and that is in the soil. She treats "pest" plants not as invaders in need of immediate chemical eradication but rather as indicators of the soil's needs. Ross' finely honed eye allows her to assess what's growing unbidden and prescribe the appropriate enrichment for the soil that will make it less hospitable for the uninvited and, at the same time, more welcoming to the grass her cattle require. "If we just kill the weeds . . . we won't know why they got there. And if you just get rid of it [without tending to the deficiencies of the soil] it just comes right back."

Betsy's sister and coworker Kathryn Chastant affirms the direction in which her sister leads. "We're moving the land back to the way it was always done until after World War II when the chemical companies had to get rid of their nitrates."

Just as Ross makes the best of the soil, she also seeks to raise the most productive grasses in each one's time. For every season there is a grass. "It's a matter of exploiting the opportunities nature presents," she explains, speaking most simply about "one grass in dry weather, another that can withstand a freeze, and one yet that is very prolific each day the temperatures and water are available."

Ross also employs a technique called "mob grazing." Calculating the weight of food the herd needs each day, she positions the cattle into areas adequate for their grazing, then moves them throughout the day. While on the restricted patch of grass, the cattle eat, trample, and leave their waste products. In a short time, they consume much of what the small area has to meet their needs. Once moved, the area will be fallow for months. And within hours, dung beetles and other arthropods, fungi, and birds move in to pull down the manure left by the crowded herd. The residual returns to nourish the healthy soil. In addition, the short duration the herd spends on each patch of ground assures it not be hardened by trampling. As a result, the cattle incur many fewer shoulder problems. The softer ground also results in naturally polished and healthier hooves.

All in all, the Betsy Ross grass-fed beef enterprise is one that demands overall health, from the roots up. Quoting Chastant again, "We consider ourselves stewards of the land." And theirs is a careful and masterful stewardship, resulting in a quality that may well be unsurpassed. One of their primary distributors is the rising star in food delivery, Whole Foods, of Austin, Texas.

No longer a start-up but in fact a leader and member of a number of standards organizations, Ross remains true to her own roots. For her standards are higher than those of the USDA and any organization to which she might belong. In Betsy Ross' words, "My name is on every piece of meat. It can't have [any flaws]."

No pesticides, no herbicides, incredible production through attention from the ground up, elegant, and effective: an anomaly in our world of chemicals, and a hope. This author's hat is off to Betsy Ross. (More information about Betsy Ross' cattle, grass-fed beef, and soil biology can be found at http://www.rossfarm.com/.)

DISTRIBUTION

Unless you are castaway on a deserted island, you probably don't grow, catch, or harvest all of the food you eat. Someone else does these things. The food products are processed, stored, transported, and then stored again, waiting your purchase either at the food store, farmers' market, or in your neighbor's kitchen. In terms of the latter case, you're probably free of the issues discussed here. But how many of us live that close to the people who make our food?

The primary risk exposure from this element of the food chain is contamination. Spoilage is number two but if something is spoiled then you don't eat it. In the case of contamination, the food can look good, smell good, taste good, and can still kill you. In 2009–2010, the United States imported agricultural products from over 170 countries [44]. Among the fastest growing categories were live animals, wine/beer, fruit/vegetable juices, wheat, coffee, snack foods, and various seafood products [45]. This is a lot of food to check! Just to give a realistic example of the enormous size of the food inspection issue, consider the following. At the Arizona border checkpoint near Nogales, in the peak winter season, almost 1,000 produce trucks arrive daily, carrying eggplants, cucumbers, tomatoes, peppers and other products destined for your food markets. How can you be sure that the inspections process works? Well, you can't.

Here again you see a situation where the potential for harm is great and the resources to protect us are finite. Food is imported through a wide variety of pathways and food safety inspection resources are no match for the volume, diversity, and geographical distribution of entry points.

In 1997, during the Clinton Administration, a risk-based set of preventive controls known as Hazard Analysis and Critical Control Points (HACCP) was introduced. These procedures focus inspection efforts on the bottle necks (or critical control points) in the food chain. The method was actually created by NASA and the Pillsbury Corporation in the 1960s for food protection during human spaceflights [46]. The HACCP method has 7 basic steps.

1. Identify the hazards
2. Determine the control points and the critical control points
3. Set critical control limits
4. Monitor critical limits
5. Take corrective action when limits are exceeded
6. Establish an effective recoding system
7. Verify the system is working as planned.

The method's philosophy is that if food is checked at its major processing points, that is, storage, defrosting, and cooking, then contamination can be identified and specific corrective actions can be taken early to minimize severity.

The government at the time used HACCP as a deregulation tool that allowed food manufacturers to do the critical point inspections themselves rather than through government inspectors. It also basically standardized food handling and food processing best practices in many different types of facilities. Regardless of the politics, the method is an important risk management tool being applied today internationally in a wide range of food production industries, including restaurants.

With food imports' growing diversity, the HACCP approach is not reliable by itself as new regions and food products enter the marketplace. The data speaks for itself. Approximately 76 million food-borne illnesses occur annually in the United States involving more than 300,000 hospitalizations and 5,000 deaths [47]. These numbers appear large but let's look at the exposure base. Assuming there are 300 million people in the United States and everyone eats three meals a day for 365 days, there are to 328.5 billion meals consumed each year. Dividing 76 million by this number estimates the probability of obtaining a food-borne illness per meal per person as 0.0002. This is a small number so your chances of getting sick from contaminated food are low, yet from a societal point of view, 76 million food poisoning illnesses is unacceptable.

It is clear that the HACCP method and other food safety measures are not working as well as required. Even though a risk-based process was being used, it had become static. Congress asked the Institute of Medicine (IOM), a nongovernmental organization under the National Academy of Science, to examine the gaps in the FDA's food safety systems since it oversees the majority (about 80%) of U.S. food supply. Their findings commended the FDA in their risk assessment approach to food safety but also said that it was not enough. They claim the present system is reactive and places too much emphasis on case-by-case analysis. The IOM recommends the FDA adopt an enhanced risk-based approach to food safety including:

- Strategic planning with performance metrics
- Public health hazard risk ranking
- Targeted information collection
- Situation analysis and intervention planning, and
- Ongoing measurement and program reviews.

The report contains the details but there are some interesting aspects of the recommendations that warrant a mention here. Often, the adoption of a risk management approach to a business or activity is viewed as primarily a change of behavior of the current workforce. Risk management can be taught, but that takes time. Their recommendation included training employees but also hiring new personnel with risk management and analytical expertise. They also identified shortcomings in the FDA's ability to collect, analyze, interpret, manage, and share data. These functions are essential for any effective risk management program.

Food safety is a complicated process in the United States, with state agencies responsible for intrastate commerce and the federal government, state, and local agencies all sharing the responsibility for interstate food commerce. The IOM's report also highlights the need for agency policy and procedural integration to ensure effective risk communication and food safety education. And, of course, since you can't change government actions without changing the laws, the report also includes a recommendation to modernize food safety laws. All of this translates to a major overhaul of a food safety system that may have worked well in previous decades, but, as of 2011, is out of date [47].

The FDA may not agree with the "out of date" part. More resources are being applied, and risk management is still the primary methodology. The FDA employs a risk-based method to identify high risk food establishments for its domestic and foreign food inspections. High risk food establishments are growers/harvesters, manufacturers/processors, packers, repackers, and holders of "high risk foods," that is, those foods that can present hazards which the FDA believes, based on scientific evidence, can pose a high potential to cause harm from their consumption [48].

High risk food commodities include, but are not limited to, modified atmosphere packaged products; acidified and low-acid canned foods; seafood; custard-filled bakery products; dairy products including soft, semisoft, and soft ripened cheese and cheese products; unpasteurized juices; sprouts; ready-to-eat fresh fruits; fresh vegetables; processed fruits; processed vegetables; spices; shelled eggs; sandwiches; prepared salads; infant formula; and medical foods. This list is not inclusive because the identification of high risk food establishments is dynamic and subject to change in response to new information. For example, in recent years, FDA has identified high risk food establishments as those that include products whose formulations do not include an allergenic ingredient but, because the products are made in a firm that also makes allergen-containing foods, may inadvertently contain an allergen which is not declared on the label. Common allergenic substances include milk, eggs, fish, crustaceans, tree nuts, peanuts, and soybeans.

In the complex multiagency bureaucracy that manages food safety, there is always room for improvement and changes are being, and will continue to be, made. Risk management is fundamental to the current and future food safety strategies, programs, and activities. As to whether the current system is adequate, let's extend the food poisoning probability calculations and see if you keep your appetite.

With a probability of 0.0002 (2.0×10^{-4}) to obtain a food-borne illness per meal per person, let's factor in how many meals you eat in a year. If you eat (3×365) meals per year, then your probability of food poisoning per year is 0.25 or 25%. You can decide for yourself if this current level of safety is acceptable or not. It is now appropriate to discuss the most important element of the food chain.

CONSUMPTION

Food poisoning is not the flu but it is defined as a flu-like illness typically characterized by nausea, vomiting, and diarrhea, due to something the victim ate or drank that contained noxious bacteria, viruses, parasites, metals, or toxins [49]. Approximately every second of every day, two people in the United States are stricken, and about 100 people die each week from food poisoning. These numbers probably underestimate the actual magnitude of the problem since doctors see only a small portion of the food-related illnesses. Many people with food poisoning assume they have a stomach bug (or the flu) and never visit the doctor. The costs? When tallied up in the United States, the consequences of foodborne illness including doctor visits, medication, lost work days, and pain and suffering amount to an estimated $152 billion annually [50]. Other researchers have cost estimates ranging from $357 billion to $1.4 trillion [51].

You can get food poisoning from eating the "wrong" type of mushrooms, the wrong parts of some fish, or food tainted with parasites. But the majority comes from eating food contaminated with different kinds of bacteria. The major pathogens are

1. *Salmonella*
2. *Campylobacter*
3. *Shigella*
4. STEC O157
5. STEC non-O157
6. *Vibrio*
7. *Listeria*
8. *Yersinia*

Even if you're not a microbiologist, you probably recognize at least one of these names, perhaps from personal experience or from media coverage of food poisoning outbreaks around the country or world. Others may be less well-known but they all are major elements of our food risk that warrant our attention and understanding. The more you know about these risks, the more you can do to prevent getting sick or worse.

Salmonella [52] causes more food-borne illness than any other pathogen. It is a genus of bacteria with similar structural characteristics that live in the intestinal tracts of humans and other animals, including birds. Reptiles are particularly likely to harbor the bacteria. Animals can have the bacteria without getting sick. It is not paranoid to have anyone, especially children, wash their hands after handling a bird or reptile. The primary transmission pathway to humans is though foods contaminated with animal feces and it only takes a small, imperceptible amount to contract the illness.

For some contamination pathways, we have little control over the risk. For example, the *Salmonella* outbreak in the summer of 2010 that forced the recall of 500 million eggs was caused by *Salmonella*-contaminated grain fed to chickens at several large egg-producing chicken farms [53]. How can you be sure the eggs you eat are safe? The answer: raise your own chickens. And then, do so carefully.

The good news is that thorough cooking kills *Salmonella*, but food may also become contaminated by an infected food handler who did not wash his or her hands with soap after using the bathroom, or from someone who handles raw eggs or meat and immediately afterward doesn't wash his or her hands or clean the work area before continuing meal preparation.

Campylobacter is also one of the most common causes of bacterial food-borne illness in the world [54]. It is found in a variety of healthy domestic and wild animals including cattle, sheep, goats, pigs, chickens, dogs, and cats. It usually lives in the intestines as part of the normal flora and is shed in feces. The bacteria can live in water troughs, stock ponds, lakes, creeks, streams, and even mud.

Because *Campylobacter* can exist in many parts of the environment, food products (especially poultry, beef, and pork) are at risk of contamination during processing. It is readily destroyed by pasteurization of dairy products and adequate cooking of meat products. As in the case of *Salmonella*, a major risk exposure to humans is cross-contamination in the preparation of food by not washing hands or sanitizing work areas after handling raw meat and other uncooked food components. An additional risk exposure for people who contract this intestinal infection is the rare occurrence of Guillain-Barré syndrome, an autoimmune condition that attacks the nervous system. Recent research has shown that exposure to the *Campylobacter* infection can cause this serious secondary illness [55].

Shigella is a genus of bacteria that can cause sudden and severe diarrhea. It can occur after ingestion of fewer than 100 bacteria [56], making *Shigella* one of the most communicable and severe forms of bacterial-induced diarrhea. The bacteria thrive in the human intestine and are commonly spread both through food and by person-to-person contact. For example, toddlers who are not completely toilet-trained can infect people they contact, people of all ages playing in public untreated shallow water areas such as fountains are at risk, and other potential sources include flies that migrate from infected sites to people and inadequately washed vegetables.

STEC O157 is an abbreviation for "Shiga toxin-producing" *Escherichia coli*. First of all, *E. coli* is a large and diverse group of bacteria, and most strains are harmless. Others can make you very sick with diarrhea, urinary tract infections, respiratory illnesses, and other illnesses. Biologically, we need *E. coli* to maintain the delicate chemical balance of our intestines. It's when the bacteria change slightly, or mutate, and have vastly different properties, that the problems begin.

The most common STEC O157 mutant in North America is O157:H7, found specifically in the intestinal tract of cattle. This strain of the *Escherichia* bacteria, discovered in 1982, was known then to cause sickness in humans but its food risk importance was not recognized for another 11 years. In 1993, after a fast-food outbreak that sickened more than 700 people and killed at least four children, both the media and food safety professionals began to treat this bacteria strain as a serious public health threat.

The bacteria's name is actually written in a detailed code. The alphanumeric labeling, in a sense, is like the bacteria's social security number, except that it refers to unique aspects of its structure. The O157 refers to a particular surface antigen marker on the outer surface of the cell. The H7 is another antigen that is located on the cell's pilae or hair-like structures. These two markers identify the particular strain of the *E. coli* bacteria. Today several new strains have been identified [57], some of which will be mentioned in the STEC non-O157 discussion.

STEC O157 has the same contamination path as the other pathogens: your mouth. You either eat contaminated food or pass small, invisible amounts of human or animal feces from your hands to your mouth. You can get infected by swallowing lake water while swimming, touching the environment in petting zoos and other animal exhibits, and by eating food picked or prepared by people who did not wash their hands well after using the toilet. The risk mitigation actions of washing your hands, not swallowing lake or stream water, and knowing what you eat at restaurants sound pretty easy. They are—but don't mistake simplicity for insignificance. The stakes are very high. STEC O157 infections are serious conditions by themselves, but the stress on other organs can induce hemorrhagic colitis and hemolytic uremic syndrome, requiring blood transfusion and kidney dialysis.

To minimize your risk exposure, stay away from unpasteurized (raw) milk, unpasteurized apple cider, and soft cheeses made from raw milk. If you work with cattle, then keeping your hands away from your face is a "no brainer," but other risk exposures like eating undercooked hamburgers or contaminated but seemingly "fresh" vegetables can be more difficult to detect.

The bacteria infection agents I've mentioned so far—*Salmonella*, *Campylobacter*, *Shigella*, and STEC O157—are the major food risk exposures we share each day. We hear about the major outbreaks, the sickness, and sometimes tragic deaths but the magnitude to the risk exposure is not adequately represented by sporadic news events. The risk exposures and the human suffering from these food-related illnesses are continuous. Figure 10.6 shows the frequency of occurrence of the four top food-borne pathogens in the United States from 1996 to 2009 [58]. The vertical axis is the number of infection cases per 100,000 people. The Centers for Disease Control and Prevention also publish national health objective targets for three of the four pathogens shown here. How these targets are selected is a subject for another book. However, to show you how the United States is actually managing our national food safety risk, the target for *Salmonella* is 6.80; the 2009 result was 15.19. The target for *Campylobacter* is 12.30; the 2009 result was 13.02. The target for STEC O157 is 1.00; the 2009 results were 0.99. As of 2010, no national health objective has been set for *Shigella*.

The next four pathogens have lower infection frequencies but are still major risk exposures. We begin with STEC non-O157. Shiga toxin-producing *E. coli*

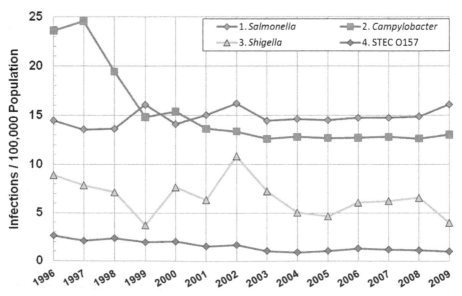

Figure 10.6   Part one: top eight food infections, United States, 1996–2009.

identification is usually done through lab testing of stool specimens. For STEC O157 strains, there are common diagnostic protocols, standards, and testing facilities, but the same cannot be said for STEC non-O157 bacteria. The testing procedures require different training and equipment and as of 2010, there are no clear guidelines for reporting, testing, and interpretation of results. Most labs can determine if an STEC is present and can identify *E. coli* O157 but to determine the O group of non-O157 STEC, strains must be sent to a state public health laboratory.

The six most common strains of non-O157 *E. coli* are O26, O45, O103, O111, O121, and O145. All of these relatively new pathogens have a track record of causing food poisoning outbreaks. Researchers believe that risk mitigation practices are the same as for the STEC O157, strains which is the good news. The food risk exposure lesson here is that basic tenets of good hygiene and food preparation practices are the best defense against this ever-changing bacteria offense.

Vibrio is a genus of bacteria that requires the salt environment of sea water for growth, so it is limited to seafood. *Vibrio* is also very sensitive to temperature and proper storage below 40°F, and sustained cooking above 140°F, will kill the bacteria. This bacteria can be problematic for countries where the diet consists primarily of seafood, for example, Japan (and any sashimi serving for that matter), where seafood is often consumed raw.

The *Listeria* [59] genus of bacteria is widespread in nature. *L. monocytogenes* is the particular strain that has caused the most illness although, evidently, not all humans are susceptible. Research has shown that this strain may actually exist in the intestines of 1%–10% of humans with no ill effects. It has also been identified in 37 species of domestic and feral mammals, at least 17 species of birds, and in some species of fish and shellfish [60].

What makes *L. monocytogenes* so dangerous is its resiliency. It can grow in refrigerated temperatures as low as 37°F, and also survive temperatures of 170°F for a short time. The bacteria's most favorable growth conditions are moist areas such as around drains and air conditioning vents that develop condensate.

Despite the small number of cases reported each year, the mortality rate (20%) is relatively high. Pregnant women and their fetuses are the most susceptible to severe illness and death. The CDC reports that pregnant women are 20 times more likely to become infected than nonpregnant healthy adults [61].

Last, but not least, in the list is the *Yersinia* bacteria family. In the United States, most sickness is caused by one particular strain called *Y. enterocolitica*. The bacteria can be found in pigs, rodents, rabbits, sheep, cattle, horses, dogs, and cats. Humans can get sick from eating contaminated food, especially raw or undercooked pork products [62]. The abdominal pain and diarrhea usually lasts from 1 to 3 weeks. While the symptoms are unpleasant, the bacteria doesn't have the life-threatening potential compared the others previously mentioned.

Figure 10.7 shows the infection rate for last four bacteria families in our list. What is particularly noteworthy about this plot are that apparent trends. Since the y-axis is computed in terms of the number of infections per 100,000 people, population growth by itself is not a factor here. Notice that the infection rates for *Listeria*, *Vibrio*, and *Yersinia* have converged and stabilized at about 0.35 infections per

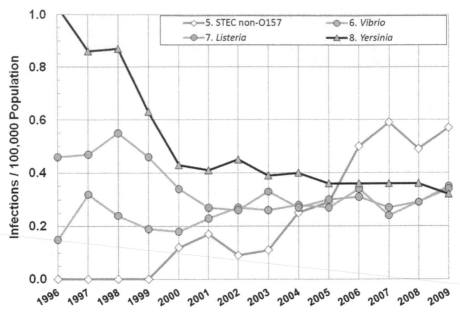

Figure 10.7   Part two: top eight food infections, United States, 1996–2009.

100,000 people. However, the infection rate for STEC non-O157 bacteria has roughly increased 300% from its 2001–2002 level. Shiga-producing toxin *E. coli* may be increasing in nature, diagnostic technology may be improving, there may be a growing awareness of the toxicity and nature of non-O157 strains, or all of these factors may have contributed to the sharp increase. The increase has been noted by the USDA's new food safety chief, Elisabeth Hagen, and additional regulations for beef processors are being planned to change the infection trend shown in Figure 10.7 [63].

However, regulations are not a guarantee that any of these eight major bacteria groups discussed here will not enter the public food supply. For right now and for the foreseeable future, it is safe to say that food safety represents, and will represent, a significant societal and personal risk exposure.

Your food safety risk management should engage both your mind and your senses. Pay attention to the sanitation conditions of the restaurants where you eat out and in the stores and markets where you buy food. Pay attention to the temperature, color, aroma, texture, and taste of the food you buy, cook, and eat. Unless you grow and cook all of your own food and watch your own food-handling habits, there is no guarantee that you won't become a food poisoning victim even if do all of the right things.

And from this section on, your view of bathroom hygiene should take on a new level of cause and effect importance. For example, why do restroom doors only have handles on the inside, requiring you to touch the handle of the door to leave? Shouldn't it be the other way around?

HEALTH

The last element of the food chain is "where the rubber meets the road," or perhaps a better metaphor is "where the food meets your mouth." At supermarkets, convenience stores, or restaurants you choose what you are going to eat. Here's where your tastes, education, training, intellect, and yes, survival instincts come into play. It might be unfair to say that "you are what you eat" but then again, maybe not. Take, for example, the amount of fruits and vegetables you consume on a daily basis. Sure, it's common knowledge that a healthy diet consists of several portions of these foods daily. But for you agnostics or ambivalent believers, recent research makes a very compelling argument. Here's a brief description of several research projects.

- An analysis compared studies involving 257,551 individuals with an average follow-up duration of 13 years. The results were based on the observed health characteristics of those individuals who ate fewer than three daily servings of fruit and vegetables. For participants who ate three to five servings, the stroke risk was reduced 11%. For those who ate more than five daily servings, stroke risk was reduced 26% [64].

- An analysis of several studies that observed the diets and health of 90,513 men and 141,536 women showed for each daily serving of fruit, stroke risk was reduced by 11%. Additional vegetable daily servings only decreased stroke risk by 3% [65].

- In a study of 72,113 women over eighteen years, those individuals whose diets consisted of more fruits, vegetables, legumes, fish, and whole grains had a 28% lower risk of cardiovascular disease and a 17% overall lower mortality risk [66].

- In a Harvard study of 91,058 men and 245,186 women, for every 10 grams of fiber consumed daily by the participants, there was a 14% decrease in heart attack risk and a 27% decrease in fatal coronary heart disease [67].

- A study of over 300,000 people in eight different European countries showed a 4% risk reduction in heart attacks caused by reduced blood supply problems for each portion of fruits and vegetables above two daily servings [68].

The large study group sizes and result consistency presents a very convincing case that it is definitely worth your time to take your diet more seriously. These studies are also interesting in that they provide a sliding scale for stroke and heart risk as a function of the daily amount of fruits and vegetables consumed. You can choose your risk reduction by the number of daily servings you decide to eat. From reading the details of these studies, it appears to me that "You 'are' (to some extent because of) what you eat."

With fruits and vegetables you can see their differences. Asparagus doesn't look like broccoli, and apples are easily distinguished from bananas. For these food groups, we can look, touch, smell, and squeeze the products before we eat them. Have you ever done this at a food store and, based on some negative feedback, decided not to buy the item?

But not all food can be examined this way. Sure we can (and should) examine the food for contamination characteristics but now it's time to use a more intellectual approach: READ THE LABELS.

Food labeling is a mixture of science, art, and politics. The science is the easiest part. Each term used on a label has a specific meaning. Once you know the language, like the meaning of the various types of fats, sugars, preservatives, oils, and alcohols, just to name a few things, you're in pretty good shape. Based on the creativity of some new courses offered in higher education, I am surprised that colleges don't have courses on food label chemistry or "how to read food labels" since everyone has a vested interest in the subject. The art form refers to what is said and shown graphically on the label, how the words and pictures are perceived and what they really mean. Food labeling content, word size, and even position are regulated in most countries. The laws appear strict when you read them but in reality, there's plenty of wiggle room for producers to present an image that's not totally accurate.

The laws require the caloric value, fat intake, and nutrient contents per serving. In addition to these items, labels will tell you all of the ingredients, including preservatives, and the relative amounts of each. But does the nutrient and ingredient data really describe what's inside the package? Let's do an experiment.

In Table 10.3, there are four label descriptions I've taken from common foods, and their four sets of ingredients. The information on these labels is impressive. But, is it useful? First, see if you can match the following nutrient descriptions to the ingredients. Then identify the food each set describes. The answers are at the end of the chapter.

Do you think we could invent a new kind of trivia game here?

In addition to the chemical contents and ingredients, products that have been irradiated contain a radura, the international symbol for irradiation, and must display the notice: "treated by irradiation" or "treated with radiation" on their packages. Consumer products that include irradiated ingredients like spices do not require any special labeling.

Here's how this radiation process works. Food is passed through an electron, x-ray, or gamma ray field for various durations, depending on the type of product, and never comes in direct contact with the radiation source. The radiation kills bacteria and other food borne pathogens like *Salmonella*, *E. coli* O157:H7, *Campylobacter*, and others, with a minimal loss of nutrients. There are some new chemical compounds formed called "radiolytic" products that are common in glucose, formic acid, and, after very extensive research, nothing has been found that poses any harm to the consumer.

Today, food irradiation is used in more than 40 countries to help reduce food contamination risks. Even though NASA has fed irradiated food to its astronauts since the Apollo 17 moon mission in 1972, there still is a concern about the safety of preserving food through radiation when it will be consumed by the general public.

Introducing the U.S. general public to irradiated food has been compared, by some public health analysts, to the controversy that arose before the initiation of pasteurization in the early 1900s. The same concerns for loss of nutrient value,

Table 10.3   Food Labels: What Are the Foods?

Part A: Nutrient Facts			
Label A		**Label B**	
Calories: 70	Fat calories: 0	Calories: 100	Fat calories: 80
Total fat:	0 g	Total fat:	9 g
Saturated fat:	0 g	Saturated fat:	3 g
Cholesterol:	0 mg	Cholesterol:	30 mg
Sodium:	1230 mg	Sodium:	340 mg
Total carbohydrate:	13 g	Total carbohydrate:	2 g
Dietary fiber:	2 g	Dietary fiber:	0 g
Sugars:	9 g	Sugars:	1 g
Protein:	2 g	Protein:	3 g

% of daily recommended daily values

		% of daily recommended daily values	
Vitamin A: 25%	Vitamin C: 60%	Vitamin A: 0%	Vitamin C: 2%
Calcium: 4%	Iron: 10%	Calcium: 0%	Iron: 2%

Label C		**Label D**	
Calories: 200	Fat calories: 45	Calories: 280	Fat calories: 130
Total fat:	5 g	Total fat:	15 g
Saturated fat:	1.5 g	Saturated fat:	2.5 g
Cholesterol:	30 mg	Cholesterol:	30 mg
Sodium:	1130 mg	Sodium:	370 mg
Total carbohydrate:	22 g	Total carbohydrate:	38 g
Dietary fiber:	4 g	Dietary fiber:	3 g
Sugars:	5 g	Sugars:	24 g
Protein:	16 g	Protein:	4 g

% of daily recommended daily values

		% of daily recommended daily values	
Vitamin A: 190%	Vitamin C: 4%	Vitamin A: 10%	Vitamin C: 0%
Calcium: 4%	Iron: 10%	Calcium: 4%	Iron: 4%

(*Continued*)

Table 10.3   *(Continued)*

Part B: Ingredients	
#1	#2
Beef stock, carrots, potatoes, cooked beef, tomato puree (water, tomato paste), peas celery, potato starch, water, contains less than 2% of the following ingredients: wheat flour, salt, vegetable oil, corn starch yeast extract and hydrolyzed yeast wheat gluten and soy protein, caramel color, spice extract, dextrose.	Tomato juice from concentrate (water, tomato concentrate), reconstituted vegetable juice blend (water and concentrated juices of carrots, celery, beets, parsley, lettuce, watercress spinach), salt, vitamin C(ascorbic acid), flavoring, and citric acid
#3	#4
Chicken, pork, water, beef, salt, contains less than 2% of: corn syrup, dextrose, flavoring, sugar, autolyzed yeast, potassium chloride, sodium phosphates, sodium erythorbate (made from sugar), sodium nitrite, extractives of paprika celery, beets, parsley, lettuce, watercress	Sugar, carrots, enriched bleached wheat flour, eggs, pineapple, margarine, walnuts, vegetable shortening, cream cheese, contains less than 2% of the following: leavening, food starch modified, salt, water, spice, natural and artificial flavors, sorbic acid mono & diglycerides, lactostearin, cellulose gum, polyglycerol esters of fatty acids

formation of harmful byproducts, and taste changes were seen. However, widespread public acceptance has been hindered by the radiation exposure's perceived danger [69].

Probably the most used and abused word on labels is "natural" or, in some cases, "100% natural." You see it on food products which are trying to appeal to the logic of "If it is natural, it must be good for you." Well, mercury, lead, arsenic, and uranium are all from nature, as well as hemlock, but I have yet to see these natural elements as part of our dietary requirements. On actual product labels, the ingredients may have natural ingredients such as apples, raisins, carrots, and oats, but if you look closely at the list you might also see very unnatural ingredients such as partially hydrogenated oils or high fructose corn syrup (HFCS). So the word is meaningless, but it is great marketing.

The FDA is starting to address the definition of "natural" in food labeling by stating that HFCS does not qualify for the "natural" label. The official statement by the FDA on their website is:

> *From a food science perspective, it is difficult to define a food product that is "natural" because the food has probably been processed and is no longer the product of the earth. That said, FDA has not developed a definition for use of the term natural or its derivatives. However, the agency has not objected to the use of the term if the food does not contain added color, artificial flavors, or synthetic substances. [70]*

So the government provides us general guidance and the individual players are left on their own to work out the details. For example, the Corn Refiners Association trade group claims that HFCS is derived from corn, so it is a natural sweetener. On the other side of the issue are the Sugar Association and consumer groups such as the Center for Science in the Public Interest. In 2007, both Cadbury Schweppes and Kraft decided to remove "natural" labels from their products containing HFCS. Why? They were threatened with lawsuits [71]. The FDA's lack of specificity on this issue creates legal and market risk exposures where the rule of law is settled by lawyers. This is one example but generally speaking, the FDA has been selective in setting food labeling policy.

As mentioned in the previous section, mandatory labeling of genetically modified foods has been proposed in the United States, but nothing has been done by Congress. The European Union requires labeling and has been resistant to even the marketing of genetically modified food in the region. Some products are approved [72, 73] but generally the Europeans have a different view of genetically engineered foods than the U.S. Congress and the food industry. To give you a perspective of the extent to which genetically modified have penetrated U.S. food products, it has been estimated that about 75% of processed food on supermarket shelves, such as soda, soup, and crackers, contain genetically engineered ingredients [74]. If Congress required genetically engineered food labels, I think we would be surprised to learn the true effect this technology has on the food we eat.

While the United States does not require labels on genetically modified foods, there are very strict regulations and operational guidelines for foods that have the "organic" label. This is one label that means what is says: only natural fertilizers, no pesticides, no herbicides, no animal antibiotics, and no animal growth hormones. In many ways, organic farms are sustainable ecosystems compared to the food factory practices of CAFOs and related farms. Organic certification is an ongoing process and the rigorous certification process helps maintain a respected level of credibility.

There are however, variations in "organic" labeling that you should know about. The first two categories are permitted to display the unique USDA Organic seal.

- "100% organic" labeling means just that. All of the ingredients qualify to be organic by the previously mentioned criteria.
- "Organic" on the label by itself signifies that at least 95% of the ingredients are organic.
- "Made with organic ingredients" means that at least 70% of the ingredients are organic.

Anything made with less than 70% organic ingredients cannot display "organic" on its label.

Another label of distinction and credibility is "grass-fed." This is the label on Betsy Ross' meat. In the past, cattle could be fed grain for the majority of their short lives except for the final few months and the meat could have this label, but not anymore. The grass-fed label standard now states:

grass and/or forage shall be consumed for the lifetime of the ruminant animal with the exception of milk consumed prior to weaning. The diet shall be derived solely from forage and animals cannot be fed grain or grain by-products and must have continuous access to pasture during the growing season. [75]

The standard also serves a second purpose. It shows that the government is capable of producing a precise, clear operational standard.

There are also tricks in labeling that give the customer one impression while meaning something quite different. For example, when a food label contains a trademark (™) or registered mark (®) after a word or phrase, the company wants you to identify the text with their product. It does not describe the product's contents or nutritional value. This subtle difference, usually only appreciated by intellectual property attorneys, now can help you be a smarter consumer. Look for some examples when you go shopping. Specific examples come and go, like most selling strategies, but the tactic is a common marketing ploy.

Other so-called label tricks have to do with words and phrases that can be easily interpreted to have a meaning that does not represent reality. To regulate the potential to mislead consumers the FDA has actually published definitions for some of them [76].

- *Fat free*: The product has less than 0.5 grams (500 mg) of fat per serving.
- *Low fat*: The product has 3 or fewer grams (3000 mg) of fat per serving.
- *Reduced or less fat*: The product has at least 25% less fat per serving than the regular version.
- *Lite or light*: This could mean several things:
  ○ Fewer calories of half the fat of the regular version.
  ○ 50% less sodium of the regular version.
  ○ The product is a not as dark as the regular version.
- *Calorie free*: The product has fewer than 5 calories per serving.
- *Low calorie*: The product has 40 calories or fewer per serving.
- *Reduced or fewer calories*: The product has at least 25% fewer calories per serving than the regular version.

Product manufacturers are constantly making up new terms that have no usage standards. Consumers are on their own to figure out what's really inside the package. Just take one look down a supermarket aisle and you'll see a plethora of colored boxes, jars, and packages with different designs, shapes, and wording. The competition is tough for commodity food products, and manufacturers use standard terminology by law and many others phrases to entice you to buy their product instead of the one right next to it. This is free enterprise at its best. Be honest, given similar pricing, what would you buy: a "low fat," "all natural," "whole grain," "fortified," "heart healthy" cereal with "less" sodium than the same basic product without this "information?"

No one can guarantee absolute food safety today. You need to be cautious, diligent, and thoughtful in your food selection and preparation activities. For example, if the meat looks old, or smells bad, don't eat it, and thoroughly wash food when appropriate. Wash cutting boards, your hands, and knives between cutting uncooked meat, eggs, and your salad ingredients. Drink pasteurized juices and don't forget to check the expiration dates! Risk management now, more than ever, is a team effort between government, producers, distributors, and us, the consumers. Obviously, food shoppers can't test for hazardous residues or find invisible pathogens. That's when we count on the HACCP inspection parts of the team. Our "negligible risk" policy allows us, as a society, to balance risks with benefits.

Now you may have the impression that food risk management is pretty basic, obvious stuff. True, for the most part a little education and some common sense are all you need. However, at the end of this chapter on food risk, I am going to introduce you to an emerging risk exposure which may have serious long-term effects—one that scientists are only just beginning to understand.

The risk is from proteins called prions. The term was coined by Stanley Prusiner who received the 1997 Nobel Prize in medicine for its discovery. Prion is short for proteinaceous infectious agent. It is a protein that does not include a nucleic acid. Scientists today believe that normal prions protect nerve cells by acting as healthy cell surface insulators [77]. Problems arise, however, when prions are deformed from their normal 3-dimensional shapes. The deformed proteins somehow aid or produce other chemicals that inhibit central nervous system and neurological functions.

The first known cases in humans were observed in the Eastern Highlands of Papua New Guinea in the 1950s. The Fore (pronounced *for-ay*) people had a ritualistic tradition of cooking and eating their dead relatives. In the ceremony, women were usually the family members that ate the brain. Over several years, mainly women and children developed a fatal neurological disease they called *kuru*, meaning "to shake from fear." People affected with this disease would suffer uncontrollable tremors and loss of muscle control before death. Since the abolition of cannibalism, the disease has virtually been eliminated [78].

Another set of prion diseases occur in cattle called bovine spongiform encephalopathy (BSE) or mad cow disease. Symptoms similar to kuru in the cattle make them appear as though they are going mad—hence the name. You might remember the mad cow epidemic in Great Britain in the 1980s and 1990s, because of which many infected herds of cattle were slaughtered.

If the disease didn't cross the "species barrier" then the implications would solely financial, but that wasn't the case. Over 300 people in Great Britain and around the world have died from a human form of the disease called Creutzfeldt-Jakob disease (CJD) acquired from eating tainted beef. It also became apparent that healthy cattle were acquiring the disease from eating food made from recycled carcasses or grazing in areas where carcasses were buried. The primary pathway to humans was through eating tainted beef although some medical-related cases were also identified. The experience in Great Britain showed, for the first time, that prion-related diseases can be transmitted from animals to humans, and to complicate the understanding of prion disease, the incubation period can be over 25 years.

In animals, prion disease appears to be self-sustaining. For example, a growing fraction of mule deer and elk in the western part of the United States are known to carry a species-specific version of a prion-related fatal neurological disease called chronic wasting disease [79]. To date, there is no statistical evidence of humans acquiring prion-induced neurological disease from these animals but with a possibly long incubation period, who knows. If you eat venison or elk, you are in effect joining a long-term scientific experiment as a lab animal.

Generally humans acquire CJD either from genetic inheritance, medical contamination or what are called "sporadic" or random events. The approximate CJD occurrence rate for the general population is very low—about 1 in a million. However, there has been some speculation that the human incidence rate has been understated since people with CJD can be mistakenly diagnosed with Alzheimers or dementia. Both diseases can have similar symptoms. The only way really to know for sure if a person has CJD is by brain autopsy. One study that did this on a group of deceased Alzheimers patients and the results showed that 13% of the people actually died from CJD [80].

Reporting the occurrence of CJD is further complicated by the fact that it is not a "reportable disease," so no central database is maintained. Autopsies are problematic and expensive just to determine actual cause of death, especially in the elderly. The expense is obvious, but the reason for the other concern is important for you to know. For autopsy and medical procedures, normal sterilization methods are designed to kill bacteria, viruses, fungi and parasites—the stuff that can cause classical disease. But with prions you can't kill something that isn't alive and normal sterilization techniques don't "inactivate" prions. For example, with the food-borne bacteria discussed earlier in the chapter, cooking at a prescribed temperature for a certain period of time normally solves the problem. With prions, you can boil them, put them in dirt for years, and put them in a fire or under high pressure, and they can be still dangerous. Even funeral directors today have prescribed procedures to eliminate prion hazards in embalming [81] and prion inactivation sterilization techniques for medical instruments have also recently been published [82].

Risk management in these areas is adapting to a new exposure. Yet since no one knows the real incidence of prion-related CJD, we don't know if there are important human prion disease pathways that need to be blocked. Our knowledge of prion functions and interspecies transportation and mutations is just beginning. We all have an interest in this research. This is food—for thought.

DISCUSSION QUESTIONS

1. Will crop productivity yields continue to match or exceed population growth over the next 20–30 years? Why?

2. Discuss your views on why the European Union has different policies towards genetic modified foods than those of the U.S. congress and the food industry.

3. As a group exercise, have each member visit a supermarket and choose a food item with the least meaningful labeling information. How you determine the best example is up to you.

4. There is a perception that eating organic and grass-fed foods is prohibitively expensive. From your local food stores and available farmers' markets, develop the meal content and costs for whatever number of people you want for a full week. Compare the results to a similar nonorganic version. What is the percentage increase in costs for the organic diet? Do you think the risk reduction of the organic diet compensates for the increase in costs?

5. The text computes the probability for a person to get food poisoning in a year as 0.25. Design and perform a sampling experiment to compute your own estimate. You might be surprised at the results.

Case Study: The Changing Farming Paradigm

Patent law is designed to promote innovation while providing financial incentives and invention protection, both of which are good things. However, new laws and regulations generally lag behind reality and, in this case, technology. When legal conflicts affect the small number of people who grow our food, there's more at stake than legal precedent. Many farmers are currently in the middle of a biotechnology farming evolution where patent law is changing the farming paradigm. I will let you judge for yourself if this is a good or bad thing.

Collecting, storing, and reapplying seeds from previous crop years have been a fundamental tenet of farming for a very, very long time. The practice is a natural part of crop production in the sense that farmers are facilitating nature's own procreation methods, but on a larger scale. Today, the combination of technology and patent law has dramatically changed this custom.

Ever since the *Diamond vs. Chakrabarty* Supreme Court decision held that living organisms were patentable subject matter, inventor companies have aggressively managed licensing their "inventions" through strict contracts (called Grower and Technology Agreements), infringement surveillance, and prosecution. The agricultural technology is in the form of genetically modified organisms (GMOs) that have desired farming traits such as resistance to specific herbicides or pesticides (which the inventor company also manufactures), desirable growth density characteristics (more plants/acre), and stronger stems. These GMOs are sold to farmers often through distributors that collect both the base cost for the seeds plus an added surcharge for the technology agreement. The agreements contain very strict wording that describes what the farmer can and cannot do with the seeds. Basically, the wording states that farmers can plant the seeds, grow the crops, and sell them commercially, but they cannot store and replant the seeds of their crops. Farmers are also restrained from doing any research or experimentation that might help them improve their yields further. These restrictions radically change framers from autonomous entrepreneurs to little more than service providers who are dependent on the GMO seed and herbicide providers.

There have been several hundred lawsuits brought on farmers by GMO inventor companies for various types of patent infringement and breach of contract situations. For example, a farmer planted all non-GMO crops in his land while surrounding farmers planted GMO crops. As the crops began to grow, the GMO inventor company discovered some GMO crops growing in his fields. They gave him three choices: pay the technology fee, remove the crops, or be sued for patent infringement. The farmer did not intentionally plot any GMO seeds and most likely they arrived on his land by the wind drift from neighboring farms, from passing seed trucks on roadways, or from cross-pollination. Nevertheless, he lost the patent infringement suit and was fined a year's profits. Even though the farmer did not actively or knowingly plant GMO crops, the simple fact that they were growing on his land was sufficient evidence for conviction. Under U.S. patent law, intent is not required for proof of patent infringement.

The farmer had alternatives. For example, the GMO inventor company agreed if he removed the GMO crops from his fields, the suit would be dismissed. However, from a practical perspective, removal of these crops would require a complete cleaning of the fields including the removal of about an inch of irreplaceable topsoil [83]. Obviously, the farmer did not view this option viable.

The non-GMO farmer can have problems from his or her GMO-planting neighbors, but these issues are small compared to the restriction of farmers to store and reuse seeds from grown crops. However, there is a new court ruling that may change this limitation in the future [84, 85]. Here is a brief description of the case.

Company A licensed its patents to company B, who included A's technology as a part of their product which was sold to company C. Company A also had a license agreement with company B that stated that B would not license its products (that included A's technology) to a third party (like company C). Company A sued company C for patent infringement and lost. If company A is replaced by the GMO seed producer; company B, the seed distributor; and company C as the farmer, you can see how this case fits into the farming industry.

The Supreme Court in this case applied, for the first time in 65 years, the Doctrine of Patent Exhaustion, also known as the first-sale doctrine that terminates all patent rights over an item sold without restriction. Before this ruling, legal precedent established that seeds obtained from crops grown with purchased seeds are under patent protection. The inventor companies claim that their seasonal licensing agreements keep seed prices low. If farmers bought seed only once, inventor companies would need to significantly raise prices since they would only have one sales opportunity for a given product. If this situation existed, they would have little financial incentive to research and develop new products since recovering the costs through single farmer sales would make product prices unrealistically high.

Now with new legal precedent established, agricultural biotech inventor companies must reconsider if the sale of patented seeds, even with contractual restrictions placed on buyers, exhausts some or all patent rights for the second-generation progeny seeds. This issue is a complex combination of patent and contract law that will take years to decide. In effect, operational risk has increased for all parties: the farmers, the distributors, the inventor companies, and for all of us, eventually, in food prices.

By the way, this new court ruling's influence is not limited to only to GMO seeds. It has implications to any patented (or patent pending) product that can make copies of itself, such as self-replicating cell lines, genetic material, and, perhaps, even software.

Questions

1. From a societal point of view, which issue is more important: farmer autonomy or seed inventor patent protection?
2. The inventor companies and farmers have reasonable arguments for their positions in the second-generation seed ownership issue. Independent of patent law, how would you construct a fair solution for both parties?

Answers to Table 10.3

1	Label A—Nutrient #2:	A 11.5 oz. can of tomato juice.
2	Label B—Nutrient #3:	One slice of bologna.
3	Label C—Nutrient #1:	A can of beef soup (with vegetables).
4	Label D—Nutrient #4:	A slice of carrot cake.

ENDNOTES

1 Biographical Directory of the United States Congress, 1774–Present, http://bioguide.congress.gov/biosearch/biosearch.asp (accessed September 7, 2010).

2 Winston Williams, "James J. Delaney, 86, a Democrat and Former Queens Congressman," *New York Times*, May 25, 1987.

3 Utah Pesticide Safety Education Program, http://home.comcast.net/~tamiwarnick. Accessed July 25, 2011.

4 Bryan Walsh. "Regulation of Toxic Chemicals Faces Tightening," *Time.com*, April 16, 2010.

5 USDA National Agricultural Statistics Service, "Farms, Land in Farms, and Livestock Operations: 2009 Summary," February 2010. Available at http://usda.mannlib.cornell.edu/usda/current/FarmLandIn/FarmLandIn-02-12-2010_new_format.pdf.

6 NRCS, "In the Time It Took to Form One Inch of Soil…" National Resources Conservation Service, USDA, 2001. Available at http://soil.gsfc.nasa.gov/inch/soiltime.htm (accessed August 22, 2010).

7 National Agriculture Statistical Service, National Statistics for Corn, 2009.

8 Mississippi Agricultural and Forestry Station, http://msucares.com/crops/soils/mgfertility.html (accessed August 11, 2010).

9 http://mygeologypage.ucdavis.edu/cowen/~gel115/115CH16fertilizer.html (accessed August 14, 2010).

10 Jaime Wisniak, "Fritz Haber: A Conflicting Chemist," *Indian Journal of History of Science*, Vol. 37, No. 2, 2002, pp. 153–173.

11 Vaclav Smil, *Enriching the Earth: Fritz Haber, Carl Bosch, and the Transformation of World Food Production*. Cambridge, MA: MIT Press, 2001.

12 http://nobelprize.org/nobel_prizes/chemistry/laureates/1918/haber.html (accessed August 11, 2010).

13 United Nations Secretariat, Population Division, Department of Economic and Social Affairs, "The World at Six Billion," ESA/P/WP.154, October 1999.

14 C. R. Darwin, *On the Origin of Species by Means of Natural Selection, or The Preservation of Favoured Races in the Struggle for Life*. London: John Murray, 1859.

15 B. F. Benz, "Archaeological Evidence of Teosinte Domestication from Guila Naquitz, Oaxaca," *Proceedings of the National Academy of Sciences of the United States of America*, Vol. 98, 2001, pp. 2104–2106.

16 Bill Penney, "Looking Ahead to Thursday's USDA Numbers," *Agriculture.com*, August 11, 2010.

17 G. H. Shull, "The Composition of a Field of Maize," Am. Breeders Assoc. Rep. 4, 1908, pp. 296–301.

18 A. R. Crabb, *The Hybrid-Corn Makers: Prophets of Plenty*. New Brunswick, NJ: Rutgers University Press, 1947.

19 J. D. Watson and F. H. C. Crick, "A Structure for Deoxyribose Nucleic Acid," *Nature*, Vol. 171, 1953, pp. 737–738.

20 http://news.bbc.co.uk/2/hi/science/nature/2804545.stm (accessed August 22, 2010).

21 Ralf Dahm, "Friedrich Miescher and the Discovery of DNA," *Developmental Biology*, Vol. 278, Issue 2, February 15, 2005, pp. 274–288.

22 http://www.ornl.gov/sci/techresources/Human_Genome/project/about.shtml (accessed August 25, 2010).

23 U.S. Patent 4,259,444.

24 Chris Braunlich, "DNA Plant Technology Unveils Second Generation Genetically-Modified Tomato," *Business Wire*. 1995. Available at http://www.thefreelibrary.com/DNA+PLANT+TECHNOLOGY+ UNVEILS+SECOND+GENERATION+GENETICALLY-MODIFIED...-a016877921 (accessed October 5, 2010).

25 http://www.isb.vt.edu/biomon/releapdf/9107901r.ea.pdf (accessed August 25, 2010).

26 Daniel Charles, *Lords of the Harvest: Biotech, Big Money, and the Future of Food*. Perseus Publishing, 2001, p. 144.

27 "Transgenic Approaches to Combat Fusarium Head Blight in Wheat and Barley Crop," *Science*, Vol. 41, No. 3, June 2001, pp. 628–627.

28 "Post-transcriptional Gene Silencing in Plum Pox Virus Resistant Transgenic European Plum Containing the Plum Pox Potyvirus Coat Protein Gene," *Transgenic Research*, Vol. 10, No. 3, June 2001, pp. 201–209.

29 http://www.ers.usda.gov/Data/BiotechCrops/, updated July 1, 2010 (accessed July 21, 2010).

30 Charles Benbrook, "Impacts of Genetically Engineered Crops on Pesticide Use in the United States: The First Thirteen Years," *The Organic Center*, www.organic-center.org, November 2009.

31 "Farmers Cope with Roundup-Resistant Weeds," *New York Times*, May 4, 2010.

32 "Gene Amplification Confers Glyphosate Resistance in *Amaranthus palmeri*," *Proceedings of the National Academy of Sciences of the United States of America*, Vol 107, 2010, pp. 955–956.

33 Genetically Modified Foods and Organisms, Human Genome Project Information, http://www.ornl.gov/ sci/techresources/Human_Genome/elsi/gmfood.shtml (accessed August 28, 2010).

34 Andres Hernandez, "Benefits and Safety Issues Concerning Genetically Modified Foods," http:// biotechpharmaceuticals.suite101.com/article.cfm/benefits-and-safety-issues-concerning-genetically-modified-foods (accessed August 29, 2010).

35 http://www.ornl.gov/sci/techresources/Human_Genome/elsi/gmfood.shtml (accessed August 29, 2010).

36 Todd Zwillich, "No Labels for Genetically Engineered Food, FDA Says Labels Won't Be Needed for Products Made from Genetically Altered Animals," *WebMD Health News*, September 18, 2008. http:// www.webmd.com/news/20080918/no-labels-for-genetically-engineered_food (accessed September 1, 2010).

37 Regulation (EC) No 1830/2003 of the European Parliament and of the Council of 22 September 2003 concerning the traceability and labeling of genetically modified organisms and the traceability of food and feed products produced from genetically modified organisms and amending Directive 2001/ 18/EC, http://eur-lex.europa.eu/smartapi/cgi/sga_doc?smartapi!celexplus!prod!DocNumber&lg=en& type_doc=Regulation&an_doc=2003&nu_doc=1830 (accessed September 2, 2010).

38 http://www.epa.gov/npdes/pubs/sector_table.pdf (accessed September 3, 2010).

39 EPA, *NPDES CAFO Rule Implementation Status—National Summary*. Washington, DC: Environmental Protection Agency, April 9, 2010.

40 William Neuman, "Heart of Iowa as Fault Line of Egg Recall," *New York Times*, August 27, 2010.

41 "Animal Feeding Operations—Compliance and Enforcement: Enforcement Cases 2009 through Present." http://www.epa.gov/agriculture/anafocom09.html (accessed September 4, 2010).

42 "Confirmed Violations/Discharges from CAFOs and Liquid-System Livestock Operations to Bean/ Tiffin Watershed and River Raisin Watershed." http://nocafos.org/violations.htm (accessed September 4, 2010).

43 Union of Concerned Scientists, "The Hidden Costs of CAFOs," *Earthwise*, spring 2009.

44 Inspections & Compliance: Import & Export, http://www.foodsafety.gov/compliance/importexport/. Accessed July 25, 2011.

45 Geoffrey S. Becker, "U.S. Food and Agricultural Imports: Safeguards and Selected Issues," March 17, Congressional Research Service, 7-5700, www.crs.gov, RL34198.

46 NASA Facts, "NASA Food Technology: Incredible Edibles from Space Dining," Lyndon B. Johnson Space Center FS-2004-08-007-JSC, August 2004.

47 Institute of Medicine, "Enhancing Food Safety: The Role of the Food and Drug Administration," report brief, June 2010.

48 Compliance Program Guidance Manual—Domestic Food Safety 7303.803, November 9, 2008.

49 "Digestive Diseases: Food Poisoning," http://www.medicinenet.com/food_poisoning/article.htm (accessed August 4, 2010).

50. Robert L. Scharff, "Health-Related Costs from Foodborne Illness in the United States," Produce Safety Project at Georgetown University, March 3, 2010.

51 T. Roberts, "WTP Estimates of the Societal Costs of U.S. Food-Borne Illness." *American Journal of Agricultural Economics*, Vol. 89, No. 5, 2007, pp. 1183–1188.

52 Centers for Disease Control and Prevention, National Center for Zoonotic, Vector-Borne, and Enteric Diseases, "Salmonellosis," November 2009. http://www.cdc.gov/nczved/divisions/dfbmd/diseases/salmonellosis/ (accessed September 15, 2010).

53 Gardiner Harris and William Neuman, "Salmonella at Egg Farm Traced to 2008," *New York Times*, September 14, 2010.

54 Centers for Disease Control and Prevention, National Center for Zoonotic, Vector-Borne, and Enteric Diseases, "Campylobacter," July 2010. http://www.cdc.gov/nczved/divisions/dfbmd/diseases/campylobacter/ (accessed September 15, 2010).

55 Ban Mishu Allos et al., "*Campylobacter jejuni* Strains from Patients with Guillain-Barré Syndrome," *Emerging Infectious Diseases*, Vol. 4, No. 2, 1998.

56 David L. Woodward. "Identification and Characterization of *Shigella boydii* 20 serovar nov., a New and Emerging *Shigella* serotype," *Journal of Medical Microbiology*, Vol. 54, 2005, pp. 741–748.

57 Carl M. Schroeder et al., "Antimicrobial Resistance of *Escherichia coli* O26, O103, O111, O128, and O145 from Animals and Humans," *Emerging Infectious Diseases*, Vol. 8, No. 12, December 2002, pp. 1409–1414.

58 http://www.cdc.gov/FoodNet/factsandfigures/2009/Table1b_all_incidence_96-09.pdf (accessed September 15, 2010).

59 J. Farber and P. Peterkin, "*Listeria monocytogenes*, a Food-Borne Pathogen," *American Society for Microbiology*, Vol. 55, No. 3, 1991, pp. 476–511.

60 Keith R. Schneider et al., "Preventing Foodborne Illness: Listeriosis," Report FSHN03-6, Food Science and Human Nutrition Department, Florida Cooperative Extension Service, Revised August 2009.

61 Centers for Disease Control and Prevention, National Center for Zoonotic, Vector-Borne, and Enteric Diseases, "Listeriosis," November 2009. http://www.cdc.gov/nczved/divisions/dfbmd/diseases/listeriosi/ (accessed September 19, 2010).

62 Centers for Disease Control and Prevention, National Center for Zoonotic, Vector-Borne, and Enteric Diseases, "Yersinia," November 2009. http://www.cdc.gov/nczved/divisions/dfbmd/diseases/yersinia/ (accessed September 19, 2010).

63 P. Brasher and D. Piller, "Green Fields: USDA Looks at Tightening Checks for *E. coli* in meat," *Des Moines Register*, September 26, 2010.

64 F. J. He, C. A. Nowson, and G. A. MacGregor, "Fruit and Vegetable Consumption and Stroke: Meta-analysis of Cohort Studies," *Lancet*, Vol. 367, 2006, pp. 320–326.

65 L. Dauchet, P. Amoyel, and J. Dallongeville, "Fruit and Vegetable Consumption and Risk of Stroke: A Meta-analysis of Cohort Studies." *Neurology*, Vol. 65, No. 8, 2005, pp. 1193–1197.

66 C. Heidemann, M. B. Schulze, O. H. Franco, R. M. van Dam, C. S. Mantzoros, and F.B. Hu, "Dietary Patterns and Risk of Mortality from Cardiovascular Disease, Cancer, and All Causes in a Prospective Cohort of Women." *Circulation*, Vol. 118, No. 3, 2008, pp. 230–237.

67 M. A. Pereira et al., "Dietary Fiber and Risk of Coronary Heart Disease," *Archives of Internal Medicine*, Vol. 164, 2004, pp. 370–376.

68 Francesca Crowe, "Cardiovascular Disease." News release, European Heart Journal, Cancer Epidemiology Unit, University of Oxford, England, British Heart Foundation.

69 Andrew Martin, "Spinach and Peanuts, with a Dash of Radiation," *New York Times*, February 1, 2009.

70 "What Is the Meaning of 'Natural' on the Label of Food?" http://www.fda.gov/AboutFDA/Transparency/Basics/ucm214868.htm (accessed September 22, 2010).

71 David Gutierrez, "FDA Says High Fructose Corn Syrup Cannot Be Considered 'Natural'," January 12, 2009. http://www.naturalnews.com/025292_corn_HFCS_food.html (accessed September 22, 2010).

72 "Processed Foods: GMOs Working Behind the Scenes," http://www.gmo-compass.org/eng/grocery_shopping/processed_foods/.

73 Leo Cendrowicz, "Is Europe Finally Ready for Genetically Modified Foods?" *Brussels Time*, March 9, 2010.

74 The Center for Food Safety, "Genetically Engineered Food," http://www.centerforfoodsafety. orggeneticall7.cfm (accessed September 23, 2010).

75 Bonnie Azab Powell, "USDA Limits 'Grass Fed' Label to Meat That Actually Is," *Ethicurean*, October 16, 2007. http://www.ethicurean.com/2007/10/16/grass-fed-label/ (accessed October 3, 2010).

76 Food and Drug Administration, Center for Food Safety and Applied Nutrition, "Guidance for Industry: A Food Labeling Guide, Appendix A: Definitions of Nutrient Content Claims," April 2008. http://www.fda.gov/Food/GuidanceComplianceRegulatoryInformation/GuidanceDocuments/Food LabelingNutrition/FoodLabelingGuide/ucm064911.htm (accessed September 25, 2010).

77 Stanley B. Prusiner, "Prions," Nobel Lecture, *PNAS*, Vol. 95, No. 23, 1998, pp. 13363–13383.

78 S. Lindenbaum, "Review. Understanding Kuru: The Contribution of Anthropology and Medicine," *Philosophical Transactions of the Royal Society B: Biological Sciences*, Vol. 363, No. 1510, 2008, pp. 3715–20.

79 John Prion Collinge, "Strain Mutation and Selection," *Science*, Vol. 328, No. 5982, 2010, pp. 1111–1112.

80 Elias E. Manuelidis and Laura Manuelidis, "Suggested Links between Different Types of Dementias: Creutzfeldt-Jakob Disease, Alzheimer Disease, and Retroviral CNS Infections," *Alzheimer Disease and Associated Disorders*, Vol. 3. 1989, pp. 100–109.

81 Centers for Disease Control, "Information on Creutzfeldt-Jakob Disease for Funeral Home, Cemetery, and Crematory Practitioners," August 23, 2010.

82 William A. Rutala et al., "Guideline for Disinfection and Sterilization of Prion-Contaminated Medical Instruments," *Infection Control and Hospital Epidemiology*, Vol. 31, No. 2, 2010, pp. 107–116.

83 Hilary Preston, "Drift of Patented Genetically Engineered Crops: Rethinking Liability Theories," *Texas 81 Tex. L. Rev.* 2002–2003, pp. 1153–1175.

84 Yuiche Watanabe, "The Doctrine of Patent Exhaustion: The Impact of *Quanta Computer, Inc. v. LG Elecs., Inc.*," *14 Va. J.L. & Tech. 273*.

85 Quanta Computer, Inc. v. LG Elecs., Inc., 128 S. Ct. 2109 (2008).

Chapter 11

Flying: The Not-So-Wild Blue Yonder

Aviation in itself is not inherently dangerous. But to an even greater degree than the sea, it is terribly unforgiving of any carelessness, incapacity or neglect.

—Captain A. G. Lamplugh, British Aviation Insurance Group, London (ca. early 1930s)

The flying public seems to demand a paradox: When it comes to the airlines, we want 100% safety, on time arrivals and departures for all passengers, *and* affordable prices. In this arena, accidents are unacceptable. Or are they? Is there an "acceptable risk" like we have seen with other areas? Can the airlines eliminate accidents completely and keep fares low or, for that matter, even reasonable?

Risk has always been a concept inherently understood by the people involved in aviation; as long as there is gravity. there is risk. The airline industry was bred with the complete understanding that theirs was a business requiring constant vigilance. It was strictly controlled by government regulations, and free of the major competitive forces that most businesses faced. But airline deregulation altered the rules. It diminished the industry by removing its insulation from typical business pressures, changing it into a dirty business of fare wars, no-frills service, tense union relations, and marketing gimmicks—just to name a few things.

Even the technology produced a new kind of risk. As airplanes got bigger, so did the number of systems, subsystems, and components that could fail. The risk exposure of placing more than 300 taxpayers (and potential plaintiffs) up in the air in a single aircraft brought new concerns. With all of those parts, the parochial maintenance design of the airlines' infancy was an expensive nightmare. So back in the 1960s, when the first wide-body jets were produced, a maintenance management strategy was invented called reliability centered maintenance (RCM). RCM revolutionized the way aircraft and other equipment were maintained. This is a "Made in the USA" engineering maintenance product that remains in use today. Maintenance

20% Chance of Rain: Exploring the Concept of Risk, First Edition. Richard B. Jones.
© 2012 John Wiley & Sons, Inc. Published 2012 by John Wiley & Sons, Inc.

tasks and the intervals of application were developed at the heart of the process, using the reliability designed into the systems. Unless you're working in commercial aviation maintenance, my guess is you haven't heard about RCM. That's probably because it's part of the nonglamorous subject of maintenance. But don't let its anonymity disguise its value; it has an impressive track record. It was the first combined airline, regulatory, and aircraft manufacturer team effort to incorporate risk into federal regulations, and was the first great re-engineering work. The result has been improved safety, reliability, and manageable costs for over 40 years. Few, if any, re-engineering projects today can top RCM's performance.

Aircraft manufacturers and airline companies freely admit that more might be done to improve safety—newer equipment, more inspections, better training, additional fire retardant materials, air bags, and so on. However, these activities would also raise fares to the point at which fewer people would fly. To make matters worse, fare increases would cause many people to travel by automobile rather than fly, greatly increasingly their chances of death or injury. The airlines are working with the resources they have. Yet, by the very fact that as a nation, as a society, and as individuals we have finite resources, there must be an "acceptable risk." We have no choice. You may not be able to get anyone to tell you what it is, but the nebulous statistic does exist.

So let's get right to the point about flying. How "safe" is it to fly? This seems like a simple question that should have a simple answer. Well, yes and no. This is a classic type of a straightforward and direct question that can have so many oblique and esoteric answers. Safety, like risk, is in the future. After all, if you've traveled by air and arrived at your destination (with or without your luggage) unscathed, then the flight was "safe." The question of safety refers to your next flight. This is the risk. No one can obviously answer this question on an individual basis so we apply statistics to measure safety on a group or population level. There are many different types of events related to safety, so a detailed definition of "aircraft accident" is required to ensure everyone collects the same data and all users can properly interpret the results. The International Civil Aviation Organization (ICAO) and the U.S. National Transportation Safety Board (NTSB) have different definitions, and NTSB's version is used here just to give you a frame of reference. According to the NTSB:

> *An aircraft accident means an occurrence associated with the operation of an aircraft which takes place between the time any person boards the aircraft with the intention of flight and all such persons have disembarked, and in which any person suffers deaths or serious injury, or in which the aircraft receives substantial damage.* [1]

Aviation safety has a plethora of statistics, so let's see how useful some of them are to address *your* flying safety.

1. *Number of fatalities per year*: If there were no fatalities, then flying would undoubtedly be deemed safe, but these numbers by themselves don't include the number of people flying or the number of flights. The number alone is impossible to relate to your safety.

2. *Fatal accidents per 100,000 flight hours*: Here we have exposure information included and a true ratio of relevant variables. The statistic is used by the NTSB, but it doesn't take into consideration the number of takeoffs and landings involved or even the actual number of deaths included in the accidents. An accident that kills one person has just as much weight to the statistic as a crash that kills 300 people.

3. *Fatal accidents per million departures*: This metric includes the number of takeoffs as the exposure variable which is good from the point of view that many accidents occur during this phase of flight. However, the numerator has the same deficiency as in the previous statistic since there is no distinction between the number of people killed in the accidents.

4. *Hull losses per million departures*: A hull loss is defined as an accident in which the aircraft is totally destroyed, damaged, or beyond repair. This is like "totaling" your car. The term also applies to situations where the airplane is missing, the search for the wreckage has been terminated, and where the wreckage is completely inaccessible. Aircraft manufacturers use this metric to measure and contrast fleet safety for different aircraft models. This is important information for their reliability engineering and marketing departments, but it doesn't provide much insight to your safety.

5. *Passenger fatalities per 100 million passenger-miles*: The numerator in this metric is important but there's no distinction between one accident that kills 300 passengers and 10 accidents that kill 30 people each. The denominator is a standard type of exposure variable often applied in transportation that combines the distance traveled with the number of people traveling. It equates the events of 100 people who each fly 1 mile to 1 person who flies 100 miles. These types of statistics are useful analysis tools for many types of aviation studies, but they don't provide information on *your* flying safety.

6. *Passenger fatalities per million passengers carried*: This one is getting closer to answering the safety question. It is the ratio of people who die in aircraft accidents to the number of people who board. This ratio is a direct measure of your probability of dying from an aircraft accident. Its weakness is from the equivalence of fatalities independent of the number of accidents. For example, this statistic treats 300 people dying in one accident the same as 30 people dying in each of 10 accidents.

7. *Passenger death risk per randomly chosen flight*: This statistic most directly quantifies your safety [2]. It measures the probability, Q, that a person who chooses a nonstop flight at random will die from an aircraft accident on that flight. It is computed from the formula:

$$Q = \frac{1}{N} \sum_{i=1}^{k} x_i$$

where N is the number of separate flights that have occurred during a given interval so $1/N$ is the probability that a passenger chooses a specific flight.

Table 11.1   Q-Statistic Example

Accident flight index	# of people onboard	# of fatalities	x_i
1	250	34	0.136
2	55	4	0.073
3	125	125	1.000
4	250	249	0.996
$\sum_{i=1}^{k} x_i$			2.205

For the number of aircraft accidents, k, the fraction of the passengers on each accident flight that died is computed as x_i. The Q statistic is precisely a passenger's death risk for a random flight. In the usual terms of frequency and severity, frequency is identified as the ratio $1/N$: the probability of selecting a given flight. Severity is the summation of passenger fractions that died on each accident flight: $\sum_{i=1}^{k} x_i$.

Here is how this statistic can be calculated in practice. Suppose, for example between 2007 and 2010 there were four domestic aircraft accidents that included fatalities with outcomes as shown in Table 11.1. If there were 2.2 million separate domestic flights (N) over this time period, then the Q statistic is about 1×10^{-6} or, stated in a more meaningful way, your chances of being killed on a domestic flight are about 1 in a million. The statistic incorporates takeoffs and landings, the number of accidents, and their survivability into a risk measurement perspective that precisely answers the question every airplane passenger has about flying. What are my chances of dying in an aircraft accident?

I think another reason why the Q statistic has been well received is its simplicity. It's easy to compute with simple arithmetic, and easily understood with small probabilities (like 1.0×10^{-6}) using the format of 1 in a million. The public is a lot more familiar with this phraseology than the use of scientific notation.

Aviation safety is a dynamic and diverse process that varies considerably between countries, the time of day, the nature of the air carrier, and the type of flying. The biggest contrast is between what is often called the developed or industrialized world and developing (or Third World) countries. It makes sense that developing countries, for example, would have poorer safety performance than industrialized countries which have established navigation, air control, and active regulatory systems. This is still the case today but safety variations also occur between air carriers in the same part of the world. Why? There is no simple answer but one aspect of this issue is clear. To blame poor safety performance on economic affordability is nonsense. Accidents are immensely more expensive than safe operations. The areas that are most likely to influence safety performance are training, maintenance, and other operational requirements such as pilot rest periods between flight duty

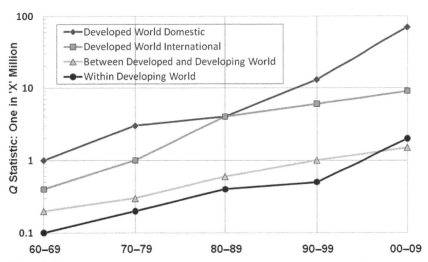

Figure 11.1   Passenger mortality risk (Q-statistic), commercial jet aviation, 1960–2009.

intervals. Yet, in spite of its shortfalls, as a global industry over the past 50 years, commercial aviation has grown and evolved as a common, reliable, and affordable means of transportation. Safety has improved, in part due to better technology but also because the aviation community has learned (and is learning) from the accidents that do occur.

It is now time to answer the aircraft accident passenger death risk question using Q-statistics. Figure 11.1 shows the Q-statistic [3, 4] for commercial jet aviation safety for flights within and between developed and developing countries. Depending where you are when you're reading this, you may feel better or worse with the following observations.

One clear conclusion is that flying within developed countries today (looking at the 2000–2009 points) is about 8 times safer than flying internationally between developed countries, 46 times safer than flying between developed and undeveloped countries and 35 times safer than flying within developing countries. Also note that flying between developed countries today is about as safe as flying within developed countries 20 years prior and that any flying involving developing countries is about as safe as flying between or within developed countries about 40–50 years ago.

These conclusions are facts. They are computational results using historical data so there is no uncertainty and no risk looking backward with Figure 11.1. What is of interest to air passengers today is the death risk for future flights. It is likely that air carrier operations are going to change significantly without new data, that is, accidents, so using the 2000–2009 numbers alone or trending the data can only produce approximations for passenger death risk statistics for the future 2010–2019 period. It may seem like forecasting the Q-statistics would be a relatively simple exercise from looking at Figure 11.1. Forecasting aviation safety is extremely unreliable due to the infrequent nature of aircraft accidents. Statistically, Q forecasts for aviation safety would have error limits so wide due to the paucity of data that they would be of little value. The best way to apply the Q statistics like in Figure 11.1 is to read the plots and make your own conclusions about flying.

The dataset in Figure 11.1 is for commercial jet aviation, but there are many other types of aviation where passengers can ask the same question. Most people fly regularly scheduled air carriers that are operated under Part 121 of the federal regulations. The data in Figure 11.1 is for these types of operations restricted to jet only aircraft (no propellers). These scheduled carriers represent the large majority of commercial aviation operations today. Yet you can also fly on commercial unscheduled air carriers called air taxi operations that make it their business to fly you where you want, when you want. This type of operator is called on demand air taxi and is governed by Part 135 of the regulations. Then there is the general aviation (noncommercial, Part 91) community where people fly their own or rented planes. If your friend is a pilot and takes you on a trip, you can pay the trip expenses, but legally the pilot can't make any money for flying you around. That would be a commercial operation and the pilot and aircraft would be subject to the higher standards of Part 135 instead of Part 91.

The location of the flight makes a big difference when it comes to safety. Not all countries have the legal and operational guidelines, operations infrastructure, training, and aircraft sophistication as the developed world, so safety performance varies a lot depending on where you are in the world as can be observed from Figure 11.1.

Up to this point, aircraft passenger death risk has been studied for accidents related to weather, engine failures, fires, bird strikes, midair collisions, and other so-called normal causes of airplane crashes. The obvious exclusion is for causes related to sabotage, suicide, terrorism, and military action. Table 11.2 shows these types of air carrier events for the developed world. The last four entries in Table 11.2 changed our view of aviation security for a long time to come.

Before September 11, 2001, aviation safety from the passenger view was primarily seen by ensuring their seat belt was fastened and the tray table was in the

Table 11.2   Air Carrier Occurrences Involving Illegal Acts in Developed World: 1986–2009 Sabotage, Suicide, or Terrorism

Date	Location	Operator	Onboard fatalities
4/2/1986	Near Athens, Greece	Trans World	4
12/7/1987	San Luis Obispo, CA	Pacific Southwest	43
12/21/1988	Lockerbie, Scotland	Pan American	259
4/7/1994	Memphis, TN	Federal Express	0
9/11/2001	New York, NY	American Airlines	92
9/11/2001	New York, NY	United Airlines	65
9/11/2001	Arlington, VA	American Airlines	64
9/11/2001	Shanksville, PA	United Airlines	44

full upright and locked position. Preboarding and passenger screening procedures were in place for scheduled air carrier flights but their apparent importance was diminished by the public's low perception of air security risk. Nevertheless, since 1999, a risk-based system was actively in operation out of public view. It was called the computer assisted passenger prescreening system (CAPPS). The system automatically cross-checked passenger information with known terror and FBI databases. Risk scores were assigned to each person with high scores requiring the airlines to perform more detailed baggage checks, personal screenings, and possibly FBI notification [5]. In fact, this system did select 9 of the 19 9/11 hijackers for additional screening of their luggage for explosives. Since they were not carrying explosives, they were allowed to board the planes [6].

Our air transportation security system today applies a "defense-in-depth" philosophy that provides a multidimensional safety net. The net consists of "20 layers of security" where each layer has the potential on its own to thwart a terrorist attack. The system is designed with and without passenger involvement. The long security lines, pat downs, and full body scans are only a small part of the security operations. And just because you can't see the activities in action doesn't mean they are not going on.

You will recognize some of the 20 layers and some you won't. They are listed in Table 11.3. Of these 20 layers, 14 are preboarding security (i.e. deterrence and apprehension of terrorists prior to boarding aircraft). The remaining six layers of security provide in-flight defense [7]. The details of each element are interesting but they are not appropriate content for this or any commercial book. This table shows that security risk is being managed. In the past, the risk management was successful, in the future, the risk remains.

The previous statement is not simply rhetoric. Air security has not diminished in importance and the tragic results speak for the need of constant vigilance for all who work or use the commercial air transportation system. Here are some examples of why air security is a major risk exposure.

Table 11.3   Functional Layers of Commercial Aviation Security

1. Intelligence	11. Checked baggage
2. Customs and border protection	12. Transportation security inspectors
3. Joint terrorism task force	13. Random employee screening
4. No-fly list and passenger prescreening	14. Bomb appraisal officers
5. Crew vetting	15. Federal Air Marshal Service
6. Visible Intermodal Protection Response Teams	16. Federal Flight Deck Officers
7. Canines	17. Trained flight crew
8. Behavioral detection officers	18. Law enforcement officers
9. Travel document checker	19. Hardened cockpit door
10. Checkpoint/transportation security officers	20. Passengers

- December 2001: An attempt to ignite explosives in the shoes of an American Airlines international passenger failed and the person was subdued by the passengers. No one is injured.
- May 2002: A China Northern Airlines jet in the People's Republic of China crashed, killing all 103 passengers and the 9 crew. The investigation revealed a passenger set fire to the cabin with gasoline.
- July 2002: A terrorist attack by a lone gunman at Los Angeles International Airport killed two and wounded four people. The gunman was also killed by a security guard.
- August 2004: Two Russian domestic passenger jets are nearly simultaneously destroyed by suicide bombers. All 88 passengers plus the two bombers were killed.
- 2006: A plot to bomb at least 10 airliners traveling from the United Kingdom to the United States and Canada was discovered and foiled by British police. The chemical explosives were planned to be made in-flight by mixing ingredients brought onboard by multiple terrorists. The demonstrated ability to manufacture bombs as an in-flight chemical mixture affected the ability of passengers to carry liquids onto commercial aircraft.
- June 2007: A car laden with propane canisters rammed Glasgow International Airport Terminal. No one was killed. It was the first terrorist attack in Scotland since the Lockerbie bombing in 1988.
- December 2009: A bomber attempted to ignite explosives hidden in his underwear as the international fight approached a U.S. destination. The crew and passengers subdued the person and no one was killed.
- October 2010: Two bombs hidden in laser printer cartridges were found in air cargo destined for the United States.

Air security is important, but most commercial aviation accidents are from other flying risks. Let's take a look at flying a domestic commercial airliner for a 1½ hour flight. We will answer three important questions:

1. What part of the flight is the most likely time for an aircraft accident?
2. In which part of the flight are the most passengers killed in aircraft accidents?
3. What part of the flight has the highest risk?

To get our answers we look to the data presented in Table 11.4, which shows a compilation of performance data for worldwide commercial jet accidents [8].

There are eight different flight modes that include accidents from takeoff to landing. We could include ground operations like taxiing in this list. The most serious type of ground operation risks involves runway incursions where aircraft have inadvertently taxied onto a runway being used for takeoffs and landings. These types of events continue to be major risk exposures for the traveling public and leaving them out of Table 11.4 is not intended to imply taxi operations are without risk. However,

Table 11.4   Fatal Accidents and Onboard Fatalities by Phase of Flight Worldwide Commercial Jet Fleet – 2000 Through 2009

Flight mode	# of accidents	# of fatalities	% Fatal accidents	% Onboard fatalities	% Duration	Time in mode (min)	Risk: deaths/ Min	Relative risk
Takeoff	11	703	14%	14%	1%	0.9	781	40
Initial Climb	8	663	10%	13%	1%	0.9	737	38
Climb	4	548	5%	11%	14%	12.6	43	2
Cruise	9	1000	12%	20%	57%	51.3	19	1
Descent	4	199	5%	4%	11%	9.9	20	1
Initial approach	10	703	13%	14%	12%	10.8	65	3
Final approach	12	632	16%	13%	3%	2.7	234	12
Landing	19	550	25%	11%	1%	0.9	611	31

This data is compiled for commercial jet airliners over 60,000 lbs. This includes some regional and all longer-distance jets. Events and fatalities excluded from the database include deaths from natural causes, suicide, stowaways, experimental tests, terrorism, sabotage, hijacking, and military action.

since the major exposure from runway incursions is with aircraft taking off and landing, we'll assume taxi risks are already contained in the other flight modes.

From Table 11.4, notice that the largest number of fatal accidents (19) occurs during landing. About 25% of the accidents occurred during this phase. This is the largest percentage in the columns so "landing" is the answer to the first question. The fact that every takeoff is "voluntary" but every landing is "mandatory" may have something to do with this flight phase having the highest fatal accident probability.

To answer question #2, look down the Percentage of Onboard Fatalities column. The largest percentage of deaths by mode of flight: 20%. The deaths occurred in the cruise flight mode. To understand, all of the reasons why the cruise segment has such a high severity requires reviewing the nine accident cases, which killed a total of 1000 passengers. This is an exercise outside this chapter's level of detail. But regardless of the causal factors, the data indicates that the cruise segment, the longest phase of flight where aircraft are at their highest altitude and speed, represents the most serious time to experience an aircraft accident situation.

Now, when it comes to risk, there is more than one way to answer question #3. One way is to identify the flight mode which has the highest product of the number of accident and number of deaths or the percentage values used to answer the preceding two questions. Notice, however, we have more information in this analysis that allows us to combine the exposure interval with the severity magnitude. Remember, the analysis in Table 11.4 is for a representative 90-minute flight. Everyone who flies is exposed to the eight phases for about the time intervals shown in Table 11.4. Since we also know the number of deaths by flight mode, we can compute risk as the number of deaths per minute of flight in each mode. The results are listed in the next to last column on the right-hand side. The actual numbers are not as important as the flight mode relative risk. Using this column as the risk metric, takeoff, followed by initial climb, are the two highest-risk flight modes for large commercial jets. Landing is ranked as third. Together takeoff and initial climb (up to 1,000 ft or the first power reduction) compose over 60% of the total death risk. Landing adds about another 25% so that the total risk contained within these three flight modes is about 85% of the total risk. Intuitively, I think these numbers make sense with most people's rational and irrational concerns of commercial aviation safety. You might see this in the white knuckles reactions of some passengers at takeoff and landing.

Accidents are the final, tragic elements in a sequence of a chain of events that may have started years before the actual crash. We're usually quick to come up with shallow descriptions of blame shortly after accidents. I say "shallow" because our forensic abilities are short-sighted, too often pointing at the most visible elements of the accident sequence rather than at the hidden and often more important factors. When the full, complex picture of the sequence is given, too often we're off pursuing other exciting news. The initial diagnosis of the problem usually sticks in our minds and we seldom bother to clear it up within the whole story. To make matters worse, the media also works to satisfy the short-term memory of the general public.

The human crew actually flying the airplanes is seldom the root cause of aircraft accidents, and yet we have seen accident database results which blame 50% of all accidents on "pilot error" [9]. It is true that, most of the time, the pilots are involved, but everyone from air traffic controllers to the Federal Aviation Administration (FAA) to the baggage handlers are fingered when someone wants to assign liability.

Since people are central to aviation safety, it shouldn't be a surprise that more research on human factors has been done in aviation than (probably) for all other areas combined [10]. The design and arrangement of the instruments in the cockpit, the colors and shapes of the controls, the workload, and a host of other factors are all designed to make the pilots and the aircraft work together as an integral unit.

Ergonomic engineers have analyzed and continue to analyze pilot eye movements to ensure that the flight instruments are arranged in such patterns that pilots gather the maximum amount of flight information per eye position [11]. The colors, shapes, and positions of the various levers are also designed to help pilots recognize the function of each by its looks and touch.

Engineers have also incorporated several audible and visual alarms that activate whenever the aircraft gets into a situation that its designers thought deserved special attention from the flight crew. Here are some examples of how modern airliners help pilots respond to possible emergency conditions.

Airfoil characteristics, or the shape of the wing, are extremely versatile to allow aircraft to fly slowly for takeoffs and landings and to allow fuel-efficient flight at cruise speeds. To accomplish these tasks, the shape of the wing is changed dramatically for the takeoff, cruise and landing phases of flight. Pilots deploy both leading edge and trailing edge flaps at various settings for both takeoffs and landings. The flying characteristics are highly dependent on the settings of these wing shape devices. Takeoff is a very unforgiving part of flight and design engineers recognize that flight crews do make mistakes. To help prevent some of these errors, engineers have developed warning systems that sounds an aural alarm if the pilots try to take off with the wing flaps not deployed in the "takeoff" configuration. The type of warning varies by airplane model, but here is one example: If engine power is increased to over 80% of maximum and the flaps are not deployed in "takeoff" configuration, a warning alarm sounds.

A similar system assists pilots for landings. If the aircraft descends too low to the ground and the aircraft is not in "landing" configuration, the ground proximity warning system (GPWS) tells the pilots to "Pull up! Pull up!" Major airline carriers around the world are installing the next generation of this system. It will give flight crews even more advanced warnings of pending collisions with terrain.

Commercial aviation has had its share of gear-up landings. Besides being expensive and dangerous, they are rather embarrassing for the flight crews. To impress new pilots I've taught with the seriousness of this risk, I gave them this piece of completely serious advice regarding gear-up landings: "There are only two kinds of pilots; those who have landed with the gear up, and those who will." (An analogous, even more potent version of this tenet is just as true for pilots of amphibious aircraft.) Flight system designers have decided that flight crews of the future will not become members of this infamous club—at least not while flying their aircrafts. If the aircraft is not configured for landing and it descends below a certain altitude, or the engine power is reduced beyond a certain point, audible alerts inform the crew that all is not right.

Engineers can't provide automated checks for all conceivable configuration errors. That would be simply impractical, expensive, and perhaps even increase risk as the added redundancies could interfere with pilot actions required for emergency situations.

One possible exception worth noting here comes from the details of a Boeing 757 crash that killed 159 out of the 163 on board: American Airlines Flight 965 from Miami, Florida, to Cali, Colombia, on December 20, 1995 [12]. There were several causal factors in this tragedy, including crew inattention, navigational errors, controller communications, and time of month. The aircraft was configured for a shallow descent with its speed brakes or wing spoilers partially deployed. These devices are large flaps located on the upper surface of the wings that disrupt airflow to reduce lift. In-flight, they are used for descending with little or no change in aircraft pitch or increase in airspeed. These types of descents are standard procedure. When the crew of Flight 965 finally realized their dangerous proximity to the ground, they applied full power and rotated the aircraft to a climb attitude. However, they never retracted the wing spoilers. In the final 13 seconds, the aircraft gained some altitude but still crashed into a mountainside, only about 100 feet below the summit.

Climbing with the spoilers deployed is like driving with your feet on the accelerator and the brake at the same time. With all of the intercoupling safety checks engineered into the Boeing 757, this particular coupling was not included. Other aircraft with different speed brake designs have included this coupling. Why not on the Boeing 757? In fact, none of the Boeing aircraft have this coupling. Here's the risk management decision behind it.

After the Cali crash, American Airlines asked Boeing to study the viability of retrofitting the Boeing 757 with an auto speed brake retraction-thrust levers coupling system. Boeing studied the detailed engineering system changes, the costs, and the aerodynamic implications and decided (in agreement with American Airlines) not to retrofit the airplanes. In fact, Boeing is not considering this coupling system on any of its aircraft. Why not? There are three reasons.

Reason #1: The retrofit system requires increasing the complexity of the already very complex aircraft control systems. Altering the control system design may reduce the risk of a particular accident scenario, but it was felt that the ongoing maintenance of this standby system plus other performance reliability issues provided little if any risk reduction benefit.

Reason #2: Since the speed brakes on the Boeing 757 and, in fact, all Boeing aircraft are located on the wings, the Safety Engineering Group observed that the aircraft could experience undesirable pitch oscillations when the speed brakes autoretracted with thrust addition. In the type of situations where a flight crew would use this safety feature, it was decided that the pitch oscillations could increase risk.

Reason #3: In studying the Cali crash, Boeing engineers asked the question: "If this particular aircraft was equipped with the auto speed brake retraction system, would the crash have been prevented?" Of course, no one knows for sure, but their results did not clearly indicate that the tragedy would have been averted.

In my research on this subject, I asked an airline pilot about his view of this incident and his opinion on the coupling of speed brakes with power applications. He saw no need for the added complexity, and said, "the best risk management action is not coupling the devices, it's 'don't get lost in the first place.'"

And today, with aircraft control coupling to high-precision global positioning system (GPS) navigation being used widely on all types of aircraft, getting lost is really hard to do. Even if you've never seen an aircraft cockpit you can get an idea of the computerization by comparing a 1970s car to a car today. Today the speedometers and many of the heating and A/C controls are electronic. The radio console no longer has the vintage vertical bar with cabling connected to knobs to identify the station. In the past, the closest thing to today's GPS maps and audio directions to the nearest restaurant was someone sitting in the passenger's seat providing you with directions. Aircraft cockpits have gone through the same kind of changes.

GPS devices today have electronic libraries of routes and navigation aids accurate to within a few feet or meters enabling pilots to identify air routes and navigational aids with simplicity, speed, and precision. The graphic displays now show aircraft position, heading, ground speed, and other variables that present pilots with an integrated view of aircraft location and performance. Before GPS, pilots relied on less precise radio navigational aids and had to integrate information from several instruments to determine location, direction, and groundspeed. Today, the basic flying instruments for attitude, altitude, airspeed, heading, turn rates, and outside air temperature have been integrated into large, colored, computer screen displays. The screens also can include navigational information, flight plan maps, positions, and even instrument approach and departure paths. This information gives pilots integrated, color-coded information in one or two side-by-side displays. For weather avoidance, these displays also can include weather radar digital maps superimposed on the aircraft's flight plan routine to help pilots (and passengers) avoid turbulence and other weather hazards. These systems can even fly the aircraft according to a programmed flight plan which can decrease fatigue on long flights and help pilots spend more time looking outside the cockpit for aircraft and bird collision avoidance at low altitudes.

All of these features are elements of flight control systems for the next generation of commercial and general aviation aircraft. They are called "glass cockpits" since the old individual gauges are replaced by one or two large computer (glass) screens. Most commercial aviation has various versions of these systems but in general aviation, the historical individual gauges for aircraft control and navigation are still being actively used.

There is no doubt that the glass cockpit's functionality can make navigation easier and increase "situational awareness," a nice term for "knowing where you are." But research on glass cockpit aircraft operations has produced a surprising finding. As of 2010, all of this great technology actually has increased fatality risk in general aviation.

The NTSB performed a study for single-engine piston airplanes, built in the interval between 2002 and 2006, with both conventional and glass cockpits [13]. The two cohort populations consisted of 2,848 conventional and 5,516 glass cockpit aircraft. The results, measured from 2002 to 2008, are clear and striking. The conventional aircraft group experienced 141 total accidents, of which 23 included fatalities. The glass cockpit aircraft had 125 accidents with 39 fatal. These are raw numbers and don't take into consideration the hours flown or the number of aircraft in each set. However, given that an accident occurs, the fraction of fatal accidents for conventional and glass cockpit-equipped aircraft are 16% and 31% respectively. In other words, if you are involved in a small aircraft accident, the chances of a fatality is almost twice as high—100% higher if the aircraft has a glass cockpit.

The real measure of exposure and aircraft safety is not the number of aircraft but the total time flown by each group. Incorporating flight hours makes glass cockpits look even more dangerous. The NTSB reports that for 2006–2007, the number of fatal accidents per 100,000 flight hours for conventional cockpit aircraft is 0.43, and for glass cockpits, 1.03. These results indicate that glass cockpit aircraft fatality risk is nearly 2.4 times higher than the old or low-tech information displays.

The NTSB report, issued in March of 2010, outlined several conclusions and recommendations. They concluded that while total aircraft accident rates were decreasing, there was an increase in fatal accident rates for the selected glass cockpit group:

Study analyses of aircraft accident and activity data showed a decrease in total accident rates but an increase in fatal accident rates for the selected group of glass cockpit aircraft when compared to similar conventionally equipped aircraft during the study period. Overall, study analyses did not show a significant improvement in safety for the glass cockpit study group. [14]

Glass cockpit issues are not just limited to general aviation. A 3-year study performed on 200 Boeing 757 pilots early in the application of new computerized flight management systems showed that over half of the pilots believed that computerization actually increased workloads [15]. Reprogramming route corrections, especially when maneuvering for landing or on departure, requires a crew member to focus his or her complete attention inside the cockpit, reducing the ability to see and avoid other aircraft. Also, the complete automation that enables the computers to fly the complete flight plan caused some interesting pilot reactions. Pilots of the 757 actually hand-flew the aircraft much more than required because they were afraid they would lose their flying skills if they allowed the computer to do all of the flying.

Now, if you are like me, after reading about all of the great benefits of glass cockpits, such as GPS pinpoint navigation, information integration, big color displays, weather updates in near real-time, and computer controlled flight plan navigation, you have to wonder how all of this good stuff can actually be more dangerous than conventional cockpit displays. The answer tragically lies with training. All of this information can be accessed but pilots need to be proficient in accessing various screen displays using the knobs and function keys, and, just as importantly, understand the equipment's limitations and warning signs for failures and malfunctions. So the answer is not that glass cockpits are inherently more dangerous, it is that the pilots flying these aircraft may not be adequately trained.

Commercial aviation today generally doesn't have these problems thanks to the rigorous and recurrent training pilots must routinely undergo to fly you around the sky. In the past, airline accidents and incidents have occurred in which the crew's mismanagement of automation has either caused or contributed to the event. This period of "learning" appears to be over, with the lessons learned incorporated into pilot training programs. Most passengers don't appreciate the rigor and intensity of commercial aviation training procedures. They are some of the best and most effective risk management practices applied today. General aviation, with its diversity of aircraft, pilots, and flight training programs, is in the process of catching up. Just like with any new technology application, it takes time and performance data to improve. Aviation is an unforgiving environment for experimentation but what's the alternative? The glass cockpit study illustrates that technology is good but unless the human using it adapts, changes behaviors, and learns, the technology can be ineffective and actually increase risk. This is a good lesson for all of us, not just for pilots.

Regardless of whether or not the aircraft has a glass or conventional type of instrument display, once the aircraft is up in the air and is being maneuvered, collision avoidance becomes a very real hazard in today's crowded skies. You might think that on clear days, pilots should look where they are going, and there shouldn't be a need for electronic collision avoidance systems. While visual references are important for collision avoidance, the high speeds of modern aircraft, coupled with the required navigational precision, make it difficult for pilots to spend a lot of time looking at the skies for other airplanes.

In the practical sense, flying requires an effective combination of navigation, aircraft control, and collision avoidance activities. To do the first two things in this list, pilots refer to the aircraft instruments. Since the pilot needs to look outside as much as possible, the placement of instruments is crucial. However, professional pilots clearly recognize that navigation and aircraft control are worthless if they do a poor job of collision avoidance. The congestion around major airports, combined with the speeds of closing aircraft, give pilots literally only seconds to take corrective action. Here's an example on how much time you need to avoid a head-on mid-air collision between two aircraft. It's taken from a U.S. Naval Aviation Safety Bulletin.

	Reaction time (seconds)
See object	0.1
Recognize aircraft	1.0
Become aware of collision	5.0
Decide on which way to turn	4.0
Muscular reaction time	0.4
Aircraft lag time	<u>2.0</u>
	Total 12.5 seconds

In practice, I believe the required response time is actually longer. Just think about yourself and how much time you need to recognize and react to hazards on the roads.

To reduce the growing collision avoidance hazard, the FAA has ruled that all transport category aircraft above a certain size must have on-board collision avoidance systems. The system, called the traffic alert and collision avoidance system or TCAS for short, continuously checks the location and flight path of all other transponder radio-equipped aircraft. If one aircraft is converging on another, an alarm is sounded with automatic instructions on which way to turn to best avoid collision. The pilot still has to turn the airplane. If the "intruder" also has TCAS, it too is warned and issued corresponding instructions on how best to avoid the pending collision. This equipment is also becoming common in private aircraft.

Even if pilots are looking outside their cockpits and correctly following procedures, there are still hazards that cannot be seen. One such threat to safety at low altitudes, namely takeoffs and landings, is wind shear produced by the localized, intense downdrafts from thunderstorms. These small, strong downdrafts are called microbursts. When an aircraft encounters a microburst area, it first experiences a head wind, followed by intense downdrafts and then a strong tailwind that can cause a rapid descent. The actual rate of descent is a function of the aircraft's speed, wind shear strength, and a host of other variables. This weather phenomenon has caused several accidents both in the takeoff and landing phases of flight. Wind shear is not new and probably has caused or contributed to other accidents in aviation's history. What has changed is the ability to measure and record the weather condition and its influence on aircraft performance. Technology has enabled engineers to record and archive flight data which aviation specialists have applied to learn wind shear's detailed characteristics. This information has, without a doubt, made flying safer. Pilot training now includes awareness of flight conditions where wind shear may occur and specific flight control procedures to follow when it is actively encountered.

Since early warning of wind shear is imperative for proactive correction, wind shear recognition systems have been installed at some airports and on some planes. Once the computer recognizes that the aircraft is entering such an area, a warning sounds and flight commands are given on the instruments to assist pilots in flying the aircraft through the downdraft and tailwind areas. At some airports, control towers monitor the wind shear systems and alert pilots when conditions are detected. This once-lethal weather condition is now being managed.

There is another expensive and sometimes lethal low-level airborne risk that only recently has gotten any attention: bird strikes. This aviation hazard has been a persistent, serious flying risk for a long, long time. However, it wasn't until the "miracle on the Hudson" event that the general public started to seriously consider birds an aviation safety risk.

On January 15, 2009, U.S. Airways Flight 1549 took off from LaGuardia Airport at 3:25 P.M. About two minutes later, while climbing through 3,200 feet, the aircraft encountered a flock of Canada geese, disabling both engines. Four minutes later the aircraft glided under control to a landing in the Hudson River, touching down at about 130 knots (150 mph). All of the 150 passengers and the crew of 5 survived [16].

Aircraft engines are tested for bird ingestion survivability but the density and size of birds that Flight 1549 encountered were beyond design limitations. The real point is that bird strikes (and near bird strikes) are very, very common and potentially deadly.

You might ask, "How can a bird bring down even a small airplane?"

The answer is kinetic energy (E). For example, a 2-lb bird striking an airplane traveling at 120 mph has about the same energy and a 1,000 weight being dropped from 1 foot. But birds can be even heavier than this. Sea gulls can weigh up to about 4 lbs (E = 2,000 lbs), vultures between 2 and 5 lbs (E = 5,000 lbs), and Canadian geese (like the type the collided with Flight 1549) between 7 and 20 lbs (E = 20,000 lbs!). As you can see, the speed of the aircraft more than makes up for the weight, turning a seemingly light, feathered animal into a deadly projectile.

For commercial and large general aviation, at least to 2010, bird strikes mainly translate to expensive engine and airframe repairs. But for small aircraft commonly flown in general aviation, bird strikes can (and have) produced fatal aircraft accidents.

Here is a top ten list of facts that shows how serious and widespread wildlife-aircraft strike risks really are from research performed by the FAA [17, 18].

1. The FAA has maintained a wildlife strike database since 1990.
2. The FAA wildlife strike database has recorded over 113,000 (civil and USAF) wildlife strikes between 1990 and 2010.

3. 92% of the bird strikes occur at or below 3,000 ft AGL (above ground level).

4. 72% of bird remains have not been identified to species level.

5. During the five years between 2004 and 2008, there was an average of 20 reported wildlife strikes per day.

6. Over 219 people have been killed worldwide as a result of bird strikes since 1988.

7. Bird and other wildlife strikes cost U.S. civil aviation over $600 million/year, measured from 1990 to 2009.

8. About 5,000 bird strikes were reported by the U.S. Air Force in 2008.

9. Over 9,000 bird and other wildlife strikes were reported for U.S. civil aircraft in 2009.

10. From 1990 to 2009, 415 different species of birds and 35 species of terrestrial mammals were involved in strikes with civil aircraft in the United States that were reported to the FAA.

Following the National Transportation Safety Board (NTSB) recommendation after the Airways Flight 1549, the FAA agreed to open its ongoing wildlife (not just birds) aircraft strike database to the general public. The database captures information about birds, deer, coyotes, and other types of animals that are involved in collisions with aircraft. It is an ongoing dynamic and systematic compilation of all reported wildlife strikes with civil aircraft and foreign carriers experiencing strikes in the United States. It contains over 113,400 strike reports from 1,599 U.S. and 561 foreign airports occurring from January 1990 to approximately the current time period. The FAA continues to collect, analyze, and support all database activities. There is some lag due to data reporting, validation, and updating, but you can download your own copy of the database if you want to do your own wildlife risk analysis [19].

Let's take a look at some interesting general facts that can be compiled from this resource. For example, Figure 11.2 shows the occurrence frequency for all wildlife strikes with aircraft as a function of the time of day in the United States from 1990 to 2010. There is a distinct peak in frequency between 7 and 8 A.M. that decreases until about 4 P.M. followed by a second peak between 5 and 6 P.M. and a relatively high strike likelihood through 11 P.M. These results are easy enough to understand on the surface but the underlying causal factors require a lot more research. For example, a wildlife–aircraft strike implies that both an animal and an airplane tried to occupy the same space at the same time. So to understand plots like Figure 11.2, both animal habits and aircraft flight patterns need to be studied. Answering these questions requires a lot more database analysis combined with bird and aircraft flying routes. Download the database yourself if you want to explore this issue. Now that we know when to expect strikes, the next question is: "Where?" Over the same period of time, Figure 11.3 shows the top 25 states for wildlife–aircraft strikes. This plot is noteworthy for no other reason than that it documents something that California and Texas have in common. They are the states with the

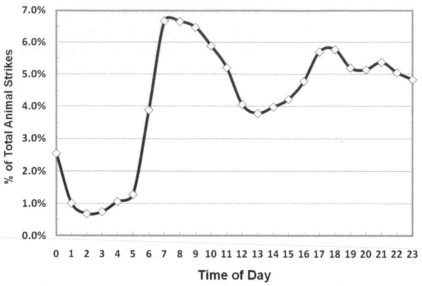

Figure 11.2   Aircraft–wildlife strike frequency by time of day.
Source: FAA Wildlife Strike Database, October 2010.

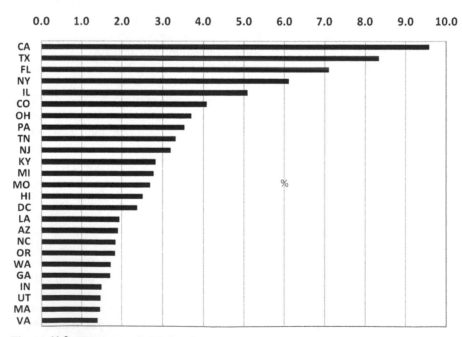

Figure 11.3   Percentage of total aircraft–wildlife strikes by state.

highest occurrence frequency, followed by Florida. The high ranking of these three states may be partially related to the fact that they are large states with the highest amount of coastline in the 48 contiguous states. In addition, this data is not normalized to include factors such as the number of flights per state.

The database is a valuable quantitative resource to identify high-risk locations and times for the purpose of developing risk mitigation actions. However, the FAA also believes that the database indicates only a small fraction, about 20%, of the total number of strikes actually occurring [20]. The ongoing support of the aviation community in providing timely and accurate strike reports is critical to making the database representative of the actual strike risk environment. Unlike wind shear, where the weather condition has warning signs, aircraft are traveling at speeds that make avoiding birds before a strike difficult, if not impossible. To help in wildlife avoidance, controllers today routinely inform pilots of local bird or wildlife activity to help them minimize the risk exposure through reduced approach speeds and increased cockpit vigilance. This risk is being managed, but to me, depending on the airport location, an aircraft–wildlife strike is a major aviation risk exposure for the first and last 3,000 feet of altitude of every flight.

But not all aviation risks are as obvious and direct as the topics mentioned so far. Some are subtle, hidden, and pervasive, with the proven potential for being just as lethal as any other type of hazard. Commercial airlines, charter operators, engine manufacturers, and certified repair shops fight the problem (or temptation) everyday. There are known and suspected aircraft accidents around the world and in the United States that were caused by the effects of this hazard. What is it? Counterfeit, bogus, or, to use the politically correct term, "suspected unapproved parts" (SUPs). The FAA defines the term as:

> **Suspected Unapproved Part (SUP).** *A part, component, or material that is suspected of not meeting the requirements of an "approved part." A part that, for any reason, a person believes is not approved. Reasons may include findings such as different finish, size, color, improper (or lack of) identification, incomplete or altered paperwork, or any other questionable indication.* [21]

The news media occasionally has covered this story with some very good investigative reports and individual accounts. They have clearly documented the hazard and that's the best they can do. We can't rely on an industry that requires some sensationalism to monitor a constant, even growing risk exposure. This is the job of the government and industry aviation community. I believe that counterfeit parts and SUPs, regardless of the media appeal, are among the top risk exposures in all of aviation today.

In 1995, when American Airlines Flight 965 crashed in a remote Colombian jungle, officials had not even removed all of the bodies before outlaw salvagers were removing parts of the engines and avionics, as well as other valuable components. The parts were quickly extracted by obviously trained people and then transported away from the crash site by helicopter. The parts found their way to a thriving counterfeit aircraft part black marketplace in Miami.

You might wonder how outlaws could steal such valuable equipment right in front of the authorities who are presiding over the wreckage. The remoteness, cover of night, and size of the wreckage pattern can make extraction of some equipment relatively easy if you have the right people and tools. I've personally observed some of their work. In the early 1980s, a twin-engine propeller aircraft crashed one wintry night on an instrument approach to Saranac Lake Airport in upstate New York. The search received national attention and went on for a considerable amount of time with no success. The family of the victims even used psychics to help locate the downed airplane. Long after hope was given up on saving the lives of the possibly injured pilot and passengers, a dog from the doomed airplane was found near the neighboring town of Lake Placid.

An air search finally found the aircraft near the top a remote mountain in the Adirondacks, many miles off the course the airplane should have taken. The site was only accessible by helicopter or by foot. Yet one night during the investigation, both engines disappeared. Apparently, an outlaw salvage crew came in under the cover of darkness and removed them. No one heard anything and no helicopters were observed in the area. The engines just disappeared from the crash site that night.

The bogus part threat was first documented in 1957 by the Flight Safety Foundation [22]. Since then, the market for counterfeit and substandard aircraft parts has been growing steadily. As airline fleets age, part replacements become frequent and today, the aircraft parts business is about a $45 billion industry worldwide. The black market industry thrives on buying substandard or salvaging used parts and then selling them to unsuspecting parts dealers, airlines, or aircraft owners at current market prices. One Colombian parts dealer told a Miami detective that she switched from drug smuggling to bogus parts dealing because the money was so much better. This is not too surprising. People can go to salvage yards and buy jet engine turbine blades for about $1 a piece and after minor refurbishment, sell them for $1,200 [23].

Why are aviation parts so special? You could take a bolt from your car and place it next to a similar one used in an airliner and probably see little, if any, difference. But there is a very big difference in operating environments. Your car weighs only about 3000 lbs, while the airliner could weigh as much as 200 tons. Your car operates at speeds from 0 to 80 mph; the airliner, at speeds from 0 to 650 mph. Your car is operated at temperatures between −20 and 100°F. On the airplane, each flight cruises where the air temperature is −50°F. It also operates at ground temperatures that can top 100°F. This temperature cycle can occur 50,000 times during the lifetime of the aircraft. Aviation parts are designed and constructed for these special operating conditions. They are special. This is why the standards and methods to accept or reject parts for restoration are so important. The combination of the two ingredients, aircraft fleets in need of parts and intense competition fueled by deregulation, sets the stage for a skyrocketing used part business.

Just how many parts are we talking about? As you might expect, airliners have a lot of them. For example, there are about six million parts on a Boeing 747 (B-747), including three million fasteners and 1.5 million rivets. The new Boeing 737 series has about 300,000 parts, the B-757 about 626,000, and the new B-777 has over three million parts, including 132,500 that are uniquely engineered. And by the way, these counts are just the for the Boeing airframes without the engines. As all of these airplanes age, sooner or later many of these parts will need to be replaced.

Not all refurbished aircraft parts are problems. When the work is performed and documented by a designated FAA repair facility using standard, approved methods, restored or overhauled parts are safe. The trouble comes when repair facilities cut corners to lower costs and sell their products under false documents. Of course, repair facilities for airlines, especially those which are not subject to FAA scrutiny, also are motivated to reduce costs.

Safety is always a concern—even for carriers who knowingly use the relatively cheap bogus parts. Usually these parts are small, detailed items and not major sections. Bolts, pins, switches, sensors, and various subassemblies are typical targets.

Identifying counterfeit or substandard parts can be a difficult task. First, there are so many different types of parts: bolts, bushings, brackets, landing gear, jet engine components, cockpit avionics, to name just a few, that it is difficult at the very least to find them. Sometimes the experienced eyes of mechanics can make the identification. At other times, it's the overall operation that gives law enforcement agents clues. An aircraft part cannot speak for itself. The documentation that accompanies each part is supposed to speak for it. The papers tell the buyer the part's life story, including the details of its manufacture and use. As you might expect, however, in a world where color copy machines have been used to make counterfeit money of reasonable quality, parts origin documentation is easily forged. While law enforcement around the world is maintaining a careful watch for us, it is a big world out there. Just to give you an indication of how widespread the problem may be, a government audit even found bogus parts in the fleet of 62 airplanes used by the FAA. This cancer is so widespread that a *Business Week* investigation found even that the fire extinguishers on Air Force One had been falsely certified by a repair station. Bogus parts have also been found in the president's helicopter and safety systems on Air Force Two [24].

Airplanes, especially the large jets used in commercial aviation today, are complex yet redundant systems. Because of the redundancies and backup systems that exist for flight critical systems, it's unlikely that a bogus part will, by itself, cause a crash. There has been only one commercial airline crash that has been related to bogus parts. A Convair 580 came apart at 22,000 feet over the North Sea in 1989, killing all 55 people on board. According to Norwegian officials, the crash was caused by counterfeit bolts that came loose, causing the tail section to violently separate from the rest of the airplane.

I mentioned that law enforcement is working on the counterfeit parts problem. However, this statement does not adequately portray the enormous effort being applied to keep our airliners and other airplanes free from these substandard parts. The government is not working alone. Aircraft and engine manufacturers are doing some very innovative, proactive things to thwart the bogus parts industry. The specifics are confidential.

Bogus part problems are not unique to commercial aviation. NASA has also expressed concern about counterfeit parts showing up in their inventories. Their concern is justified; counterfeit parts have been found on the shuttle vehicles. Some people have expressed doubt that even some bolts and fittings you or I might buy at our local hardware store may not meet their advertised tensile strengths. You can't tell if a bolt meets its labeled properties or not just by examining it. The bottom line is that bogus parts—parts advertised as one strength but actually made to a lower standard—may be more widespread than anyone can imagine.

In the aviation industry, an estimate made in 2002 by the chairman of the Italian aviation agency was that 1% of all aviation spares in the international market are SUPs. In 2008, the FAA stated that they believe that about 2% of the 26 million airline parts installed each year are counterfeit, or approximately 520,000 valves, gauges, flanges, gears, radar, and other parts [25].

Although a great deal of progress has been made, all parties still agree that more could be done. The chronology of most commercial air carrier accidents shows that crashes are usually the end result of a cascade of events. It makes discomforting but logical sense that as more counterfeit parts infiltrate our airliners' airframes and systems, the more likely it becomes that someday they will be a domino in an accident sequence. This is the risk exposure.

This is not what you want to hear if you fly on commercial planes. The main risk issues here are safety and affordability—the familiar cost–benefit equation. But now there are four related (or covariant) variables: safety, affordability, airline profitability, and government regulations. Whether it's the airline's Q statistic analysis, takeoff risk, landing risk, bird strike risk, wind shear risk, pilot training risk, airline security risk, or a plethora of other risks, flying can be stressful and to some, even fearful.

Risk perception plays a strong role for the paying passenger. For people who need to fly and, for whatever the reason, possess high anxiety or abject fear about the experience, the minutes or few hours confined within a flying tube can seem like torturous days. If you're one of these white knuckled (or worse) folks, there's something productive you can do about it. Some airlines and private companies actively help fearful and stress-laden flyers manage, minimize, or, in the best of cases, eliminate their flying discomfort. These organizations provide a valuable service in helping passengers modify their risk perception of commercial aviation through education. Here's a list of some of the companies around the world that help people manage their flying risk perception.

MySky Program: [26] This private company originated from Northwest Airlines' in-house curriculum. Licensed psychologists and airline pilots direct a rigorous 3-day course. Seminars are held in Northwest's two principal hub cities and usually include a short flight. With advance registration, you can save a little money and the prices are listed on their website.

The SOAR Program [27]: This program was founded in 1982 by Tom Bunn, an airline captain and licensed therapist. He has helped over of 7,000 individuals conquer fear of flying. It's the only program proven highly effective by impartial academic research.

Fear of Flying Clinic [28]: This company is based in San Mateo, California. It was founded in 1976 by two female pilots and they operate workshops throughout the year in northern California. Pricing information is offered through the website.

Online Self-Help Course [29]: This is a free online self-help course, but you can buy the DVD or download the video if you want a more extensive course.

Flying Without Fear (U.K.) [30]: Virgin Atlantic Airways conducts these seminars at regular intervals at airports all around Great Britain. Of particular note is that the syllabus offers a seminar adapted just for children, which lasts half a day and includes virtual flight technology.

Stress-Free Flying (France) [31]: Sponsored by Air France, this program is best described as an "anti-stress workshop." It includes an interview with a flight stress expert and a session in a flight simulator.

The Valk Foundation (The Netherlands) [32]: This program is supported in part by KLM Royal Dutch Airlines and Amsterdam's Schiphol Airport, with oversight provided by Leiden University. The curriculum includes visits to static aircraft, flight simulators, and real flights within Europe. If you are timid about admitting your flying fears in a group, private training activities are available.

Fearless Flyers Inc. (Australia) [33]: The course is sponsored with Qantas Airways and Airservice and has been in operation for more than 30 years. It's a nonprofit, volunteer organization run by women pilots. The curriculum includes relaxation and stress management techniques.

After reading this chapter you may feel like you need one of these courses but relax (with the 100+ other people around you), sit back (maybe 2 inches), stretch out (in your dreams!), and enjoy flying. But seriously, the aviation industry is constantly examining new designs, design improvements, new training tools, and better procedures to ensure on-time arrivals and departures for all passengers, at prices you can afford. And commercial aviation, in particular, is doing a lot more behind the scenes, out of public view, not just to maintain, but to improve your safety. If you really want to experience a high risk, albeit less dramatic way of traveling, get in your car and read Chapter 12.

DISCUSSION QUESTIONS

1. Compute the previous ten-year Q statistic for specific airlines. What are your conclusions?
2. Repeat the calculations done in Question 1, except use nonfatal injuries instead of fatalities. What are your conclusions?
3. For the previous ten years, or other suitable time period, compute three of the six safety statistics discussed in this chapter and interpret the results. Do the results from all three metrics together give you a good picture of aviation safety?
4. Using the FAA wildlife strike database, estimate the bird strike risk for your state or city. How does that compare to the national average? What other factors (environmental, human demographic trends, etc.) might relate to the differing rates of strikes between the states?
5. From recent historical research, identify a serious example of aircraft counterfeit parts.

5 June 20, 1972 (T −1.68 years)

In the summer of 1972, both Lockheed and Douglas were starting a major worldwide marketing tour with their new aircraft in hopes of securing sale orders. The last thing that Douglas needed at this time was any bad news on safety, maintenance, or reliability issues with the DC-10. After the June 12th incident, the FAA official in charge of the western region where the DC-10 was being built drafted an Airworthiness Directive or "AD" to send to all DC-10 owners regarding the hazards and mitigation actions that should be taken. An Airworthiness Directive is the highest priority notice the FAA can distribute and it requires actions be taken, usually by a certain date. Compliance with an AD has the legal power of federal law.

No AD was ever issued. The FAA officials who had drafted the AD were informed that the FAA Administrator, John Shafer, made a "gentleman's agreement" with Jackson McGowen, the President of Douglas, that corrective measures would be taken and no AD needed to be issued. Here is an excerpt of a memo written by the FAA Western Region Director whose department drafted the AD. The memo was written "to file" for the record rather than to anyone in particular.

> *Mr. McGowan [sic] said he had reviewed with the Administrator the facts developed which included the need to beef up the electrical wiring and related factors that had been developed by the Douglas Company working with the FAA.*
> *He indicated that Mr. Shaffer had expressed pleasure in the finding of reasonable corrective actions and had told Mr. McGowen that the corrective measures could be undertaken as a product of a "Gentleman's Agreement" thereby not requiring the issuance of an FAA Airworthiness Directive.*

6 June 27, 1972 (T −1.67 years)

It was clear, however, that not everyone was comfortable with this type of arrangement. Daniel Applegate, the Director of Product Engineering at Convair, the contractor who made the fuselage (and cargo door), wrote a memorandum detailing his concerns to his immediate supervisor. The subject title he used was "DC-10 Future Accident Liability." Here is the beginning, which summarizes his primary concerns:

The potential for long-term Convair liability on the DC-10 has caused me concern for several reasons.

1. The fundamental safety of the cargo door latching system has been progressively degraded since the program began in 1968.
2. The airplane demonstrated an inherent susceptibility to catastrophic failure when exposed to explosive decompression of the cargo compartment in the 1970 ground test.
3. Douglas has taken an increasingly "hard-line" with regards to the relative division of design responsibility between Douglas and Convair during change cost negotiations.
4. The growing "consumerism" environment indicates increasing Convair exposure to accident liability claims in the years ahead.

His memo goes on to discuss the technical problems in more detail and also speculates on Douglas' motivation to delay aggressively fixing the problems:

My only criticism of Douglas in this regard is that once this inherent weakness was determined by the July 1970 test failure, they did not take immediate steps to correct it. It seems to me inevitable that, in the twenty years ahead of us, DC-10 cargo doors will come open and I would expect this to usually result in the loss of the airplane. This fundamental failure mode has been discussed in the past and is being discussed again in the bowels of Douglas and Convair organizations. It appears however that Douglas is waiting and hoping for government direction or regulations in the hope of passing costs on to us or their consumers.

This memorandum was read by executives Convair but, due to contractual obligations, the contents of this prophetic document never reached the FAA.

7   July 3, 1972 (T −1.66 years)

Dan Applegate's superior, J. B. Hurt, program manager for the DC-10 Support Program, was the primary relationship contact between Douglas and Convair. He responded to the Applegate memo with comments that "look at the other side of the coin" as far as Convair is concerned. The points he makes basically say that Convair is in agreement with what Douglas is doing and has done in the past. However, in his planned recommendations to Douglas he is concerned about how they will aside blame and costs. In his last paragraph Mr. Hurt states:

We have an interesting legal and moral problem, and I feel that any direct conversation on this subject with Douglas should be based on the assumption that as a result, Convair may subsequently find itself in a position where it must assume all or a significant portion of the costs that are involved.

8   March 3, 1974, 2:20 PM (T = 0)

The 346 people who perished on Flight 981 do not represent the largest tragedy in aviation history, but that's not the point. In each tragedy, the direct loss of human life represents a minute part of the total loss. Here's just a very small portion of the societal loss from this tragedy.

The people on Turkish Airlines Flight 981 included a nine-month old baby, 30 college students, 21 engineers, 3 architects, 1 priest, 4 doctors, 2 nurses, 8 teachers, 3 accountants, and 3 lawyers. Fifteen entire families were wiped out, 80 women and 29 men lost their spouses, 53 children lost both of their parents, and 207 children lost one parent.

9   April 14, 1974 (T +42 days)

Seldom does a failure of this magnitude occur without precursor events, and sure enough, the FAA found in a post accident investigation that in the preceding 6 months, 1,000 Maintenance Reliability Reports (MRRs) were made for cargo doors failing to close properly. Douglas Aircraft Company had to be aware of these events since the airlines routinely submit MRRs to the aircraft manufacturers for these types of problems. To think the airlines were careless in filing these reports is unrealistic. The cargo door problems had a direct safety connection and of course, after paying $40 million (1970$) for an airplane, you would have a clear motivation to let the manufacturer know about its problems.

After reading this timeline, it is hard to believe that definitive corrective actions were not taken to solve the cargo door problems. From the FMEAs developed by Convair, the static tests of the first fuselage, and the cargo door problems documented during the 6 months prior to the Paris crash, the signs were obvious. There was also a reliable paper trail of actions or, rather, inactions, that identified the people and organizations involved in the apparent negligence. As you might expect, the legal actions that followed were enormous, complex, and expensive. After all, here you have an American built aircraft, owned and operated by a Turkish airline, crashing in France. Just these facts alone complicated the adjudication process.

All of this leads up to a single basic question in my mind. Here we have a well-documented series of events wherein several people could have taken actions to prevent 346 people from losing their lives, yet no one was held accountable. Today corporate executives either have been sent to, or are still in jail, for not what they have done, but for simply lying about their actions. I am not suggesting the executives are in any way being treated unfairly. Perhaps legal standards have changed, perhaps not. Regardless, the ethical dichomity between other so-called crimes and the causes of the Flight 981 tragedy are part of our history forever.

The events of this case study reveal how normal people, albeit in positions of authority, made decisions in the environment of uncertainty. In hindsight, we know they exercised extremely poor judgment. But at the time, the men considered the highly competitive business environment, the technical fixes that were being applied, and the risk of aircraft loss of control, and made a decision. There are always failure modes that can cause an aircraft to crash. Engineering never guarantees complete reliability. So safety becomes an economic decision as to how much should be spent to mitigate risk. In aviation, risk is especially significant. As Captain A. G. Lamplugh states in this chapter's beginning quotation: "…it is terribly unforgiving of any carelessness, incapacity or neglect."

Discussion Questions

1. Based on the ethical, risk acceptance, and operational standards we operate commercial aviation under today, do you think this scenario could happen again, albeit for another part of an aircraft?
2. Identify the people and organizations responsible for crash of Flight 981 and assign comparative negligence percentages.
3. Do you think criminal charges should have been made against some of the people? Defend your answer.

ENDNOTES

1 National Transportation Safety Board, NTSB Form 6120.1, Pilot/Operator Aircraft Accident/Incident Report, NTSB Form 6120.1 (rev. 10/2006).
2 Arnold Barnett, "Air Safety: End of the Golden Age?" Year 2000 Blackett Memorial Lecture, Royal Aeronautical Society, November 27, 2000.
3 Arnold Barnett, "Measure for Measure," Flight Safety Foundation, Aerosafetyworld, November 2007.
4 Ben Sherwood, "The Sky Is Falling: How Safe Is Flying in the U.S. vs. Abroad?" *The Huffington Post*, January 25, 2010, www.huffingtonpost.com (accessed October 22, 2010).
5 FAA Report: Air Carrier Standard Security Program, May 2001, pp. 75–76.
6 9-11 Commission Report, The Aviation Security System and the 9/11 Attacks, Staff Statement No. 3.
7 Mark G. Stewart and John Mueller, "A Risk and Cost-Benefit Assessment of United States Aviation Security Measures," *Journal of Transportation Security*, Vol. 1, 2008, pp. 143–159.

8 "Statistical Summary of Commercial Jet Airplane Accidents Worldwide Operations, 1959–2009," Aviation Safety, Boeing Commercial Airplanes http://www.boeing.com/news/techissues/pdf/statsum.pdf (accessed July 2010).

9 Naveeta Singh, "Pilot Error No. 1 Reason behind Plane Crashes, 50-Year Data shows," *Agency: DNA*, May 25, 2010. http://www.dnaindia.com/mumbai/report_pilot-error-no-1-reason-behind-plane-crashes-50-year-data-shows_1387357 (accessed October 27, 2010).

10 Earl L. Wiener and David C. Nagel, eds., *Human Factors in Aviation*. Academic Press, 1988.

11 J. W. Senders, "A Reanalysis of the Pilot Eye-Movement Data (Pilot Eye Movement Data by Recording Eye Fixations on Flight Instruments)," *IEEE Transactions on Human Factors in Electronics*, Vol. HFE-7, June 1966, pp. 103-106.

12 Peter Ladkin, "AA965 Cali Accident Report Near Buga, Colombia, Dec. 20, 1995," Peter Ladkin Universität Bielefeld Germany, November 6, 1996. Available at http://www.skybrary.aero/bookshelf/books/1056.pdf. Accessed July 25, 2011.

13 News Release, "NTSB Study Shows Introduction of 'Glass Cockpits' in General Aviation Airplanes Has Not Led to Expected Safety Improvements," SB-10-07, March 9, 2010.

14 National Transportation Safety Board, "Introduction of Glass Cockpit Avionics into Light Aircraft," Safety Study, NTSB/SS-01/10, March 2010.

15 David Hughes, "Pilot React to the Automated Cockpit Glass Cockpit Study Reveals Human Factors," *Aviation Week & Space Technology*, Vol. 131, No. 6, August 7, 1989, p. 32.

16 Matthew L. Wald and Al Baker, "1549 to Tower: 'We're Gonna End Up in the Hudson,'" *New York Times*, January 18, 2009, http://www.nytimes.com/2009/01/18/nyregion/18plane.html?_r=1 (accessed October 28, 2010).

17 FAA Wildlife Strike Database, http://wildlife-mitigation.tc.faa.gov/wildlife/default.aspx (accessed October 28, 2010.

18 Bird Strike Committee USA, http://www.birdstrike.org/ (accessed November 3, 2010).

19 FAA Wildlife Strike Database.

20 Alexis Madrigal, "Open Data: FAA Releases Bird Strike Database," *Wired Science*, April 24, 2009. http://www.wired.com/wiredscience/2009/04/open-data-faa-releases-bird-strike-database/ (accessed October 29, 2010).

21 U.S. Department of Transportation, Federal Aviation Administration, Advisory Circular, "Detecting and Reporting Suspected Unapproved Parts, AC No: 21-29C," July 22, 2008.

22 Joseph Chase, *The Problem of Bogus Parts*. New York: United States Flight Safety Foundation.

23 "Warning! Bogus Parts Have Turned Up in Commercial Jets. Where's the FAA?" *Business Week*, June 10, 1996.

24 Jack Hessburg, *Air Carrier MRO Handbook: Maintenance, Repair, and Overhaul*. New York: McGraw-Hill, 2001, p. 25.

25 ICC Counterfeiting Intelligence Bureau, *The International Anti-Counterfeiting Directory: A Key Information Resource to Combat the Global Plague of Counterfeiting*. London: ICC Counterfeiting Intelligence Bureau, 2009.

26 http://www.myflyingfear.com/index.php/recommends/MySky_Program/9/ (accessed October 15, 2010).

27 http://www.fearofflying.com/store/ (accessed October 14, 2010).

28 www.myflyingfear.com/index.php/recommends/Fear_of_Flying_Clinic_/10/ (accessed October 14, 2010).

29 http://www.fearofflyinghelp.com/ (accessed October 14, 2010).

30 http://www.myflyingfear.com/index.php/recommends/Flying_Without_Fear_/12/ (accessed October 16, 2010).

31 http://www.myflyingfear.com/index.php/recommends/_Stress_Free_Flying_/13/ (accessed October 16, 2010).

32 http://www.myflyingfear.com/index.php/recommends/The_VALK_Foundation_/14 (accessed October 16, 2010).

33 http://www.fearlessflyers.com.au/ (accessed November 11, 2010).

34 Paul Eddy, Elaine Potter, and Bruce Page, *Destination Disaster: From the Tri-Motor to the DC-10, the Risk of Flying*. Quadrangle/New York Times Book Co., 1976.

Chapter 12

Risk on the Roads

Better late in this world than early in the next one.

—Signpost on a rock above a winding road on the way to the Vale of Kashmir

Imagine for a moment that you developed a new, revolutionary technology that would dramatically improve, without precedent, the quality of life in today's society. It has the potential to transform lifestyles and world economies, and offers a plethora of other benefits. There is only one problem: just in the United States alone, it would kill over 33,000 people a year, year after year. Do you think you could find investors? How do you think the government would react to such an invention? Such a technology exists. In fact, just in the United States, the industry is responsible for at least 5.8 million jobs [1, 2], about 2% of the workforce, and brings in about $76 billion of revenue each year [3]. The 33,000-plus deaths are not the only visible downside. Each year, this industry's products cause about 2,200,000 injuries whose overall costs exceed $99 billion [4, 5]. These numbers give the technology a statistical value of life of about $3 million. The product? Your car.

As a society and as individuals, we accept motor vehicle risk. After all, it's familiar, and we feel that we have some degree of control over our destiny. For most people, work, pleasure, and shopping destinations are located beyond walking distance and driving or riding in motor vehicles is viewed as a necessary, manageable risk. It's one of the most common risks in countries with sufficient road, service, and fuel supply infrastructures.

Motor vehicles eliminate some travel risks. No longer are people required to walk or ride animals long distances to travel. Adverse weather, wild animals, robbers, and remote health issues are some of the hazards virtually eliminated by the modern automobile. But other hazards, like accidental collisions, take their place. Also, road conditions, driver behavior, car types, regulations, and regulatory enforcement vary by country so it's not surprising that motor vehicle death rates are variable as well. Figure 12.1 shows motor vehicle death rates for a selected set of countries, with Malaysia and Argentina being the most hazardous and Sweden and the United

20% Chance of Rain: Exploring the Concept of Risk, First Edition. Richard B. Jones.
© 2012 John Wiley & Sons, Inc. Published 2012 by John Wiley & Sons, Inc.

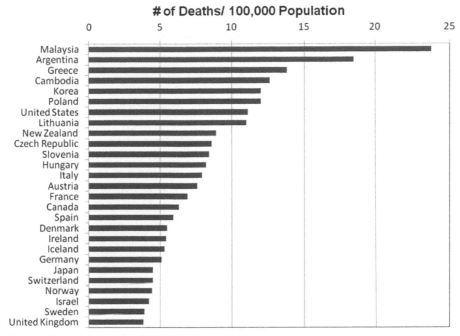

Figure 12.1   Motor vehicle death rate by country.

Kingdom being the safest [6]. The number of deaths per 100,000 population range from about 4 to 24, placing the United States just below the average value of 14 and above the median value of 8. This result is a little surprising since the United States generally has an excellent road infrastructure, access to high quality motor vehicles, and an active law enforcement system. With these advantages you might expect the United States to be a safer driving country, but there is more to consider here. For example, since the denominator is a country's population, the fraction of people who drive or ride in automobiles is another factor to consider in interpreting these results. In the United States, more people drive than ride trains or other modes of public transportation than in some of the other countries in Figure 12.1. Also, the time and distances traveled by car in each country can influence the results. Figure 12.1 represents one international safety benchmark but pertinent issues like the aforementioned need to be considered to properly interpret the results.

Looking at the United States alone, Table 12.1 shows the top ten safest and most hazardous states using the same statistic [7]. This table shows that driving in Mississippi is even more dangerous that driving in Malaysia. Even the safest state, Massachusetts, is more dangerous than several countries in Figure 12.1. The point of these tables is that even though the process is basically the same, that is, a human behind a steering wheel in a car, the safety results vary considerably. True, some of the factors I've mentioned in the preceding paragraph need to be included depending on the application of the statistics. Yet, there is no doubt that motor vehicle risk is

Table 12.1   State Driving Safety: Top Ten Lists (# deaths due to motor vehicle accidents per 100,000 population, 2007)

	Most dangerous			Safest	
1	Mississippi	31.6	Massachusetts		6.7
2	Montana	27.6	New York		7.4
3	Alabama	25.9	Rhode Island		7.6
4	Wyoming	25.3	New Jersey		8.2
5	South Carolina	24.2	Connecticut		8.7
6	Louisiana	24.0	District of Columbia		8.9
7	Arkansas	23.7	Washington		9.9
8	West Virginia	23.6	New Hampshire		10.3
9	Tennessee	21.0	Hawaii		10.3
10	Oklahoma	20.4	Illinois		10.6

a global risk management issue involving you, other drivers on the roads, car manufacturers, and government agencies.

Unless your occupation is extremely hazardous, driving is the highest risk activity of your day, no matter where you are. The danger, of course, is due to the sudden, uncontrolled stop from the collision of your car with other cars or objects. The car collision by itself is not what is dangerous. It's not even your collision with the dashboard or other objects inside or outside the car. Injuries are caused by your organs suddenly crushed or otherwise damaged from their sudden deceleration. It's this third collision where the damage is done.

A car's metal frame provides some protection from harm. In fact, auto designers perform thousands of crash tests to study how passenger compartments and the passengers are damaged in collisions from all sides and at various speeds. Actually, the entire motor vehicle structure is designed to provide some risk reduction. How much? It depends on the car.

As a rule, the bigger and heavier the car, the safer it is. When you're shopping for a car, do you relate its size to risk reduction or safety? For the most part, car size and price are related until you enter the high performance and specialty car markets. While most small cars are relatively inexpensive and have lower operating costs, their owners and occupants implicitly accept considerable risk. In effect, small car owners are trading the risk-reduction benefits of large cars for lower purchase prices and operating costs. In a sense, smaller car and truck owners are self-insuring. They have accepted risk, and in return they receive a premium of reduced purchase prices and operating costs.

While the insurance analogy is technically correct, I doubt that many look at buying a car this way. When you go to a car dealership and look at the sticker details, there are no standard survivability statistics for consumers to compare. You'll see miles per gallon data and a list of features with their costs. The fact that two and a half more people are killed in small cars than in large cars isn't listed on the sticker

[8]. Car dealers and the automakers don't want you thinking about this kind of risk when you're buying a car.

This raises the issue of how motor vehicle risk is regulated, perceived, and managed. Here is a clear instrument of death, and yet there are no standards or government warning labels to inform the consumer of the risks associated with using this product. Take cold medicines for a comparison, mentioned in Chapter 8. Look at the warning labels on the packages. They take up most of the back side of the box, and how many are killed from their use? On the other hand, we have the common, very familiar motor vehicle, which kills thousands at enormous societal costs each year, and the consumer receives no risk information during the sales process. Perhaps the reason is that our society and our government are still adolescent in their understanding of risk. Another explanation could be deduced from the phrase, "Follow the money."

What's the change in risk with car size? According to the Insurance Institute for Highway Safety, small and light cars do in fact have a higher death rate than large and heavier cars; about a factor of 2 for multivehicle accidents and about a factor of 3.3 for single car crashes [9]. This is not surprising when you analyze the physics involved in a collision between a light and heavy car. For example, a testing agency staged a head-on 40 mph crash (80 mph closing speed) between cars weighing 1,800 kg and 3,600 kg. The force of the collision caused the lighter car to reverse direction at 13 mph and the heavier car's forward speed was only reduced to 27 mph. The lighter car's velocity change was 53 mph, twice that of the heavier car [10].

Intuitively it's apparent that the heavier car is exerting more force, but how much more, and what are the forces involved in car collisions? No two collisions are the same but the following example will give you an appreciation of the high force magnitudes involved in car collisions. First, some basic physics of motion: Force (F) is defined as the product of a body's mass (m) and its acceleration (a). If a body, like a car, has a force applied to it and is moved a distance, s, then the work (W) done on the car by the force is defined as $F \times s$. In summary:

$$\text{Force } (F) = m \times a \tag{12.1}$$
$$\text{Work } (W) = F \times s \tag{12.2}$$

The kinetic energy (E) involves a car's mass and velocity (v):

$$E = \tfrac{1}{2}\, mv^2 \tag{12.3}$$

The energy equation shows that doubling a car's speed increases its kinetic energy by a factor of 4. A more subtle example is that increasing your speed from 30 mph to 42 mph also corresponds to a doubling of the car's energy.

Work and energy are related by a theorem that states if an external force acts upon a rigid object, causing its kinetic energy to change from E_1 to E_2, then the work done on the object is given by:

$$W = F * d = \Delta E = \frac{1}{2} m \left(v_2^2 - v_1^2 \right)$$

Solving for F:

$$F = \frac{1}{2d} m \left(v_2^2 - v_1^2 \right) \tag{12.4}$$

Now we have the math required to compute the forces involved in collisions. Let's use the two cars that the testing agency did for the head-on collision experiment. Suppose two cars, weighing 1,800 kg and 3,600 kg (3,969 lb and 7,938 lb) respectively, traveling at 30 mph (44 ft/sec), hit an immovable object such as a large rock or a tree. And further, suppose the cars come to a complete stop within a distance of 1 foot. We apply Equation 12.4 and substitute the values stated in the previous sentence, with F_L and F_H representing the lighter and heavier car forces.

$$F_L = \frac{1}{2} \frac{m}{d} \left(v_2^2 - v_1^2 \right) = \frac{1}{2} \frac{3969 lb / 32 \text{ ft} / \text{sec}^2}{1 \text{ ft}} \left((44 \text{ ft} / \text{sec})^2 - 0 \right) \tag{12.5a}$$

$$F_H = \frac{1}{2} \frac{m}{d} \left(v_2^2 - v_1^2 \right) = \frac{1}{2} \frac{7938 \, lb / 32 \text{ ft} / \text{sec}^2}{1 \text{ ft}} \left((44 \text{ ft} / \text{sec})^2 - 0 \right) \tag{12.5b}$$

English units are used in Equation 12.4 to compute force in terms of pounds. The term 32 ft/sec^2 is the acceleration due to gravity, also known as "1 g" when used in reference to forces on the body. For example, you sometimes hear of fighter pilots "pulling 9 gs" in combat. This refers to forces 9 times their body weight.

Equations 12.5a and b compute forces that are so large that you might want to check my math:

$$F_L = 120{,}062 \text{ lb} \approx 60 \text{ tons}$$
$$F_H = 240{,}124 \text{ lb} \approx 120 \text{ tons}$$

From the size of these forces, it's not surprising that collisions cause so much physical damage and injuries. A critical variable in these force computations is the distance required to come to a complete stop: the term d in Equations 12.4 and 12.5. If the stopping distance is increased to 2 feet, the forces are reduced by a factor of 2.

Car manufacturers now make their vehicles of all sizes safer by designing a structure that distributes the collision forces over the entire frame. This allows the structure to absorb the impact forces, essentially shrinking it like an accordion during a collision. The net result is the stopping distance is increased reducing the impact forces.

Vehicle structural designs are hidden so you can't see their benefits but the one benefit you can see is the car's size and relative weight compared to other cars. Increasing weight corresponds to increasing m in Equation 4 and the car with the larger weight transfers force to the lighter car. When it comes to car collisions it is better (and safer) to be the "transferor" than the "transferee." A clearer picture of your risk as a function of the car you drive is presented in Figure 12.2 [11]. Driver

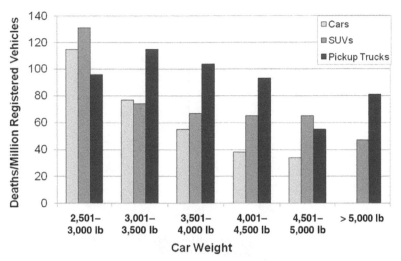

Figure 12.2   Effect of car weight on driver death rates, 2001–2004 models during 2002–2005.

Table 12.2   Car Size Classification Definitions

Class	Passenger and cargo volume (ft^3)
Minicompact	<85
Subcompact	85–99
Compact	100–109
Midsize	110–119
Large	>120

deaths are normalized by the number of registered cars, making the results more pertinent to the assessment of your death risk as a function of the car you drive. The lightest cars have a death risk 3.4 times that of the heaviest cars. The lightest SUVs have a death risk 2.8 times that of the heaviest SUVs. You can also see from this plot that even small versus light pickup trucks share similar safety characteristics. These results, again, show that small cars are dangerous. Over the past decades there have been several studies that have come to the same conclusion [12, 13]. And "small" does mean having less weight than "large" cars. Automakers try to keep the weight down as a function of car size by using more plastics, aluminum, and titanium, but in practice it hasn't worked. The materials cost too much.

Car size is also measured by the car's roominess or volumetric capacity as listed in Table 12.2 [14]. You may not routinely use this type of language but it will be necessary for talking with any rental car agent. The driver death risk, measured in the same terms, is shown in Figure 12.3. For cars, the safety risk of "large" and "very large" cars is about the same as the result measured in terms of gross weight. Yet there are some interesting additional safety characteristics that are seen here for

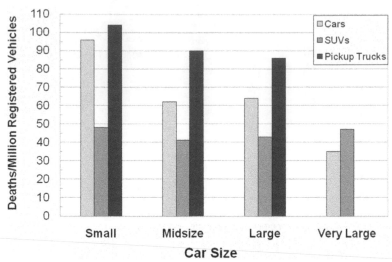

Figure 12.3 Effect of car size on driver death rates, vehicle age: 1–3 years, 2007.

midsize cars. They have about the same death risk as large cars. This result suggests that you can enjoy the safety benefits of "large" cars without the additional operating costs. Also notice that for SUVs, the death risk is about the same across all car sizes. Additional factors need to be studied to explain why the apparent differences in weight are not very helpful in mitigating the risk. But one possible explanation is that the deaths in these types of vehicles may be due to rollovers rather than from blunt force trauma usually associated with collisions. The extremely high death risk values for trucks of all sizes may be due to the combined rollover and collision force effects. However, regardless of the causality explanations, pickup trucks represent the highest death risk followed by small, midsize and large cars. There is no doubt that size (and weight) matters when it pertains to motor vehicle safety.

It is also interesting and particularly noteworthy that car safety ratings are published for cars by category, but they are not comparable. For example, a small car and a large car can both have good crash test ratings but that does not mean the cars are equally safe. A "good" small car rating means that it has good crash test performance relative only to other cars in its same size and weight category—not relative to large cars. It is entirely (and most likely) probable that a poor "large" car rating would be safer than a "good" small car rating. This fact is not a secret and you can look up actual driver death rates for many different car models in the online archives of the Insurance Institute of Highway Safety (www.iihs.org). Check out the type of car you drive. To me this type of information is just as important, if not more important, than the mpg rating on the sales sticker.

Since so many people are killed and injured every year from auto accidents you might ask why absolute (compared to relative) crash test data and driver death risk information is not included on the pricing stickers along with the city and highway miles per gallon for new cars. Sure, new models with enhanced safety features need time to develop statistics, but there are additional factors that are motivating the automakers to sell more small, fuel efficient cars. The answer involves reducing the nation's dependency on foreign oil imports.

Let's go back to 1973 to understand how events caused the automakers to change their manufacturing strategy and begin making cars lighter and smaller to achieve better gas mileage.

The United States' support of Israel during the October 1973 Arab–Israeli War angered many Arab nations. As a result, the Arab members of the Organization of Petroleum Exporting Countries (OPEC) formed a very effective oil embargo against the United States. Gasoline rationing and long lines at service stations were commonplace and the price of oil skyrocketed, leading to increased costs for all countries that were supplied by OPEC. The high prices and reduced supply further added to the threat of a global recession if countries' oil reserves were depleted before the embargo ended [15]. The United States negotiated an end to the embargo and life returned to normal. However, the demonstration of OPEC's effective world power caused President Nixon to develop a plan to reduce the country's dependence on foreign oil. He called the initiative "Project Independence" [16]. It included several energy conservation and alternative energy initiatives such as reducing the national speed limit to 55 mph, converting oil power plants to coal plants, and funding more mass transit. Subsequently in 1975, Congress passed the "Energy Policy Conservation Act," which included a segment on miles per gallon targets for all new cars. In 1975, cars were getting under 15 mpg and the first stage targets were 18 mpg for cars and 17.2 mpg for passenger trucks, to be achieved by 1985 [17]. The government presented automakers with a formula that is still used today to measure their cars' fuel efficiency performance. The formula, called Corporate Average Fuel Economy (CAFE), is computed as a weighted average over a manufacturer's fleet of cars. Here is how it works. Suppose an automaker produces four models of cars, as shown in Table 12.3. The best way to show the CAFE formula is to apply it to the data in Table 12.3 and compute this automaker's overall fuel economy performance. For the statistically-oriented readers, you will notice the application of the harmonic mean.

$$\frac{200,000 + 80,000 + 60,000 + 20,000}{\dfrac{200,000}{40} + \dfrac{80,000}{27} + \dfrac{60,000}{20} + \dfrac{20,000}{9}} = 27.3 \text{ mpg}$$

Table 12.3   Fleet Models and Sales Volume

Model name	MPG	Production volume
Electron	40	200,000
Commuter	27	80,000
Grand Puba	20	60,000
Rocket	9	20,000

So selling 200,000 highly fuel efficient (and small) "Electrons" along with the slightly larger (and heavier) "Commuter" cars enables the automaker to also sell the low mileage large and muscle cars where weight and horsepower are the drivers' benefits.

This example shows that automakers, responding to government pressure to make fuel efficient cars, need to make, and therefore sell, many small cars to offset the poor mileage of their larger (and safer) car models. The benefit is lower fuel consumption for the country. But as long as there are cars on the roads with large differences in vehicle weights, smaller car drivers' operating savings is offset by an increase in death risk.

The 2010 CAFE standards are 27.5 mpg for cars and 22.2 mpg for light trucks, and there is speculation of target increases in the future. However, if the past is any indication of the future, if you look at the actual targets by year over the 1980s and 1990s, you will see that the government has not been aggressive in changing the fuel economy targets.

Even though the standard methodology encourages automakers to make small cars, not all manufacturers comply and, as a result, these manufacturers pay the associated fines. For example, Table 12.4 summarizes some of the more notable auto manufacturers and the total CAFE fines paid to the U.S. treasury since the inception of the fuel economy law.

The total fine amount paid from all automakers between 1983 and 2007 as of January, 2009, was $772,850,459 [18]. To these auto manufacturers, paying the fines is simply a cost of doing business and no doubt the penalty "expenses" were passed on to car buyers. However, times are changing; Mercedes-Benz, the historical leader in mileage under performance, expected to pay no fines in 2010 thanks to its new hybrid cars. Other manufacturers in the list are also seeing the benefits of hybrid car development [19]. The message is apparently getting through to even the high-end manufacturers. Fuel economy is more than a fad. Buyers want more miles per gallon, even for high-performance and even for large (and heavier) cars.

Table 12.4   Total CAFE Fines (1983–2007)

Automaker	Total CAFE fines
Mercedes-Benz of North America, Inc.	$255,433,521
BMW of North America	$244,572,552
Porsche Cars North America, Inc.	$58,274,621
Volvo Cars of North America	$56,421,280
DaimlerChrysler Corp.	$55,690,756
Jaguar Cars, Inc.	$40,069,650
Ferrari Maserati North America, Inc.	$5,077,248
Maserati Automobiles of America, Inc.	$2,737,735
Ferrari North America, Inc.	$1,968,318
Total	**$720,245,681**

Cars are being designed and made safer each new model-year in all size-weight categories. Yet the safety measures do not negate the laws of conservation of momentum and energy. One thing is clear when it comes to auto safety: the more your car resembles a tank, the safer it probably is. Even with the reduced gas mileage, the risk reduction advantages of lots of metal around you may make increases in gas costs a bargain. Are the lives of your loved ones worth more than a few miles per gallon in fuel costs?

Here is a simple example that compares the fuel savings to the safety. Suppose we look at two vehicles; Car A, which gets 30 mpg, and Car B, which gets a gas-guzzling 15 mpg. You can use your imagination or experience to identify makes and models that fit these profiles. The 30 mpg vehicle is smaller and much lighter than the 15 mpg "tank" car. Now to see what the annual fuel costs are, let's assume you drive 20,000 miles each year with a gas price of $3.00 per gallon:

$$\text{Car A:} \quad \frac{20,000 \text{ miles/yr}}{30 \text{ miles/gal}} \times 3.00 \text{ \$/gal} = \$2,000$$

$$\text{Car B:} \quad \frac{20,000 \text{ miles/yr}}{15 \text{ miles/gal}} \times 3.00 \text{ \$/gal} = \$4,000$$

$$\text{Difference} = \$2,000$$

You save about $2,000 a year with Car A on gas or, over a 5-year car life, about $10,000. When you compare this amount to the typical medical costs associated with car accidents, the cost of fuel becomes trivial. To place the Car B "costs" or Car A "savings" in some perspective, according to the National Safety Council, the average economic cost (including medical expenses) per nonfatal disabling car accident injury in 2008 was $63,500 [20].

If we take the ratio of $63,500 to $2,000, we see that the average annual injury costs are about 32 times the fuel savings from driving smaller, more injury prone vehicles. Sure, these calculations include several assumptions, but nevertheless, the savings obtained from gas mileage differences can go away very quickly if an accident occurs.

The lesson from this exercise is that the large print miles per gallon numbers on the car stickers provides you with only a part of the information you really need to decide which car to buy. It's hard to argue with the math, yet practically, from the proliferation of small cars on the road, many people either believe the decrease in operating costs offset the increase in death risk or they simply are unaware of the inherent risks associated with light cars. The automakers and the government are encouraging the purchase of small cars so I doubt you will see any absolute safety information on these stickers. Yet from a risk management perspective, here is the question: Are the dollars saved in gas consumption really a benefit or a cost? What good is an extra $2,000 a year if you're not around to spend it? (Author's note: I do have an environmental conscience. My well-maintained, full-sized car averages about 36 mpg.)

Table 12.5   Contributing Factors to U.S. Motor Vehicle
Deaths: 2008

Contributing factor	Occurrence (%)
Occupant protection	33
Speeding	31
Alcohol	31
Distractions	17
Young drivers	16
Large trucks	11

Car size is related to motor vehicle fatalities, but what really causes fatal accidents to occur? The National Safety Council's analysis shows there are six contributing factors. They are listing in Table 12.5 along with the occurrence percentages. The percentages do not add up to 100 since more than one contributing factor can be involved in an accident. Let's take look at each of these factors.

OCCUPANT PROTECTION

Some of the major risk exposures mitigated by seat belts and child seat restraints are the tremendous forces that act on the human body during collisions. Clearly the speed at which the collision occurs plays a major role in determining the forces, but there is more to the equation than just speed. Here is an illustration that shows how you and your children's risk increases with speed and other factors of collision.

We start with Equation 12.4 by expanding F as the product of mass (m) and acceleration (a) and dividing by the acceleration due to gravity (g).

$$F = \frac{1}{2d}\frac{m}{g}\left(v_2^2 - v_1^2\right) = m\frac{a}{g} \tag{12.5}$$

The ratio a/g measures acceleration and force relative to the size of the force of gravity. Simplifying gives us the equation which translates speed changes to g force

$$\left(\frac{a}{g}\right) = \frac{\left(v_2^2 - v_1^2\right)}{2dg} \tag{12.6}$$

Let's apply Equation 12.6 to the collision scenarios we discussed in reference to the amount of impact force created during collisions. A car traveling at 30 mph (44 ft/sec) strikes an immovable object and comes to a complete stop ($v_2 = 0$) in a distance of 1 foot. Substituting these values and $g = 32$ ft/sec^2 into the right-hand side show that the stopping acceleration is $-30g$. The negative sign signifies deceleration. If the stopping distance is increased to 2 feet, the g force is reduced to $-15g$.

The point here is that your body's internal organs may not withstand acceleration forces of these magnitudes. I use the word "may" because the human tolerance to high acceleration forces is dependent on several variables [21]. Here are some of them:

- Magnitude: number of g forces.
- Direction: Most tolerable: forward direction ("eyes in").
    Least tolerable: z direction ("eyes down").
- Duration: number of milliseconds of sustained acceleration.
- Rate of onset: how rapidly the acceleration is applied.
- Position/restraint/support: g-load distribution over the body.

For these reasons and others, there is no specific maximum g-load that the human body can survive and the many variables involved in auto collisions make any number practically meaningless anyway. But this analysis does show that we are defenseless to cope with such high forces. You and your passengers have a much better chance of crash survival being properly supported and restrained than the alternative of becoming an involuntary projectile inside (and maybe outside) of the car. The auto industry is constantly working on better vehicle structures to absorb high deceleration forces [22], but for their engineering to work, you must become a part of the structure instead of a projectile inside it. Seat belts keep you fixed to the structure, helping your body share the distributed deceleration forces with the surrounding structure.

In the case of children, the high g-forces can be even more dangerous. Child restraint seats distribute the high g-forces better over their bodies and also keep them secure so the high deceleration forces can be distributed around them through the car's frame. Preventing all accidents is out of your control but you can control how your passengers are restrained in your car. Proper restraints, whether booster seats, or, when applicable, seat belts, are excellent risk management decisions, and just plain good sense.

Knowledge of the risk reduction value of seat belts is probably the most commonplace safety fact about cars. The reduction in fatalities and injuries due to seat belts is indeed impressive. The National Safety Council reports [23] that wearing lap/shoulder belts reduces death risk by 45% for front seat occupants and serious injury risk by 50%. It's easy to ignore statistics like these, but there are other facts that are indisputable. In 2007, for example, 42% of the passenger vehicle occupants killed were unbelted.

Buckling a seat belt is clearly a simple and very effective risk reduction activity. Yet with over 30,000 motor vehicle deaths and $255 billion in costs each year, how serious are the state governments in enforcing a behavior that has indisputable societal and individual benefits? The record looks good on the surface. Most states have passed mandatory seat belt laws but when you look at the law details more closely, you see a tapestry sewn of politics and rhetoric. In most cases, the laws are not a deterrent. Tables 12.6A and B present the laws and fines by state [24]. Primary

Table 12.6A   U.S. Seat Belt Laws—Primary Enforcement

State	Who is covered (yrs.)	In what seat	Maximum fine, 1st offense
Alabama	≥15	Front	$25
Alaska	≥16	All	$15
Arkansas	≥15	Front	$25
California	≥16	All	$20
Connecticut	≥7	Front	$92
Delaware	≥16	All	$25
Washington, DC	≥16	All	$50
Florida	>6	Front	$30
	6–17	All	
Georgia	6–17	All	$15
	≥18	Front	
Hawaii	8–17	All	$92
	≥18	Front	(incl. admin.fees)
Illinois	≤18, if driver <19	All	$25 plus court
	≥16	Front	costs
Indiana	≥16	All	$25
Iowa	≥18	Front	$25
Kansas	14–17	All	$60
	≥18	Front	$10 ($5 until
		(other positions are secondary enforcement)	7/1/11)
Kentucky	≤6 and >50 in	All	$25
	≥7		
Louisiana	≥13	All	$25
Maine	≥18	All	$50
Maryland	≥16	Front outboard	$25
Michigan	≥16	Front	$25
Minnesota	Anyone not covered by child pass. safety law	All	$25
Mississippi	≥7	Front	$25
New Jersey	≤7 and >80 lb	All	$46
	≥8	(secondary: rear	
	(*effective 1/20/11*)	seats)	
New Mexico	≥18	All	$25
New York	<16	Rear	$50
	All	Front	
North Carolina	≥16	All	$25.50 + court costs
		(secondary: rear seats)	$100.50 (front seat)

Table 12.6A *(Continued)*

State	Who is covered (yrs.)	In what seat	Maximum fine, 1st offense
Oklahoma	≥13	Front	$20
Oregon	≥8 or 4'9"	All	$97
South Carolina	≥6	All	$25
Tennessee	≥16	Front	$50
Texas	≤7 (and ≥57 in)-17	All	$200
	≥17		$50
Washington	>8 or >4'9"	All	$124
Wisconsin	>8	All	$10

Table 12.6B U.S. Seat Belt Laws—Secondary Enforcement

State	Who is covered (yrs.)	In what seat	Maximum fine, 1st offense
Arizona	5–15	All	$10
	≥5	Front	
Colorado	≥18	Front	$72
Idaho	≥7	All	$10
			(drivers <18 pay $51.50, including court costs)
Massachusetts	≥13	All	$25
Missouri	8–15 (law is primary)	All	$25
	≥16	Front	$10 (drivers and and passengers)
Montana	≥6	All	$20
Nebraska	≥18	Front	$25
Nevada	≥6	All	$25
North Dakota	≥18	Front	$20
Ohio	8–14	All	$30 drivers; $20 passengers
	≥15	Front	
Pennsylvania	8–14	All	$10
	≥18	Front	
Rhode Island	≥18 (primary for <18)	All	$75
South Dakota	≥18	Front	$25
Utah	≥16 (primary for <19)	All	$45
Vermont	≥18 (primary for <18)	All	$25
Virginia	≥18	Front	$25
West Virginia	8–17	All	$25
	≥18	Front	
Wyoming	≥9	All	$25 drivers; $10 passengers

enforcement allows police to ticket a driver for not wearing a seat belt, without any other traffic offense taking place. Secondary enforcement means that police may issue a ticket for not wearing a seat belt only when there is another citable traffic infraction. After reviewing this information, it's apparent most states have seatbelt laws, but the "fines" fall far short of the penalty that is justified from the carnage statistics.

In 1997, the average U.S. percentage of seat belt wearers was 64%. Actual seat belt use for 2009 was measured at 84%. The understanding that seat belts increase safety is now a generally accepted fact. Can you remember a time when, if you got into someone else's car and buckled the seat belt, they would say, "What's wrong? Don't you trust my driving?"

It's apparent, independent of the lackluster enforcement penalties, that most car occupants today understand the device's risk reduction value. Yet for the remaining 16%, the National Highway Traffic Safety Administration (NHTSA) estimates [25] that if seat belt use was 90% in every state, more than 1,600 lives could be saved and 22,000 injuries prevented.

Why do people choose not to use these simple, passive devices? This question has many answers but let's start with the only state that is missing from Tables 12.6A and B. New Hampshire's citizens have always taken pride in their laissez-faire approach to government. This would explain its decision not to order its adult constituents to buckle up in their cars. The state does have seat belt laws for minors, but New Hampshire's law makers believe adults are old enough to make their own decisions. The state, however, does recommend the use of seat belts for all occupants. But it's not the law.

New Hampshire's "Live Free or Die" citizens are practicing a sentiment common to many people who don't like government telling them what to do. There is some validity to their fear of government intrusion into personal freedoms. After all, the government doesn't control your diet or how much you exercise and that can have an influence on your health, so why should government tell you to wear a seat belt?

New Hampshire does, however, have laws for some truck occupants. On January 1, 1997, a law went into effect banning dogs from traveling in the back of pickups unrestrained. In this state, truck sales outnumber cars sales 3 to 1 and having your dog in the truck bed is a long-standing tradition. However, dogs falling out of pickup trucks are a common accident throughout the country. Today, at least eight states, in addition to New Hampshire, have laws requiring owners to secure their dogs or other animals that ride in open areas. The fines range from $50 to $200. Just for your information, the laws exclude hunters and farmers. Also, if you are in riding in the truck bed, you're not required to be secured, just your dog is.

Moving beyond dogs and animals, another group that doesn't have sufficient life experience to understand motor vehicle risk is children. The good news is that child restraints have an even more impressive risk-reduction benefit than do seat belts for adults. Child seats reduce the risk of fatal injury by 71% for infants less than one year old. For children between the ages of 1 and 4, the special seats reduce death risk by 54% [26]. In 1983, child restraint usage was at 41% and in 2009 the usage had grown to an impressive 86%.

One of the major questions with child seats is their compatibility with the seat belt system of the car. The problem is that several of the more than 50 varieties of child seats currently on the market can't be safely or easily secured to the approximately 30 different seat belt systems. There are websites today where you can go to determine the seat brand, name, and type that is compatible with your car. The usage rate for children younger than 4 years old is high but as children become older and larger, parents allow children to use adult seat belts, thinking they have outgrown booster seats. The appropriate restraint system for 4- to 7-year-old children is either a forward-facing safety seat or a booster seat, depending on the child's height and weight. A 2009 national survey of booster seats use found that 45% of children ages 4 to 7 in the United States were not being properly protected. The studied observed 32% of their sample in seat belts and 13% completely unrestrained [27].

Tragic accident stories of children killed because the child seat was improperly secured are all too prevalent. While the toll from this risk exposure is not precisely known, an estimated 600 children under five years of age die in car crashes each year because their child seat was improperly connected to seat belt systems

According to the U.S. Transportation Secretary,

NHTSA's 2009 child fatality data found that, last year [2009], motor vehicle crashes were the leading cause of death for young people ages 3 to 14. In 2009, an average of four children age 14 and younger were killed and 490 were injured every day. [28]

All states have child restraint laws but there a great deal more that can be done for children.

Properly installing a child restraint system or a booster seat in your car is not a simple task. All cars are different and there are a several companies that manufacture these products at prices ranging from $20 (yes $20) to $620. This is a perfect case for the development of new standards. One industry must interface with another. The companies that make light bulbs don't make the sockets, so standard light bulb sizes were developed and accepted by all interested parties. Our civilization would be really chaotic if we didn't have these simple agreements among manufacturers that require that everyone conform to certain manufacturing specifications. The good news is that today, several countries [29–31] including the United States have adopted product standards for the $2 billion global child restraint seat industry. Now people just have to use them.

Another part of occupant protection is the passive supplemental restraining system (SRS) in your car. The term is a long way of saying "air bags." The standard versions are stored in the steering wheel or dashboard and are activated in frontal collisions, typically at speeds in excess of 12 mph. The objective of the bag is to cushion your contact with the steering wheel, dashboard, or other car structure by distributing your body's collision forces across the air bag surface. In order for this to occur, the air bag must be activated after the collision and then inflated before you get there. To be sure this happens, the bags expand at nearly 200 mph.

In 1984, the Secretary of Transportation, Elizabeth Dole, announced that all newly produced cars sold in the United States would be equipped with air bags by September 1989, unless two-thirds of the country's population was covered by state mandatory seat belt laws (*Federal Register*, 1984, 28962). The 1984 rule that governed air bags required that they inflate with enough force to save a 169-lb man who wasn't wearing a seat belt [32]. When the rule was written, few people wore seat belts.

Chrysler became the first U.S. company to offer air bag restraint systems as standard equipment. In 1994, TRW began production of the first gas-inflated air bags that are now mandatory in all cars since 1998. By the NHTSA's own estimates, air bags have saved nearly 30,000 lives since 1998. That's more than 2,300 people a year. NHTSA calculates that using a seat belt and having an air bag reduces death risk by 61% [33].

The early years of air bag applications and perceived benefits were tainted with design and usage issues. Auto deaths in minor collisions, where the occupants would have otherwise survived, were determined to be caused by the air bag's violent deployment. The small size and low weight of the victims were identified as the primary causal factors. Mechanics were regularly asked, against the wishes of manu-facturers, to disable passenger-side bags. This led to the introduction of controversial on/off switches—which are still permitted until 2012. Figure 12.4 shows NHTSA estimates of the lives saved by these safety devices [34].

Since September 1, 2006, NHTSA has required that new vehicles come with advanced frontal air bag systems. Advanced front passenger air bag systems can also—based on your weight and position—turn the air bag off. These new models are configured to render moot the public outcry over the first-generation air bags—that is, the belief that the air bags crushed children and small adults. Today's air bags deploy with less force than the earlier versions with mixed results. Child deaths are down but the lower deployment forces appear to have actually decreased the life-saving benefits for belted drivers [35]. This research is continuing. In any case, today you can buy cars with up to 10 air bags, and some automakers even include additional air bags in their option lists [36].

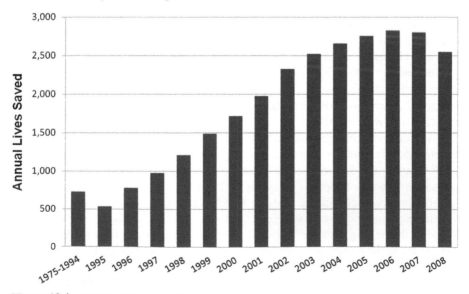

Figure 12.4   Estimated lives saved by air bags.

While reading this information about the lifesaving value of air bags in crashes you might have thought about why the technology isn't used in airplanes. Arguments for not including them on aircraft state the added weight or lower payloads would raise airline fares and that aircraft crash forces can far exceed survivable limits of air bag deployment. Yet the National Transportation and Safety Board adopted a study that concluded that at least in general aviation airplanes, occupants in accidents involving survivable forward impacts do benefit from air bags especially when used with lap or shoulder harness restraints. As of 2010, there are nearly 18,000 air bag-equipped seats in over 7,000 of the 224,000 general aircraft in the United States and about 30 aircraft manufacturers offer air bags as standard or optional equipment [37]. Air bags are not just for cars anymore.

SPEEDING

I want to start this discussion with the good news. While motor vehicle deaths have varied up and down over the past 35 years in the United States, overall, motor vehicle deaths are trending lower. In 2009, the actual number of motor vehicle deaths was at its lowest since about 1950. And on both population and motor vehicle-mile bases, death rates are lower today than when the statistics were first recorded in 1925. But there are additional important facts that need to be mentioned.

Equations 12.4 and 12.6 share two key risk factors: speed (v) and stopping distance (d). In both equations, acceleration and force both increase with the square of speed. Stopping distance is also a mathematical variable but in practice, for low speed collisions, stopping distance is usually not a concern. It is mainly for high speed crashes that deceleration and stopping distance play a key role. This is one reason that some high speed collisions where the cars travel large distances in the collision event can be survivable. Speed control is voluntary, but usually, after a collision, drivers become involuntary passengers subjected to whatever deceleration forces occur due to the nature of the collision. What's the lesson in this physics exercise? Drivers control a major risk factor in producing fatal accidents: speed.

With this in mind, let's look at how our federal and state governments have responded to this primary risk factor. Until 1973, highway speed limits were set by the states. In 1974, due to national oil shortages and for safety considerations, the maximum speed limit was reduced by the U.S. Congress' Emergency Highway Conservation Act. This was another part of Project Independence that, along with the Corporate Average Fuel Economy (CAFE) mile per gallon targets for cars, sought to help reduce the country's dependence foreign oil imports. The expectation was that nearly 200,000 barrels/day of fuel would be saved by the speed reduction [38]. The act created a national maximum speed limit of 55 mph on all high-speed roads in the nation. This included 40,000 miles of interstates and about 500,000 miles of state highways. Did traffic fatalities go down because of this? Yes. In 1973 there were 54,052 traffic fatalities. In 1974, with the 55 mph speed limit in place, there were 45,196 traffic-related deaths. We cannot say that the speed reduction saved all of these lives. Due to the gas shortages during this time period, people were also driving less and conserving fuel use by a nationally imposed gas rationing scheme, and perhaps also by the really long lines at the gas stations. It is interesting to note, however, that this reduction in traffic deaths was the largest single year decline since World War II.

There is more evidence that the speed limit reduction has saved lives. In 1984, a National Academy of Sciences study concluded that the 55 mph speed limit law saved between 3,000 and 5,000 lives in its first year and between 2,000 and 4,000 each year the speed limit was in place [39]. There have been several laws generated since that time selectively removing the 55 mph cap. Finally, in 1995, with the oil crisis over and in recognition of the fact that very few people were obeying the speed limit anyway, Congress repealed the national maximum speed limit altogether.

The results have been impressive—in a tragic way. In reaction to the final repeal, here's how the states' speeds have changed: four states have kept 55 mph, 21 states have kept their previously raised limit of 65 mph, seven states raised the limit to 65 mph, nine states raised the limit to 70 mph, 10 states raised the limit to 75 mph, and one state briefly had no speed limit (only "reasonable and prudent"). The latter didn't last long. As far as increases in fatalities, the death data speaks for itself. According to the 1996 National Safety Council report [40] describing 1995 accident experience, 68% of all fatal accidents involved excessive or unsafe speeds—an increase of 5% from the previous year. For all accidents, speed is a factor in 75% of all crashes—an increase of 8% over the previous year. In 2010, excessive speed was still the primary factor in fatal accidents [41].

The risk reduction of lowering speed limits has also been measured in other countries [42] and the results are shown in Table 12.7A. These scientific studies

Table 12.7A Effects Lowering Speed Limits

Country	Change	Results
Sweden	110 kmph to 90 kmph (68 mph to 56 mph)	Speeds declined by 14 kmph Fatal crashes declined by 21%
Denmark	60 kmph to 50 kmph (37 mph to 31 mph)	Fatal crashes declined by 24% Injury crashes declined by 9%
U.K.	100 kmph to 80 kmph (62 mph to 50 mph)	Speeds declined by 4 kmph Crashes declined by 14%
Australia	110 kmph to 100 kmph (68 mph to 62 mph)	Injury crashes declined by 19%
Switzerland	130 kmph to 120 kmph (81 mph to 75 mph)	Speeds declined by 5 kmph Fatal crashes declined by 12%
Germany	60 kmph to 50 kmph (37 mph to 31 mph)	Crashes declined by 20%
Australia	5–20 kmph decreases (3–12 mph decreases)	No significant change (4% increase relative to sites not changed)
U.S. 22 states	5–20 mph decreases (8–32 kmph decreases)	No significant changes

Table 12.7B   Effects of Raising Speed Limits

Country	Change	Results
U.S.	55 mph to 65 mph (89 kmph to 105 kmph)	Fatal crashes increased by 21%
U.S.	55 mph to 65 mph (89 kmph to 105 kmph)	Fatal crashes increased by 22% Speeding increased by 48%
U.S. (40 states)	55 mph to 65 mph (89 kmph to 105 kmph)	Fatalities increased by 15% Decrease or no effect in 12 States
U.S. (Michigan)	55 mph to 65 mph (89 kmph to 105 kmph)	Fatal and injury crashes increased on rural freeways
U.S. (Ohio)	55 mph to 65 mph (89 kmph to 105 kmph)	Injury and property damage crashes but not fatal crashes
Australia	100 kmph to 110 kmph (62 mph to 68 mph)	Injury crashes increased by 25%
U.S. (40 states)	55 mph to 65 mph (89 kmph to 105 kmph)	Statewide fatality rates decreased 3–5% (Significant in 14 of 40 States)
U.S. (Iowa)	55 mph to 65 mph (89 kmph to 105 kmph)	Fatal crashes increased by 36%
U.S. (Michigan)	Various	No significant changes
Australia (Victoria)	5–20 kmph increases (3–12 mph increases)	Crashes increased overall by 8% 35% decline in zones raised from 60–80
U.S. 22 states	5–15 mph (8–24 kmph)	No significant changes

document the actual changes in death risk as a function of the speed reduction and the results reinforce the tremendous life-saving value of lowering speed limits. As some states and countries raised the posted speed limits for various reasons, scientific studies were also conducted to examine changes in fatality risk. Table 12.7B shows that in most cases, the increase in speed limits was associated with an increase in fatalities. Notice that not all studies measured an increase in death risk. This information has been used to support increasing speed limits. The decrease in travel time and the associated increase in commerce are nonstatistical realities. These facts, along with the statistical uncertainty in death risk, have been used effectively by politicians and others to either support increasing speed limits or support maintaining the current values.

Nevertheless, based on Tables 12.7A and B, the laws of physics, and the overwhelming documentation accumulated over the past 20 years, my conclusion is that speed does kill. The actual speed limits that countries and states impose show the diversity of risk management decisions balancing death risk and its associated costs with societal benefits.

Another way to assess the risk management decisions associated with speed is to examine the penalties associated with speeding-related fines. A common belief today is that the speed limit is the most broken law in the United States. I am not going to argue either way on this issue but there are some classic examples of speeding fines that illustrate how violating this law is treated around the world.

In Helsinki, Finland, a man received a speeding ticket for traveling 80 km/hr (50 mph) in a 40 km/hr (25 mph) zone. His fine was £116,000 (~$180,000). In Finland, traffic fines are not solely fixed by the actual speeding values. They include a component associated with the offender's income. The person was an heir to a large, family-owned business. His previous year's salary was £7 million (~$11M) [43]. Fines linked to income are common in Europe, although some countries do impose caps so the cash amounts never become "extraordinarily" high.

It appears the largest fine ever imposed on a speeder was levied in Switzerland when a speeder driving a £140,000, 6-liter Mercedes was clocked at 180 mph (288 km/hr) traveling 2.5 times the speed limit. He was charged with 300 days of fines at 3,600 Swiss Francs per day—about £727,000 or $1.12M. In Switzerland, fines are also dependent on personal income and apparently, from the type of car this person was driving, the court believes the amount is appropriate. The police actually did not pull the car over while speeding. The driver was arrested when he stopped for a rest break. The case was not closed at the time this chapter was written. The car was being checked to verify the driver's claim that the "speedometer was malfunctioning."

Other countries' financial penalties for speeding are listing in Table 12.8. These laws are in a constant state of change, so the information is historical [44]. The fine amounts are considerably more than for the penalties associated with not using seatbelt and child restraints. The considerable differences in this table show that countries have different approaches to speed risk management. While the underlying

Table 12.8   Maximum Speeding Fines

Country	Fine amount
Finland	(unlimited)
Australia	$857
Canada	$25,000
Denmark	$200
France and Italy	$2,100
Germany	$623
Iceland	$2,700
Norway	10% of annual income + jail time
Portugal	$1,800
The Netherlands	$800
United Kingdom	$8,000
United States	$2,500

reasons are too detailed to be discussed here, the fact that they are different should be no surprise. It's not uncommon for penalties for the same law violation, even in the United States, to vary by local jurisdiction.

ALCOHOL

Drinking and driving form a very deadly combination. What more can be said? The effects of alcohol on the human body are well-documented, yet in spite of the well-known facts people still voluntarily drink alcohol and then drive their motor vehicles. Alcohol impairs judgment, lowers situational awareness, instills a false sense of wellness, and reduces motor skills. These facts are indisputable, yet people gamble their lives, their passengers' lives, and the lives of everyone else on the roads because they think a crash won't happen to them. Maybe they are lulled into this behavior like gaming addicts who replay their winnings at the poker table in the hope of the big win. The odds are against the gambler and the alcohol impaired drivers. In fact from a relative risk perspective, it's probably a better (and clearly a safer) bet to gamble your money than to drink any amount of alcohol and drive. The relative risk of death for drivers in single car accidents with a high blood alcohol content (BAC) is 385 times that of a zero-BAC driver. For male drivers with high BAC, the risk of death is 707 times that of a sober driver [45]. I doubt if any gambler would take this kind of bet, yet people get in their cars and assume this risk every day.

From a regulatory perspective, in the United States, all states and the District of Columbia have laws in place defining driving with BAC levels greater than 0.08% (0.08g alcohol per 100 ml blood) as a crime. Other countries have different values and the BAC values defining this type of criminal behavior can change. Table 12.9 presents other countries' risk tolerance for drinking and driving [46].

Since BAC levels determining criminal motor vehicle-related behavior can change over time, the current values can be different from the thresholds shown in Table 12.9. And if you're thinking about drinking and driving in these (or other) countries, it would be prudent to check the current laws. Notice the large range of risk tolerance for drinking and driving shown in Table 12.9. Countries in the first column have zero tolerance. To me this is the correct level, given the potential consequences, yet countries in the other two columns show much higher thresholds or tolerances. And the United States, known for its high lawyer density and litigation propensity, is in the highest BAC level column. Why? There are no simple answers to this question.

Even though the United States has a common threshold for determining criminal behavior related to driving under the influence of alcohol, the penalties by state vary considerably. Table 12.10 shows the U.S. criminal penalty variation for violating the BAC level laws [47].

Generally, a driver loses his or her license following a conviction for alcohol-impaired driving. But even before this judgment occurs, under a procedure called "administrative license suspension," licenses are taken before conviction when a driver refuses to take or fails a chemical test. Administrative license suspension laws

Table 12.9   International Blood Alcohol Limits

Country	BAC limit (%)	Country	BAC limit (%)	Country	BAC limit (%)
Armenia	0	Argentina	0.05	Brazil	0.08
Azerbaijan	0	Australia	0.05	Canada	0.08
Bahrain	0	Austria	0.05	Chile	0.08
Czech Republic	0	Belgium	0.05	Ecuador	0.08
Hungary	0	Denmark	0.05	Ireland	0.08
Jordan	0	Finland	0.05	Jamaica	0.08
Kyrgyzstan	0	France	0.05	Luxembourg	0.08
Mali	0	Germany	0.05	Malaysia	0.08
Pakistan	0	Israel	0.05	New Zealand	0.08
Romania	0	Italy	0.05	Puerto Rico	0.08
Saudi Arabia	0	Namibia	0.05	Singapore	0.08
Slovak Republic / Slovakia	0	South Africa	0.05	Uganda	0.08
United Arab Emirates (UAE)	0	Switzerland	0.05	United Kingdom	0.08
Uzbekistan	0	Turkey	0.05	United States	0.08

are independent of criminal procedures and drivers who refuse to be tested lose their driving privileges at the time of their arrest. For some convicted offenders, states can require the installation of ignition interlocks on their vehicles that analyze a driver's breath. If the driver has been drinking, he or she can't start the car.

Notice in Georgia, the duration of the administrative license suspension can be up to one year, while in other states, such as Pennsylvania and South Dakota, there is no suspension penalty for failing to take a chemical test. There are other significant penalty variations in Table 12.10 that you can see in each of the columns. The regulatory variations show, again, the diversity of risk tolerance attitudes that are present between the states. Driving regulations are the responsibility of the state governments. There are no differences in how people respond to alcohol by state. The people who live in South Dakota are just as susceptible to alcohol's effects as the residents of Louisiana. Yet these and the other states have developed significantly different penalties for the same offense. The range of BAC criminal thresholds in Tables 12.9 and penalties in Table 12.10 demonstrates the diversity of risk management decisionmaking in practice. This subject is much more complex and esoteric than the mathematics associated with accident frequency and severity calculations.

Thanks to education and enforcement actions, drunk-driving fatalities appear to be declining, but there is an emerging related behavior that deserves discussion here. This is drivers operating vehicles under the influence of drugs other than alcohol. In a study of drivers admitted to a Maryland Trauma Center, 34% tested positive for drugs while only 16% tested positive for only alcohol [48]. This is a single data

Table 12.10   State Penalties for Driving with Alcohol Impairments

State	Administrative license suspension 1st offense	Restore driving privileges during suspension?	Mandatory ignition interlocks?	
			First offenders	Repeat offenders
Alabama	90 days	No	No state interlock laws	
Alaska	90 days	After 30 days[1]	All offenders	yes
Arizona	90 days	After 30 days[1]	All offenders	yes
Arkansas	6 months	Yes[1]	All offenders	yes
California	4 months	After 30 days[1]	All *offenders (in 4 counties)*[2]	no
Colorado	3 months	Yes[1]	All offenders	yes
Connecticut	90 days	Yes[1]	No	no
Delaware	3 months	No	No	no
DC	2–90 days	Yes[1]	No	no
Florida	6 months	After 30 days[1]	High-BAC offenders	yes
Georgia	1 year	Yes[1]	No	no
Hawaii	3 months	After 30 days[1]	All offenders	yes
Idaho	90 days	After 30 days[1]	No	no
Illinois	3 months	After 30 days[1]	All offenders	yes
Indiana	180 days	After 30 days[1]	No	no
Iowa	180 days	After 90 days[1]	No	no
Kansas	30 days	No	High-BAC offenders	yes
Kentucky	no	Not applicable	No	no
Louisiana	90 days	After 30 days[1]	All offenders	yes
Maine	90 days	Yes[1]	No	no
Maryland	45 days	Yes[1]	No	no
Massachusetts	90 days	No	No	yes
Michigan	no	Not applicable	No	no
Minnesota	90 days	After 15 days[1]	No	no
Mississippi	90 days	No	No	no
Missouri	30 days	No	No	yes
Montana	no	Not applicable	No	yes
Nebraska	90 days	After 30 days[1]	All offenders	yes
Nevada	90 days	After 45 days[1]	No	no
New Hampshire	6 months	No	High-BAC offenders	yes
New Jersey	no	Not applicable	High-BAC offenders	yes

(Continued)

Table 12.10   (*Continued*)

State	Administrative license suspension 1st offense	Restore driving privileges during suspension?	Mandatory ignition interlocks?	
			First offenders	Repeat offenders
New Mexico	90 days	After 30 days[1]	All offenders	yes
New York	prosecution duration	Yes[1]	All offenders	yes
North Carolina	30 days	After 10 days[1]	High-BAC offenders	yes
North Dakota	91 days	After 30 days[1]	No	no
Ohio	90 days	After 15 days[1]	No	no
Oklahoma	180 days	Yes[1]	No	yes
Oregon	90 days	After 30 days[1]	All offenders	yes
Pennsylvania	No	Not applicable	No	no
Rhode Island	No	Not applicable	No	no
South Carolina	No	Not applicable	No	yes
South Dakota	No	Not applicable	No state interlock laws	
Tennessee	No	Not applicable	No	no
Texas	90 days	Yes[1]	No	yes
Utah	120 days	No	All offenders	yes
Vermont	90 days	No	No state interlock laws	
Virginia	7 days	No	High-BAC offenders	yes
Washington	90 days	yes[1]	All offenders	yes
West Virginia	6 months	After 30 days[1]	High-BAC offenders	yes
Wisconsin	6 months	Yes[1]	High-BAC offenders	yes
Wyoming	90 days	Yes[1]	High-BAC offenders	yes

[1]Drivers usually must demonstrate special hardship to justify restoring privileges during suspension, and then privileges often are restricted.

[2]First offender pilot program in four counties: Alameda, Los Angeles, Sacramento, and Tulare.

[3]In New York, administrative license suspension lasts until prosecution is complete.

point but even in 2004 when this study was published, drug use was a risk factor in motor vehicle accidents.

In the United States, drugs other than alcohol are classified as narcotics, depressants, stimulants, hallucinogens, cannabinoids, phencyclidines (PCP), anabolic steroids, and inhalants. When drug testing is done in fatal accidents, any positive test results are classified in the Fatality Analysis Reporting System (FARS), a publicly available database that is actively used in safety studies. The NHTSA used this database to perform a drug occurrence study. The analysis reviews drug presence in the majority of fatally injured drivers over several years when the testing was done [49]. Out of the tested subjects, a significant percentage was found to test positive for drugs. Table 12.11 shows some of the research results. The far right column shows that from the tested sample each year, there is a significant fraction of people who were under the influence of drugs at the time of the accident. In simple terms, the trend appears to be about a 1% increase in fatal driver drug use per year.

In 2009, 3952 out of the total of 21,798 drivers, or 18%, tested positive for drugs. However, this conclusion assumes without justification that the testing sample contains all of the drug-related deaths. A more realistic assumption is that testing was performed randomly, relative to the drug use, and that a more representative ratio of the total population (21,798 in 2009) is the number of positive tests divided by the number of tests: 3,952/12,005, or about 33%. This suggests that about 1 in 3 drivers killed on the roads in 2009 were under the influence of a known drug other than alcohol. This is a national average and the results by state vary over a wide range. For example, the two states with the highest fraction of drugged driver fatalities in 2009 were Montana and Connecticut (77%, 57%). The two states with the lowest were Maine and New Mexico (0%, 1%).

The study findings only show that drugs were in the drivers' system. They could be prescriptions, over-the counter-medications, or illegal substances. Also, there's no causal evidence relating drug effects to the accidents. The study only measures drug presence frequency—not the effects. Under current laws, the amount of drug present is not measured nor are there strict standards on drug testing practices for laboratories. Consequently, the type of drug, its quality, quantity, and effects on the human body are not available to further refine this analysis. Based on the surprising and tragic trend shown in Table 12.11, perhaps we should develop drug tests and testing standards similar to alcohol BAC to reduce this emerging risk exposure.

As an individual, you can choose not to use alcohol, to buckle your seat belt, to drive at reasonable speeds, and not to use drugs behind the steering wheel. That's good—but what about the other drivers around you?

Table 12.11   Drug Test Results for Fatally Injured Drivers: 2005–2009

	Total drivers	Drivers tested with known results	Drugs reported in tested drivers	% of tested drivers with drugs
2005	27,491	13,324	3,710	**28**
2006	27,348	14,325	4,018	**28**
2007	26,570	14,893	4,214	**28**
2008	24,254	14,381	4,267	**30**
2009	21,798	12,055	3,952	**33**

DISTRACTIONS

Historically, the most common driver distractions have been reading, eating, smoking, conversations with passengers, or putting on makeup, just to name a few. These "risk exposures" have not gone away, but the newest and fastest growing driving distraction risk from comes from cell phone use and texting. These facts are common knowledge today, yet people still seem to accept the risk exposure rather than change their behavior. In the United States, about 25% of all accidents each year are estimated to be caused by driver distractions and inattentiveness [50]. Several U.S. states have cell phone restrictions and some countries have also developed special laws restricting cell phone use while driving.

Cell phone distractions have been reduced by local, state, and international laws and "hands free" technologies clearly help to minimize driver distractions. This is good news for the rest of the people on the roads since research has shown that dialing a cell phone in a car or light truck increases crash risk by a factor of 5.9, or 590%. There is a relatively small increase in risk (1.3 or 130%) for talking on cell phones while driving, but these distractions pale to the hazards associated with text messaging. Sending and receiving text messages requires your hands, your eyes, and, yes, your attention. People are falsely confident they can multitask and that there is no loss in driving awareness during this activity. However, the science on this subject is clear. For example, a study of 18- to 21-year-old students with low driving experience showed a 400% increase in drivers' eyes not looking at the road while retrieving and sending messages. If you have done any texting I think you can appreciate these results. People involved in texting usually are using both hands and both eyes. Now imagine this behavior behind the wheel of a moving car in traffic. The point is that you, or the texting drivers around you, cannot respond to emergencies or situations requiring quick actions if attention is focused in any other place than the road. Text messaging requires many visual and manual steps. For this reason, texting while driving is estimated to increase crash risk by a dangerous 23 times compared to normal driving [51]. This research was performed in commercial trucks but nevertheless, the findings give some indication of the increased risk we face on the roads from this behavior.

YOUNG DRIVERS

It seems that in regions where cars are the primary means of transportation, in addition to the ramifications of puberty, getting a driver's license is a major rite of passage along the way to adulthood for young people. Independent of the dramatic increase in insurance costs, most parents I know have mixed feelings about allowing their son or daughter to drive. And it's not because they are simply growing up. It's because of the large increase in risk associated with young drivers. Auto-related deaths and injuries are common enough that most parents either have direct or secondary knowledge of teenagers seriously injured or killed in auto accidents. So when their child takes to the roads, to many parents, the best words in the entire language

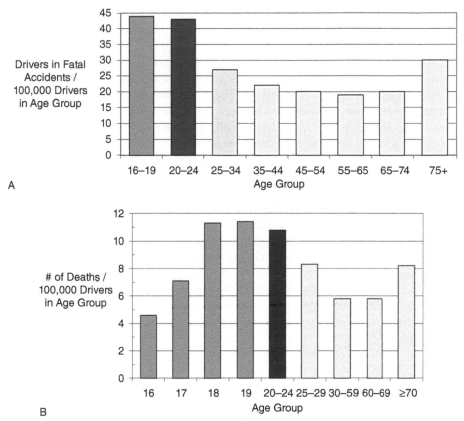

Figure 12.5   (A) Rate of fatal accidents by driver age. (B) Driver death rate by age group, United States, 2009.

are: "I'm home." Parents' concern is justified. In the United States, on a vehicle mile basis, teenagers drive less than all but the oldest people, but their crash numbers and death rates deaths are disproportionately high [52].

There are several ways to look at motor vehicle deaths by age, and two are shown here. Figure 12.5A shows the ratio of the number of driver deaths by people in each age group and Figure 12.5B measures driver fatality rates using the number of drivers in each age group as the statistic divisor. There are other measures but they basically show consistent results with these plots. Teenage driving is a high-risk activity.

Yet there is some actually good news if we contrast the recent and historical data. In 2009, a total of 3466 teenagers died in motor vehicle accidents. Yes, each of these deaths is a tragedy, but this number is 60% lower than in 1975.

Why? Today, cars are designed to higher crash impact standards. There are fewer young teens today receiving their drivers' licenses at the youngest ages possible. In addition, we have seat belt laws, air bags, and many other features that make cars more survivable in accidents. But these latter features don't address the root cause of many accidents involving this age group: immaturity and low driving experience.

To help decrease driving risk for teenage drivers, U.S. state governments and some countries have started implementing "graduated licensing systems" which employ a stepwise approach to driving. There are many program versions but for example, a program could include between 30 to 50 hours of supervised driving with strong restrictions on night driving and on driving with other teenage passengers. And it should be no surprise to you that some programs expressly prohibit the use of cell phones and texting. Driving restrictions are relaxed usually as the drivers age; for example, night driving and teenage passenger limits can be lifted at the age of eighteen [53].

These programs have reduced teenage crash rates by between 10 and 30% [54] and the more rigorous the program, the better the safety results. The insurance companies may even reduce teenage driver premiums more if they pass a formal, approved driver education course. Giving young drivers a structured approach to driving and providing them with as much training as possible does reduce their crash and death risk. If we are to allow young people to become productive members of society, driving can be a necessary activity. But there are societal costs, and for some families, a very high price: about 10 young people die on the roads every day.

LARGE TRUCKS

Any discussion of risk on the roads cannot conclude without the mention of road vehicles that have almost 4 times the kinetic energy of a M1 battle-ready army tank at full speed. Heavy trucks, the 18-wheeler variety, can have a gross weight of 80,000 pounds or 40 tons. The M1 tank is heavier, at about 63 tons, can travel only at about 30 mph, and is seldom on the same roads as your car. Tractor-trailer rigs, while lighter than M1 tanks, routinely travel at highway speeds in excess of 70 mph and often drive so close to cars that all you can see in your rear view mirror is part of the engine grill. You don't need to do the stopping distance mathematics to realize that these tank-like vehicles have the potential to crush your car, whatever its size, like a paper bag.

In 1932 the maximum rig length was 42 feet [55] and in 2011 the length can be 80 feet for a single trailer, 113 feet for a double, and larger, depending on the state or country. In addition to the growing length of these monsters, there's legislation called the "Safe and Efficient Transportation Act (SETA) of 2010," S. 3705, introduced August 4, 2010, by U.S. Senators Mike Crapo (R-Idaho), Susan Collins (R-Maine), and Herb Kohl (D-Wisconsin). The proposed law would raise the maximum gross weight from 80,000 to 97,000 pounds for interstate highways. The lawmakers claim that raising the weight limit, according to this organization, is both safe and efficient. They claim that four trucks at the higher weights would accomplish the work of five trucks today. The transport, fuel, and emissions efficiency arguments make sense, but the safety argument is not as convincing. Having fewer trucks on the roads is good but the reduced maneuverability and increased kinetic energy of the road-borne goliaths make car-truck collisions even more dangerous. You can decide for yourself about whether increasing the weight limit is "safe and efficient." One fact is clear, however; in collisions between passenger cars and tractor trailers, cars lose. In 2009, 3163 people died in large truck crashes. Only 14% of these deaths were the truck occupants.

There is some good news in the recent carnage statistics that is worth noting here. In 2009, fewer people died in large truck crashes than in any year since data on fatal crashes began to be collected in 1975 [56]. This fact is the result of stricter commercial vehicles law enforcement, improved stability controls, antilock brakes, GPS data recording, good driver training, and probably several other factors. Regardless of the reduced truck crash fatality statistics causes, what has not changed are your survival chances if your car collides with a large truck. Large trucks are more likely to be involved in a multiple-vehicle fatal crash than passenger vehicles—82% versus 58%, respectively [57].

So even though large trucks can be aggressive road adversaries, the implications from a collision with these large, heavy (20 to 30 times a car's weight), and poorly maneuverable vehicles gives car drivers little to gain from being in their close proximity. You should treat their kinetic energy with a great deal of respect. The risk reduction strategy for drivers of smaller vehicles is to actively and aggressively stay out of their way.

Whether it's the law of large numbers, bad luck (if you're not a statistician), fate, or countless other factors, accidents do occur. And when the severity, measured in terms of lives lost, medical injury costs, suffering, and consequential damages is high, there can be a great deal of effort involved in determining who is to blame. From a civil law perspective, blame can be shared among several parties where each participates in the financial settlement to the victims or victims' families. From a criminal point of view, the investigation can determine who is telling the truth.

In the past, this work was done by crash forensic experts who would piece together evidence with models and analysis to support one or more crash scenario theories. Today, these experts and even you have much of the data needed to solve motor vehicle accident mysteries. Just like airliners have flight data recorders or "black boxes" that record flight parameters for the final minutes of flight, many cars today (and all new cars in the near future) will have similar devices called event data recorders (EDRs).

In the 1990s, the NHTSA began testing the efficacy of EDRs with car manufacturers. The technical community developed the technical standards and the car industry started voluntarily installing EDRs in new cars by 2005. For example about 64% of the 2005 model passenger vehicles had the device; General Motors, Ford, Isuzu, Mazda, Mitsubishi, Subaru, and Suzuki were voluntarily equipping all of their vehicles with EDRs [58]. In 2010, about 85% of all new cars will have EDRs installed. The current law has a stepwise implementation schedule but from model year 2013 going forward, all EDRs must record the following [59]:

- Change in forward crash speed
- Maximum change in forward crash speed
- Time from beginning of crash at which the maximum change in forward crash speed occurs
- Speed vehicle was traveling
- Percentage of engine throttle, percentage full (how far the accelerator pedal was pressed)
- Whether or not brake was applied
- Ignition cycle (number of power cycles applied to the EDR) at the time of the crash
- Ignition cycle (number of power cycles applied to the EDR) when the EDR data were downloaded
- Whether or not driver was using safety belt
- Whether or not frontal air bag warning lamp was on
- Driver frontal air bag deployment: time to deploy for a single stage air bag, or time to first stage deployment for a multistage air bag
- Right front passenger frontal air bag deployment: time to deploy for a single stage air bag, or time to first stage deployment for a multistage air bag
- Number of crash events
- Time between first two crash events, if applicable
- Whether or not EDR completed recording

This is a fairly extensive list of details and you can see that having this information would be helpful in determining the real cause of accidents. Advanced EDR data is also available for some cars. This information includes things like:

- Sideways, forward, or rearward acceleration
- Engine speed
- Driver steering input
- Right front passenger safety belt status
- Engagement of electronic stability control system
- Antilock brake activity
- Side air bag deployment time for driver and right front passenger,
- Seat track positions for both the driver and right front passenger.

Flight data recorders can record about 30 minutes of data. This is usually plenty of time to study how the aircraft was flying in the minutes that lead up to a crash. Event data recorders capture up to 5 seconds of data. This may not seem like a very long amount of time, but most serious accidents involve rapid deceleration forces measured in terms of milliseconds. From this perspective, 5 seconds is a very long time.

So the car you're driving today likely has one of these devices constantly gathering data when the ignition is turned on. A common question is who owns the information in your EDR? In the United States as of 2010, 12 states have laws related to EDR standards and use, with most other states following this trend. The general precedent for data ownership is that the owner of the car is the owner of the data. But depending on the situation, courts can subpoena the information and your insurance company may require it as per your policy statement. Many policies state that you must cooperate with the insurance company as part of your duties in the administration of a claim. This can mean that you need give them this information if they are handling the claim settlement.

Earlier in this discussion I mentioned that the automakers voluntarily installed EDRs on their cars. You might wonder why would automakers actively absorb these costs. There are two reasons. The automakers have historically been the deep pockets in civil lawsuits where plaintiffs have argued that the injuries or deaths were caused in some way by the design, manufacture, or some functional failure of the car. It is significantly easier to successfully apply the civil standard of "preponderance of the evidence," implying there is enough evidence to conclude that it is more likely than not that the plaintiff's claims are true than the "beyond a reasonable doubt" standard of a criminal case. Automakers see the EDRs as the unbiased, third party, factual expert witnesses to reduce the frequency and severity of civil ligation.

Car manufacturers also use this information to measure the strengths and weaknesses of their car designs in different types of crash scenarios. This information helps design safer cars. Both of these reasons are good for us. The elimination of excessive civil settlements helps keep car prices down and the accident data applied to design changes improves safety. Here is a low-cost technology application that is transparent to the users. It provides irrefutable and valuable data on many critical aspects of motor vehicle accidents that can reduce costs and reduce risk. It is rare when you find such a unique combination.

The list of road risk issues does not end with these illustrations. I hope you have seen how unbalanced and out-of-perspective some road transportation risk management decisions are with respect to public safety, as well as how some are right on the mark. We could go for several chapters and discuss similar risk contradictions regarding bicycle safety, railroad transportation, all-terrain vehicles, pedestrians, and, of course, motorcycles. You should be able to apply the same thought processes used in this chapter to look at these risks more effectively. For the most part, risk management on the roads is your decision.

DISCUSSION QUESTIONS

1. In this chapter, the phrase, "follow the money" was used in reference to the omission safety statistics beside the fuel efficiency information presented for new motor vehicle sales. Do you believe they should be included? How in practice could this be achieved? What are the additional costs?

2. Suggest a new motor vehicle fuel economy strategy to replace CAFE that would not penalize automakers for manufacturing larger, less fuel efficient cars. How are other countries managing this issue?

3. Suppose that proven motor vehicle ejection seat technology was available for new passenger cars at a nominal cost. Discuss the risk management issues regarding how the seat would be activated? Do you think the technology should be implemented?

4. Several tables in this chapter show the diversity in state penalties for speeding, seat belt use, and alcohol impairment. Does it make sense to align the penalties across the country? From a risk management perspective, what do you think the penalties should be?

5. The use of graduated drivers licenses and structured driver training courses has greatly reduced young driver crash risk. Discuss a risk or cost–benefit analysis of having all drivers undergo such a program every five years. In particular, what is your estimate of the cost benefit of mandatory annual testing for all drivers over 70 years of age? Should there be a yearly test and perhaps restrictions for drivers over the age of 70?

Case Study: The Electronic Witness

Legal systems are generally designed to determine guilt or innocence in a systematic manner utilizing available credible information. The assignment of blame in many cases is similar to statistical inference where researchers show Effect A is different than Effect B at a prescribed level of significance. In the legal community, for criminal cases, defendants and plaintiffs each present their side of the case, and the jury or judge then decides the binary outcome to the standard of "beyond a reasonable doubt."

Doubt can be increased by providing direct information that conflicts with the other side's argument and can also be enhanced by showing the evidence is flawed. For example, witnesses may testify that the defendant was with them at the time of the crime or they saw the defendant at the crime scene, but it was a dark, rainy night. In an attempt to reduce uncertainty, expert witnesses, called by both sides, add to the mix by providing technical insights. Now consider the entrance of technology as an unbiased third party into this mix.

A good place to start is the admission of polygraph "lie detector" evidence in court cases in the landmark case *Frye v. United States* [60] in 1923. The ruling created a legal standard that an expert's testimony, regardless of its quality, was an insufficient basis for admission into legal proceedings. The particular scientific techniques used in the analysis must be "generally accepted" by the relevant scientific community. Also, as discussed in the Chapter 9 case study, since 1993, the "Frye Standard" has been replaced by the "Daubert Standard" which rejects the need for "general acceptance." Under Daubert [61], judges can perform independent assessments of the technology or scientific methods used to gather the information intended to be added into testimony. A key element that judges use to allow this type of evidence is peer review and validation. Today there is court precedence for allowing evidence gathered through technology to reduce uncertainty and to provide more clarity in the assignment of guilt or innocence beyond a reasonable doubt.

In the case of car crashes, event data recording (EDR) technology has provided these insights for both criminal and civil courts with success. The following example shows how this information can be used to reduce uncertainty or, in plain language, determine which story is closer to actually what happened.

A woman put on her seat belt and left her home at 7:15 A.M. in her newly acquired General Motors car. The posted speed limit was 35 mph and she claimed her speed was in the 35–45 mph range. About a half a mile from her house the two lane road had a rough section leading to a curve and as she entered the curve, her car crossed over into the oncoming lane and collided head-on with a delivery truck. She received permanently disabling injuries and the crash was determined to be due to her loss of control. The driver of the truck saw the oncoming car fishtail and then swerve into his lane. He tried to avoid a collision by slowing down (almost to a complete stop at the point of collision) and turning to the right in an attempt to avoid the oncoming car.

There is no doubt that her car crossed the road and hit the truck. The legal issue is why. The driver filed a law suit against General Motors, the car manufacturer; the car dealer which sold her the car less than a month before the accident; Delphi Automotive Systems; and Delco Electronic Systems [62]. Delphi and Delco were the designers and manufacturers of the air bag systems. She claimed that as the car entered the rough part of the road, the air bag spontaneously deployed ("out of the blue") rendering her unconscious. She stated that due to her short height (5 feet), she needed to adjust the seat close to the steering wheel, placing her so close that the air bag deployment pushed her hands off the steering wheel and knocked her unconscious.

To add to the veracity of her position was the fact that earlier in the year, General Motors began receiving reports of inadvertent air bag deployment in the type of car she was driving. A causal scenario that was suggested speculated that small stones or debris hitting a specific area of the floor pan under the car would cause a resonance frequency that would sometimes trigger air bag deployment. No data was ever found, however, to support this idea.

The data in her car's EDR had the air bag deployment time plus many other speed and force variables. And without this information, assuming it could be deemed "reliable," the exact sequence of events would really never be known. The court held a "Frye Hearing" and ruled that the EDR information was an accurate and reliable record of crash events and was admissible evidence into this case.

Highly qualified research engineers testified that the inadvertent air bag deployments actually occurred at a very low stopping Δv range—in the order of 1–3 mph. In the accident case, the recorded stopping Δv was 16.2 mph, consistent with normal deployment. Additional timing evidence showed that the air bag did not malfunction and in fact performed correctly. The 16.2 mph Δv was consistent with the truck collision and the acceleration vector showed a direction (front to back) consistent with the deployment algorithm.

Based on this evidence explained by experts, the question of "why" was answered: driver error and excessive speed. The trial court judged in favor of the defendants. The verdict was upheld on appeal.

Discussion Questions

1. The discussion of EDR applications in litigation is just one example of how technology influences our courts. Make a list of technologies and scientific methods now routinely used. What future technologies and methods do you think will be added to this list in the next decade? What are the privacy implications?

2. Under what situations would you disapprove of electronic witness applications?

ENDNOTES

1 Bureau of Labor Statistics, "Automotive Industry: Employment, Earnings, and Hours," November 2010, http://www.bls.gov/iag/tgs/iagauto.htm (accessed November 27, 2010).

2 Bureau of Labor Statistics, "Table A-1, Employment Status of the Civilian Population by Sex and Age," October 2010, http://www.bls.gov/news.release/empsit.t01.htm (accessed November 27, 2010).

3 "Car and Automobile Manufacturing in the U.S.," October 26, 2009, NAICS 33611a, http://www.techventuresource.org/wp-content/uploads/2010/01/usautoindustry.pdf (accessed November 27, 2010).

4 Centers for Disease Control and Prevention, "CDC Study Finds Annual Cost of Motor Vehicle Crashes Exceeds $99 Billion," press release, August 25, 2010,

5 National Highway Traffic Safety Administration, "An Analysis of the Significant Decline in Motor Vehicle Traffic Fatalities in 2008," Report DOT HS 811 346, June 2010.

6 International Transport Forum Paris, "A Record Decade for Road Safety," press release, September 15, 2010.

7 Centers for Disease Control and Prevention, National Center for Health Statistics, Division of Vital Statistics, National Vital Statistics Report Volume 58, No. 19, May 2010, Table 29.

8 Institute of Highway Safety, "New Crash Tests Demonstrate the Influence of Vehicle Size and Weight on Safety Crashes; Results are Relevant to Fuel Economy Policies," news release, April 14, 2009.

9 Insurance Institute for Highway Safety, Status Report, Vol. 42, No. 4, April 19, 2007.

10 Insurance Institute for Highway Safety, Status Report, Vol. 44, No. 4, April 14, 2009.

11 Institute of Highway Safety, "New Crash Tests."

12 Institute of Highway Safety, "Bigger Is Safer but Costs Fuel," status report, Vol. 25, No. 8, September 8, 1990.

13 Institute of Highway Safety, "Vehicle Size, Weight, and Fuel Economy Interrelate to Affect Occupant Safety," status report, Vol. 37, No. 4, April 6, 2002.

14 40 Code of Federal Regulations (CFR), Title 40: Protection of Environment, 600.315–82.

15 John M. Carland, ed., *Foreign Relations of the United States, 1969–1976, Volume 8: Vietnam, January–October 1972*. Washington, DC: United States Government Printing Office, Office of the Historian Bureau of Public Affairs.

16 Richard Nixon, "The President's Address to the Nation Outlining Steps to Deal with the Emergency," November 7, 1973. Available at http://www.ena.lu/address_given_richard_nixon_november_1973-2 -11710 (accessed June 15, 2011).

17 National Highway Traffic Safety Administration, CAFE Overview, http://www/nhtsa.gov/cars/rules/cafe/overview (accessed December 14, 2010).

18 National Highway Traffic Safety Administration, Summary of CAFE Fines Collected, January 16, 2009, http://www.nhtsa.gov/DOT/NHTSA/Rulemaking/Articles/Associated%20Files/CAFE_Fines.pdf (accessed December 14, 2010).

19 Matthias Krust, "Mercedes Expects No CAFE Penalty in U.S. This Year," *Automotive New Europe*, March 12, 2010.

20 National Safety Council, *Injury Facts*®, 2010, p. 95.

21 Dennis F. Shanahan, "Human Tolerance and Crash Survivability," NATO Research and Technology Organization, Human Factors and Medicine, Lecture Series, Pathological Aspects and Associated Biodynamics in Aircraft Accident Investigation, Madrid, Spain, October 2004, RTO-EN-HFM-113.

22 Damon Lavrinc, "Energy-Absorbing Gel Destined for Your Next Car," January 19, 2011, http://translogic.aolautos.com/2011/01/19/mazda-investigating-use-of-energy-absorbing-gel/ (accessed January 20, 2011).

23 National Safety Council, *Injury Facts*®, 2010, p. 98.

24 Governors Highway Safety Association, "Seatbelt Laws," December 2010 http://www.ghsa.org/html/stateinfo/laws/seatbelt_laws.html (accessed December 19, 2010).

25 Traffic Safety Facts, "The Increase in Lives Saved, Injuries Prevented, and Cost Savings If Seat Belt Use Rose to at Least 90 Percent in All States," Research Note, DOT HS 811 140, May 2009.

26 National Safety Council, *Injury Facts*®, 2010, p. 99.

27 National Highway Traffic Safety Administration, "The 2009 National Survey of the Use of Booster Seats," DOT HS 811 377, September 2010.

28 National Highway Traffic Safety Administration, "U.S. Transportation Secretary Ray LaHood Unveils New Data and Urges Parents to Install Proper Safety Seats During Child Passenger Safety Week," NHTSA 11-10, September 23, 2010.

29 U.S. Code of Federal Regulations, Part 571.213: Standard No. 213; Child restraint systems.

30 Vehicle Standard (Australian Design Rule 34/01–Child Restraint Anchorages and Child Restraint Anchor Fittings) 2005, subsection 7 (1) of the Motor Vehicle Standards Act 1989, November 29, 2005.

31 Restraining devices for child occupants of power-driven vehicles (Child restraint system), UNECE Vehicle Regulations, Addendum 43: Regulation No. 44, March 2010.

32 "Shaky Statistics Are Driving the Air Bag Debate," *Wall Street Journal*, January 22, 1997.

33 Donna Glassbrenner, "Estimating the Lives Saved by Safety Belts and Air," National Center for Statistics and Analysis, National Highway Traffic Safety Administration, Proceedings 18th International Technical Conference on the Enhances safety of Vehicles, DOT HS 809 543, May 2003.

34 U.S. Department of Transportation, National Highway Traffic Safety Administration, *Traffic Safety Facts 2008 (Final Edition)*. Washington, DC, 2009, http://www-nrd.nhtsa.dot.gov/cats/listpublications.aspx?Id=E&ShowBy=DocType (accessed July 6, 2010).

35 Institute of Highway Safety, Status Report, Vol. 45, No. 1, February 6, 2010 (accessed January 2, 2010).

36 Peter Valdes Dapena, "What's Important Isn't the Number of Air bags, It's the Type of Air bags and Their Effect on Crash Scores," *CNN*, July 3, 2007; cnn.com (accessed January 2, 2011).

37 NTSB, "NTSB Study Shows That Air bags Can Provide Occupant Protection in General Aviation Accidents," press release SB-11-03, January 11, 2011.

38 President Nixon's Statement on Signing the Emergency Highway Energy Conservation Act, January 2, 1974, http://www.presidency.ucsb.edu/ws/index.php?pid=4332 (accessed December 23, 2010).

39 Transportation Research Board, TRB Special Report 204: 55, "A Decade of Experience Evaluates the Benefits and Costs of the 55 mph Speed Limit and Assesses the effectiveness of State Laws in Inducing Compliance." National Academy of Sciences, January 1984.

40 National Safety Council, *Accident Facts*®, 1996.

41 National Safety Council, *Injury Facts*®, 2010, p. 107.

42 Jack Stuster et al., "Synthesis of Safety Research Related to Speed and Speed Management," Anacapa Sciences and Safety Design Division, Federal Highway Administration, 1998. http://www.tfhrc.gov/safety/speed/spdtoc.htm (accessed January 17, 2011).

43 "Finn's Speed Fine Is a Bit Rich," *BBC News*, February 10, 2004, http://news.bbc.co.uk/2/hi/business/3477285.stm (accessed December 26, 2010).

44 Craig Howie, "The World's Highest Speeding Fines, Don't Speed Here Unless You've Got the Cash," http://autos.aol.com/article/highest-speeding-fines/ (accessed December 26, 2010).

45 Testimony of Richard F. Healing, Member, National Transportation Safety Board before the House Judiciary Committee State of Maryland Regarding House Bill 763, February 12, 2004. http://www.ntsb.gov/speeches/healing/rfh040212.htm (accessed January 4, 2011).

46 http://www.driveandstayalive.com/articles%20and%20topics/drunk%20driving/artcl–drunk-driving-0005–global-BAC-limits.htm (accessed January 4, 2011).

47 Highway Loss Data Institute, Highway Safety Research and Communications, Insurance Institute of Highway Safety, "DUI/DWI Laws," January 2011.

48 J. M. Walsh et al., "Epidemiology of Alcohol and Other Drug Use among Motor Vehicle Crash Victims Admitted to a Trauma Center," *Traffic Injury Prevention*, Vol. 5, No. 3, 2004, pp. 254–260.

49 "Drug Involvement of Fatally Injured Drivers," Traffic Safety Facts, DOT HS 811 415, November 2010.

50 K. Young and M. Regan, "Driver Distraction: A Review of the Literature." In: I. J. Faulks, M. Regan, M. Stevenson, J. Brown, A. Porter, and J. D. Irwin (eds.), *Distracted Driving*. Sydney, NSW: Australasian College of Road Safety, 2007, pp. 379–405.

51 Virginia Tech Transportation Institute, "New Data from Virginia Tech Transportation Institute Provides Insight into Cell Phone Use and Driving Distraction," press release, July 27, 2009.

52 Insurance Institute for Highway Safety, "Unpublished Analysis of Data from the US Department of Transportation's National Household Travel Survey, General Estimates System, and Fatality Analysis Reporting System," Arlington, VA.

53 http://www.iihs.org/brochures/pdf/gdl_brochure.pdf (accessed January 11, 2011).

54 R. E. Trempel, *Graduated Driver Licensing Laws and Insurance Collision Claim Frequencies of Teenage Drivers*. Arlington, VA: Highway Loss Data Institute, 2009.

55 Rob Wagner, "History of Cabover Trucks," http://www.ehow.com/about_5387387_history-cabover-trucks.html#ixzz1AsFhg02n (accessed January 12, 2011).

56 Institute of Highway Safety, Highway Loss Data Institute, "Fatality Facts 2009."

57 National Safety Council, *Injury Facts*®, 2010, p. 94.

58 National Highway Traffic Safety Administration, "Final Regulatory Evaluation—Event Data Recorders. Table III-1: Estimate of the Number of EDRs in Light Vehicles with a GVWR of 3,855 Kilograms (8,500 Pounds) or Less," July 2006. Washington, DC: U.S. Department of Transportation, pp. 111–112.

59 National Highway Traffic Safety Administration—Final Rule. Docket no. NHTSA-2006-25666; 49 CFR Part 563 Event Data Recorders. *Federal Register*, vol. 71, no. 166, 2006, pp. 50998–51048. Washington, DC: National Archives and Records Administration.

60 *Frye v. United States*, 293 F. 1013 (D.C. Cir. 1923).

61 *Daubert v. Merrell Dow Pharmaceuticals*, 509 U.S. 579 (1993).

62 *Bachman et al. v. General Motors Corp., Uftring Chevrolet-Oldsmobile, Delphi Automotive Systems and Delco Electronics Systems*, Ill 98L21 (1998).

Chapter 13

An Incomplete List of Life's Risks

. . . though we know not what the day will bring, what course after nightfall destiny has written that we must run to the end.

—Pindar, *Sixth Nemean Ode*

Risk is an inherent part of life. There is no way to eliminate it. You can follow every safety and health rule in the world, and you still won't be able to avoid risk. This life, by definition, ends; we have yet to find a way around that. But there are things that you can do to better manage day-to-day life risks. Even though we all know that our journey through life is finite, you can increase the odds that the trip is a long, pleasant one.

Life is a series of choices. And we are always weighing the benefits of whatever we do with the not-so-beneficial consequences. If you want to manage your risks, make conscious, direct decisions about the choices you have in your control. This doesn't mean you're supposed to sit and fret over every decision or choice you make, but do think about your general lifestyle. Admonitions about not smoking, wearing seatbelts, exercising and maintaining a healthy weight, having a good diet, and general healthcare tips really do have tremendous value. While making wise choices cannot assure you a long and prosperous life, ignoring healthy directions certainly stacks the deck against it.

Risk management in our lives is about what can go wrong—all of the different ways we can and do "run to the end." Life is an experience whose hazards can be unforgiving, irreversible, and final. Some of these risks are known and accepted. The choice to continue the behavior is a conscious one. Others are deliberately avoided or reduced. Whatever our plan, sooner or later we all become risk statistics. The final severity act of our lives will occur with a probability equal to one. All that is unknown is the frequency—when will we breathe our last breath. The conditions of that moment end up in lists like the ones shown here.

The majority of information in this chapter came from the 2010 edition of *Injury Facts*® published by the National Safety Council (www.nsc.org) and reports published by the World Health Organization (www.who.org). In some tables, the presentation of data has been modified to make it easier to interpret. Instead of citing death and injury rates, or number of cases of each, I translated the risks to a 1 in *N* basis. For example, the risk of injury in playing football is 1 in 42. This means that on the average, 1 out of every 42 people who play football suffer injuries. The "1 in . . . " applies to the group involved in the activity, age group, or sex (male or female), and does not mean 1 in the general population. As in every compilation of statistics, there are assumptions, errors, and limitations of how the numbers should be interpreted. Don't get tied up in the absolute precision here. The value of these tables is to show you the relative risk of death or injury between different categories, diseases, health conditions, or activities.

At the intellectual or clinical level, the information in this chapter shows a variety of aspects of our global, societal, and personal risks. The numbers show what life's hazards were in the past. After reading *Injury Facts*® and other similar publications for several years, I've seen that while these numbers are changing, about the same number of people die or are injured each year in the listed activities. The element we don't know is the all-important "who." We should all learn from the past. Here, in this incomplete list of life's risks, are some of its lessons. Behind each of these silent numbers is a person—a mother, a father, a son, or a daughter. Let their past serve as a guide to help you reduce your risk, today and in the future.

AN INCOMPLETE LIST OF LIFE'S RISKS

Risk of Injury While Playing Sports

Football (touch and tackle)	1 in 42
Wrestling	1 in 56
Basketball	1 in 61
Skateboarding	1 in 66
Soccer	1 in 78
Bicycling, excluding mountain biking	1 in 89
Ice hockey	1 in 90
Baseball	1 in 95
Cheerleading	1 in 99
Snowboarding	1 in 99
Softball	1 in 112
Volleyball	1 in 217
Swimming including diving, slides, and pool equipment	1 in 357
Weight lifting	1 in 475
Tennis	1 in 533
Fishing	1 in 582
Water skiing	1 in 642

Golf (excluding cart injuries)	1 in 676
Mountain biking	1 in 820
Mountain climbing	1 in 1,359
Archery	1 in 1,583
Bowling	1 in 2,340
Billiards, pool	1 in 7,615

Motor Vehicle Risks

Based on U.S. population estimate: 305,690,698, December 2008				
Hazard	Injury risk		Death risk	
Overall	1 in	146	1 in	7,838
(This means 1 out of every 146 people who ride in a motor vehicle will be injured in a given year.)				
Motorcycles	1 in	69	1 in	1,380
(per registered motorcycle)				
Collision type:				
Pedestrian	1 in	4,076	1 in	54,588
Angle collision	1 in	200	1 in	20,244
Head-on collision	1 in	429	1 in	40,222
Rear-end collision	1 in	2,729	1 in	76,423
Sideswipe and other two-vehicle collisions	1 in	495	1 in	127,371
Railroad train	1 in	3,474	1 in	277,901
Pedalcycle	1 in	305,691	1 in	1,528,453
Animal, animal-drawn vehicle	1 in	5,181	1 in	339,656
Fixed or other object	1 in	61,138	1 in	3,056,907
Noncollision, e.g., roll over	1 in	926	1 in	24,261

Should You Ride or Walk?

	Pedestrian death risk		Auto death risk	
Age group	(in age group)		(in age group)	
0–4	1 in	42,305	1 in	62,500
5-14	1 in	40,417	1 in	80,000
15–24	1 in	4,881	1 in	62,500
25–44	1 in	6,876	1 in	55,556
45–64	1 in	7,559	1 in	40,323
65–74	1 in	7,459	1 in	45,455
75+	1 in	5,184	1 in	17,241

Age group	Car accident death risk (in age group)		Pedestrian death risk (in age group)	
0–4	1 in	18,519	1 in	62,500
5–14	1 in	19,608	1 in	80,000
15–24	1 in	3,704	1 in	62,500
25–44	1 in	5,952	1 in	55,556
45–64	1 in	6,757	1 in	40,323
65–74	1 in	5,714	1 in	45,455
75+	1 in	3,058	1 in	17,241

Bus, Train, or Plane?

	Death risk	
Commuter rail	1 in	14,303,030
Light rail	1 in	32,428,571
Bus	1 in	88,460,317
Heavy rail	1 in	114,419,355
Air carrier	1 in	247,768,667
School buses (passengers)	1 in	624,000,000

	Deaths per passenger	
Railroad passenger trains	1 in	30,250,000
Intercity buses	1 in	180,000,000
School buses	1 in	614,571,429
Transit buses	1 in	1,000,000,000
Scheduled airlines	1 in	1,613,981,763

Actual Number of Recorded Deaths (2006)

Transportation	
Motor-vehicle accidents	45,316
Pedestrian	6,162
Pedalcyclist	926
Motorcycle rider	4,787
Car occupant	14,119
Occupant of pick-up truck or van	3,411
Occupant of heavy transport vehicle	432
Bus occupant36Animal rider or occupant of animal-drawn vehicle	126
Occupant of railway train or railway vehicle	17
Water transport accidents	514

Complications of medical and surgical care	2,521
Falls	20,823
Exposure to smoke, fire, and flames	3,109

Drowning & choking

Drowning and submersion while in or falling into bath-tub	413
Drowning and submersion while in or falling into swimming-pool	698
Drowning and submersion while in or falling into natural water	1,611
Other and unspecified drowning and submersion	857
Inhalation of gastric contents	351
Inhalation and ingestion of food causing obstruction of respiratory tract	872
Inhalation and ingestion of other objects causing obstruction of respiratory tract	3,109

Contact animals and plants

Bitten or struck by dog	32
Bitten or struck by other mammals	73
Bitten or stung by nonvenomous insect and other arthropods	7
Bitten or crushed by other reptiles	1
Contact with venomous snakes and lizards	8
Contact with venomous spiders	4
Contact with hornets	61
Contact with other and unspecified venomous animal or plant	10

Exposure to forces of nature — **1,340**

Exposure to excessive natural heat	622
Exposure to excessive natural cold	519
Lightning	47
Earthquake and other earth movements	25
Cataclysmic storm	75
Flood	10

Intentional self-harm

Poisoning	6,109
Hanging	7,491
Firearm discharge	16,883
Other and unspecified means	2,817

Event of undetermined intent

Poisoning	3,541
Hanging	152
Drowning and submersion	255
Firearm discharge	220
Exposure to smoke	128
Falling	80
Other and unspecified means	755

Risks at Home: Overall

Hazard	Death risk	
Poisoning	1 in	12,179
Falls	1 in	20,164
Fire, flames, smoke	1 in	112,766
Choking	1 in	121,788
Mechanical suffocation drowning	1 in	169,150
Firearms	1 in	380,587
All other home	1 in	46,132

All Home Hazards

Age group	Death risk	
0–4	1 in	9,524
5–14	1 in	100,000
15–24	1 in	14,286
25–44	1 in	6,667
45–64	1 in	5,236
65–74	1 in	5,291
75+	1 in	1,072

Poisoning: Is Your Kitchen Cabinet Locked?

Hazard	Death risk	
0–4	1 in	454,545
5–14	1 in	833,333
15–24	1 in	16,949
25–44	1 in	7,519
45–64	1 in	7,576
65–74	1 in	33,333
75+	1 in	42,017

Slips and Falls: Are Your Floors and Stairs Slippery?

Hazard	Death risk	
0–4	1 in	333,333
5–14	1 in	10,000,000
15–24	1 in	833,333
25–44	1 in	243,902
45–64	1 in	41,152
65–74	1 in	11,236
75+	1 in	1,689

Fires, Flames, and Smoke: Do You Have
Smoke Detectors? Are the Batteries Working?

Hazard	Death risk	
0–4	1 in	109,890
5–14	1 in	227,273
15–24	1 in	227,273
25–44	1 in	312,500
45–64	1 in	105,263
65–74	1 in	44,643
75+	1 in	25,000

Choking: Do You Know the Heimlich
Maneuver?

Hazard	Death risk	
0–4	1 in	250,000
5–14	1 in	10,000,000
15–24	1 in	1,000,000
25–44	1 in	500,000
45–64	1 in	142,857
65–74	1 in	58,824
75+	1 in	14,493

Mechanical Suffocation: Hanging,
Strangulation, Smothering

Hazard	Death risk	
0–4	1 in	20,833
5–14	1 in	1,000,000
15–24	1 in	500,000
25–44	1 in	500,000
45–64	1 in	333,333
65–74	1 in	1,000,000
75+	1 in	333,333

Drowning

Hazard	Death risk	
0–4	1 in	83,333
5–14	1 in	1,000,000
15–24	1 in	1,000,000
25–44	1 in	1,000,000
45–64	1 in	500,000
65–74	1 in	250,000
75+	1 in	100,000

Firearms: Are Your Guns Unloaded and
Locked?

Hazard	Death risk	
0–4	1 in	10,000,000
5–14	1 in	10,000,000
15–24	1 in	333,333
25–44	1 in	500,000
45–64	1 in	1,000,000
65–74	1 in	333,333
75+	1 in	500,000

Home Injury Risks

Hazard	Injury risk	
Stairs or steps	1 in	252
Floors or flooring materials	1 in	253
Beds	1 in	542
Knives	1 in	736
Chairs	1 in	902
Tables	1 in	962
Ceilings and walls	1 in	976
Doors (not glass)	1 in	976
Household cabinets, racks, and shelves	1 in	1,106
Bathtubs and showers	1 in	1,173
Household containers and packaging	1 in	1,403
Ladders	1 in	1,699
Sofas, couches, davenports, divans, etc.	1 in	1,798
Saws, hammers, tools	1 in	2,089
Porches, balconies, open-side oors	1 in	2,167
Nails, screws, tacks or bolts	1 in	2,229
Rugs and carpets	1 in	2,335
Windows	1 in	2,587
Fences or fence posts	1 in	2,746
Tableware (excluding knives)	1 in	3,105
Toilets	1 in	3,814
Bottles and jars	1 in	3,850
Drinking glasses	1 in	3,955
Door sills or frames	1 in	5,186
Stools	1 in	5,555
Counters or countertops	1 in	6,165
Handrails, railings or banisters	1 in	6,857
Bags	1 in	7,264
Poles	1 in	8,094
Waste containers, trash baskets, etc.	1 in	8,972
Benches	1 in	9,392
Glass doors	1 in	10,052
Scissors	1 in	10,265
Cookware, bowls and canisters	1 in	10,425
Mirrors or mirror glass	1 in	11,871
Manual cleaning equipment (excl. buckets)	1 in	12,302
Sinks	1 in	12,363
Paper products	1 in	12,542
Other kitchen gadgets	1 in	13,909

Based on U.S. population estimate: 305,690,698, December 2008.

TOP TEN CAUSES OF DEATH: 2006 STATISTICS

All Ages

All causes	Total		Male		Female	
	1 in	123	1 in	122	1 in	124
From:						
1) Heart disease	1 in	472	1 in	466	1 in	479
2) Cancer (malignant neoplasms)	1 in	533	1 in	507	1 in	561
3) Stroke (cerebrovascular disease)	1 in	2,174	1 in	2,695	1 in	1,835
4) Chronic lower respiratory diseases	1 in	2,392	1 in	2,481	1 in	2,320
5) Diabetes mellitus	1 in	4,115	1 in	4,082	1 in	4,149
6) Alzheimer's disease	1 in	4,115	1 in	6,944	1 in	2,950
7) Influenza and pneumonia	1 in	5,291	1 in	5,714	1 in	4,926
8) Nephritis and nephrosis	1 in	6,579	1 in	6,667	1 in	6,494
9) Motor vehicles	1 in	6,579	1 in	4,651	1 in	11,111
10) Septicemia	1 in	8,696	1 in	9,434	1 in	8,065

Younger Than 1 Year Old

All causes	Total		Male		Female	
	1 in	146	1 in	133	1 in	162
From:						
1) Congenital anomalies	1 in	715	1 in	687	1 in	747
2) Short gestation, low birth weight	1 in	859	1 in	795	1 in	939
3) Sudden infant death syndrome	1 in	1,792	1 in	1,567	1 in	2,105
4) Maternal complications: pregnancy	1 in	2,469	1 in	2,262	1 in	2,732
5) Placenta, cord, membrane Complications	1 in	3,650	1 in	3,311	1 in	4,082
6) Respiratory distress	1 in	5,051	1 in	4,525	1 in	5,714
7) Bacterial sepsis	1 in	5,155	1 in	4,902	1 in	5,464
8) Mechanical suffocation	1 in	5,319	1 in	4,545	1 in	6,494
9) Neonatal hemorrhage	1 in	6,711	1 in	5,917	1 in	7,874
10) Circulatory system disease	1 in	7,634	1 in	7,092	1 in	8,403

Between 1 and 4 Years Old

All causes		Total		Male		Female	
		1 in	3,509	1 in	3,279	1 in	3,802
From:							
1)	Motor vehicles	1 in	27,778	1 in	26,316	1 in	29,412
2)	Congenital anomalies	1 in	31,250	1 in	34,483	1 in	29,412
3)	Drowning	1 in	35,714	1 in	28,571	1 in	47,619
4)	Cancer	1 in	43,478	1 in	40,000	1 in	47,619
5)	Homicide	1 in	45,455	1 in	40,000	1 in	50,000
6)	All other unintentional injuries	1 in	71,429	1 in	62,500	1 in	83,333
7)	Fires and flames	1 in	83,333	1 in	71,429	1 in	100,000
8)	Heart disease	1 in	100,000	1 in	90,909	1 in	111,111
9)	Influenza/pneumonia	1 in	125,000	1 in	125,000	1 in	125,000
10)	Septicemia	1 in	200,000	1 in	200,000	1 in	166,667

Between 5 and 14 Years Old

All causes		Total		Male		Female	
		1 in	6,536	1 in	5,682	1 in	7,752
From:							
1)	Motor vehicles	1 in	30,303	1 in	27,027	1 in	34,483
2)	Cancer	1 in	43,478	1 in	40,000	1 in	50,000
3)	Homicide	1 in	100,000	1 in	83,333	1 in	142,857
4)	All other unintentional injuries	1 in	111,111	1 in	83,333	1 in	200,000
5)	Congenital anomalies	1 in	111,111	1 in	111,111	1 in	125,000
6)	Drowning	1 in	166,667	1 in	111,111	1 in	333,333
7)	Heart disease	1 in	166,667	1 in	142,857	1 in	166,667
8)	Fires and flames	1 in	200,000	1 in	250,000	1 in	200,000
9)	Suicide	1 in	200,000	1 in	142,857	1 in	333,333
10)	Chronic lower respiratory diseases	1 in	333,333	1 in	250,000	1 in	500,000

Between 15 and 24 Years Old

All causes	Total		Male		Female	
	1 in	1,209	1 in	832	1 in	2,326
From:						
1) Motor vehicles	1 in	3,831	1 in	2,717	1 in	6,757
2) Homicide	1 in	7,353	1 in	4,348	1 in	28,571
3) Suicide	1 in	10,101	1 in	6,135	1 in	31,250
4) Poisoning	1 in	14,286	1 in	9,346	1 in	33,333
5) Cancer	1 in	25,641	1 in	21,739	1 in	32,258
6) All other unintentional injuries	1 in	34,483	1 in	22,222	1 in	83,333
7) Heart disease	1 in	38,462	1 in	30,303	1 in	55,556
8) Drowning	1 in	66,667	1 in	40,000	1 in	333,333
9) Congenital anomalies	1 in	90,909	1 in	76,923	1 in	111,111
10) Falls	1 in	166,667	1 in	100,000	1 in	1,000,000

Between 25 and 34 Years Old

All causes	Total		Male		Female	
	1 in	932	1 in	674	1 in	1,541
From:						
1) Motor vehicles	1 in	5,435	1 in	3,610	1 in	11,494
2) Poisoning	1 in	7,576	1 in	5,236	1 in	14,286
3) Suicide	1 in	8,000	1 in	5,025	1 in	21,277
4) Homicide	1 in	8,475	1 in	5,102	1 in	26,316
5) Cancer	1 in	10,989	1 in	11,494	1 in	10,417
6) Heart disease	1 in	12,048	1 in	8,850	1 in	19,608
7) All other unintentional injuries	1 in	28,571	1 in	17,857	1 in	83,333
8) HIV	1 in	33,333	1 in	27,778	1 in	43,478
9) Diabetes mellitus	1 in	58,824	1 in	55,556	1 in	66,667
10) Stroke	1 in	76,923	1 in	71,429	1 in	83,333

Between 35 and 44 Years Old

All causes		Total		Male		Female	
		1 in	524	1 in	417	1 in	703
From:							
1)	Cancer	1 in	3,125	1 in	3,636	1 in	2,740
2)	Heart disease	1 in	3,521	1 in	2,519	1 in	5,848
3)	Poisoning	1 in	5,780	1 in	4,329	1 in	8,621
4)	Motor vehicles	1 in	6,494	1 in	4,566	1 in	11,364
5)	Suicide	1 in	6,579	1 in	4,292	1 in	14,286
6)	HIV	1 in	10,870	1 in	7,692	1 in	18,519
7)	Homicide	1 in	14,493	1 in	9,615	1 in	28,571
8)	Chronic liver disease and cirrhosis	1 in	16,949	1 in	12,987	1 in	24,390
9)	Stroke	1 in	19,608	1 in	18,868	1 in	20,408
10)	Diabetes mellitus	1 in	20,833	1 in	16,949	1 in	26,316

Between 45 and 54 Years Old

All causes		Total		Male		Female	
		1 in	234	1 in	185	1 in	314
From:							
1)	Cancer	1 in	858	1 in	839	1 in	879
2)	Heart disease	1 in	1,134	1 in	775	1 in	2,058
3)	Poisoning	1 in	5,236	1 in	4,016	1 in	7,463
4)	Chronic liver disease	1 in	5,618	1 in	3,846	1 in	10,000
5)	Suicide	1 in	5,814	1 in	3,817	1 in	11,905
6)	Motor vehicles	1 in	6,536	1 in	4,405	1 in	12,195
7)	Stroke	1 in	6,803	1 in	6,098	1 in	7,692
8)	Diabetes mellitus	1 in	7,576	1 in	6,061	1 in	10,101
9)	HIV	1 in	9,901	1 in	6,494	1 in	19,608
10)	Chronic lower respiratory diseases	1 in	10,989	1 in	10,870	1 in	11,111

Between 55 and 64 Years Old

All	Total		Male		Female	
causes	1 in	112	1 in	90	1 in	145
From:						
1) Cancer	1 in	311	1 in	275	1 in	355
2) Heart disease	1 in	482	1 in	337	1 in	805
3) Chronic lower respiratory diseases	1 in	2,551	1 in	2,381	1 in	2,725
4) Diabetes mellitus	1 in	2,762	1 in	2,278	1 in	3,436
5) Stroke	1 in	3,003	1 in	2,577	1 in	3,534
6) Chronic liver disease	1 in	4,367	1 in	2,985	1 in	7,692
7) Suicide	1 in	6,897	1 in	4,386	1 in	14,493
8) Motor vehicles	1 in	6,993	1 in	4,878	1 in	11,628
9) Nephritis and nephrosis	1 in	7,246	1 in	6,211	1 in	8,475
10) Septicemia	1 in	7,813	1 in	7,092	1 in	8,696

Between 65 and 74 Years Old

All	Total		Male		Female	
causes	1 in	48	1 in	40	1 in	60
From:						
1) Cancer	1 in	138	1 in	115	1 in	165
2) Heart disease	1 in	204	1 in	151	1 in	289
3) Chronic lower respiratory diseases	1 in	670	1 in	613	1 in	727
4) Stroke	1 in	1,038	1 in	926	1 in	1,156
5) Diabetes mellitus	1 in	1,221	1 in	1,040	1 in	1,435
6) Nephritis and nephrosis	1 in	2,538	1 in	2,217	1 in	2,890
7) Septicemia	1 in	3,115	1 in	2,874	1 in	3,367
8) Influenza/pneumonia	1 in	3,125	1 in	2,571	1 in	3,802
9) Chronic liver disease	1 in	3,846	1 in	2,882	1 in	5,376
10) Motor vehicles	1 in	6,494	1 in	4,739	1 in	9,434

75 Years and Older

All	Total		Male		Female	
causes	1 in	14	1 in	15	1 in	13

From:

1)	Cancer	1 in	13	1 in	50	1 in	42
2)	Heart disease	1 in	76	1 in	70	1 in	82
3)	Stroke	1 in	106	1 in	241	1 in	240
4)	Chronic lower respiratory diseases	1 in	191	1 in	255	1 in	158
5)	Diabetes mellitus	1 in	279	1 in	455	1 in	210
6)	Motor vehicles	1 in	514	1 in	553	1 in	483
7)	Falls	1 in	634	1 in	637	1 in	627
8)	Choking	1 in	668	1 in	662	1 in	652
9)	Chronic liver disease	1 in	939	1 in	1,057	1 in	860
10)	Poisoning	1 in	1,303	1 in	1,383	1 in	1,263

Risk as a Function of Time

Accidental injury	1 every	2 seconds
Home: accidental injury	1 every	5 seconds
Public non-motor vehicle injury	1 every	5 seconds
Workers: off the job injury	1 every	5 seconds
Workers: on the job injury	1 every	8 seconds
Accidental death	1 every	6 minutes
Motor Vehicle death	1 every	12 minutes
Workers: off the job death	1 every	15 minutes
Home: accidental death	1 every	20 minutes
Public non-motor vehicle death	1 every	27 minutes
Workers: on the job death	1 every	103 minutes

Leading Causes of Death: United States

Cause of death	Death frequency	
Heart disease	70	every hour
Cancer	64	every hour
Stroke (cerebrovascular diseases)	16	every hour
Chronic lower respiratory diseases	15	every hour
Accidents (unintentional injuries)	14	every hour
Alzheimer's disease	9	every hour
Diabetes	8	every hour
Influenza and pneumonia	6	every hour
Nephritis, nephrotic syndrome, and nephrosis	5	every hour
Septicemia	4	every hour

Leading Causes of Death: Worldwide

Cause of death	Death frequency	
Heart disease	822	every hour
Stroke, cerebrovascular diseases	652	every hour
Lower respiratory infections	477	every hour
Chronic obstructive pulmonary disease	345	every hour
Diarrheal diseases	247	every hour
HIV/AIDS	233	every hour
Tuberculosis	167	every hour
Trachea, bronchus, lung cancers	151	every hour
Road traffic accidents	145	every hour
Prematurity and low birth weight	135	every hour

WORLD FOOD RISK

Hunger: the want or scarcity of food in a country.

—*Oxford English Dictionary (1971)*

925 million people are undernourished.

—United Nations Food and Agriculture Organization (October 2010)

World Poverty Statistics

Region	% in $1.25 a day poverty	Population (billions)	Pop. in $1 a day poverty (billions)
East Asia and Pacific	16.8	1.884	0.316
Latin America and the Caribbean	8.2	0.550	0.045
South Asia	40.4	1.476	0.596
Sub-Saharan Africa	50.9	0.763	0.388
Total Developing countries	28,8	4,673	1.345
Europe and Central Asia	0.04	0.473	0.017
Middle East and North Africa	0.04	0.305	0.011
Total		5.451	1.372

World Water Risk

World Health Organization region	Population using improved drinking-water sources (%)	
	1990	2008
Africa	50	61
The Americas	91	96
Southeast Asia	73	86
European	96	98
Eastern Mediterranean	85	83
Western Pacific	71	90
Income group		
Low	57	67
Lower middle	71	86
Upper middle	89	95
High	99	100
Global	77	87

Child Mortality Risk

World Health Organization Region	Infant mortality rate[1]		Under-five mortality rate[2]	
	2000	2008	2000	2008
Africa	98	85	165	142
The Americas	22	15	27	18
Southeast Asia	63	48	87	63
European	18	12	22	14
Eastern Mediterranean	66	57	90	78
Western Pacific	28	18	34	21
Income group				
Low	88	76	137	118
Lower middle	55	44	78	63
Upper middle	26	19	32	23
High	7	6	8	7
Global	54	35	78	65

[1]Probability of dying by age 1 per 1000 live births.
[2]Probability of dying by age 5 per 1000 live births.

World Heath Risk, Part 1 of 4

Death classification	Number of deaths		
Communicable, maternal, perinatal and nutritional conditions	**6,634,951**		
Infectious and parasitic diseases	4,320,933		
Tuberculosis		494,404	
STDs excluding HIV		56,847	
Syphilis			38,846
Chlamydia			8,889
Gonorrhea			492
Other STDs			9,112
HIV/AIDS		1,012,939	
Diarrheal diseases		1,036,506	
Childhood-cluster diseases		389,197	
Pertussis			125,077
Poliomyelitis			529
Diphtheria			2,523
Measles			203,806
Tetanus			57,791
Meningitis		159,377	
Hepatitis B		31,096	
Hepatitis C		18,089	
Malaria		433,167	
Tropical-cluster diseases		57,608	
Leprosy		1,445	
Dengue		8,985	
Japanese encephalitis		5,911	
Trachoma		1	
Intestinal nematode infections		2,837	
Other intestinal infections		862	
Other infectious diseases		611,502	
Lower respiratory infections		2,013,782	
Upper respiratory infections		35,804	
Otitis media		2,258	
Maternal conditions	526,625		
Maternal hemorrhage		139,696	
Maternal sepsis		61,956	

Death classification	Number of deaths	
Hypertensive disorders		62,018
Obstructed labor		33,927
Abortion		67,628
Other maternal conditions		161,400
Perinatal conditions (e)	1,523,189	
Prematurity and low birth weight		567,049
Birth asphyxia and birth trauma		410,635
Neonatal infections and other conditi		545,505
Nutritional deficiencies	264,204	
Protein-energy malnutrition		124,014
Iodine deficiency		2,168
Vitamin A deficiency		7,818
Iron-deficiency anemia		97,801
Other nutritional disorders		32,403

World Heath Risk, Part 2 of 4

Death classification	Number of deaths	
Noncommunicable diseases	**17,027,016**	
Malignant neoplasms	3,270,458	
Mouth and oropharynx cancers		96,038
Esophagus cancer		177,296
Stomach cancer		304,193
Colon and rectum cancers		303,463
Liver cancer		191,992
Pancreas cancer		128,359
Trachea, bronchus, lung cancers		380,625
Melanoma and other skin cancers		31,196
Breast cancer		516,644
Cervix uteri cancer		268,245
Corpus uteri cancer		54,926
Ovary cancer		144,266
Bladder cancer		49,152
Lymphomas, multiple myeloma		141,572

(Continued)

Death classification	Number of deaths	
Leukemia		122,709
Other malignant neoplasms		359,782
Other neoplasms	80,177	
Diabetes mellitus	632,941	
Endocrine disorders	161,825	
Neuropsychiatric conditions	615,455	
Unipolar depressive disorders		8,040
Bipolar disorder		595
Schizophrenia		14,424
Epilepsy		60,074
Alcohol use disorders		12,717
Alzheimers and other dementias		311,785
Parkinson disease		52,420
Multiple sclerosis		10,694
Drug use disorders		16,467
Post-traumatic stress disorder		43
Mental retardation, lead-caused		2,018
Other neuropsychiatric disorders		126,816
Sense organ diseases	2,087	
Glaucoma		31
Macular degeneration and other		2,056
Cardiovascular diseases	8,735,181	
Rheumatic heart disease		171,142
Hypertensive heart disease		529,714
Ischemic heart disease		3,371,347
Cerebrovascular disease		3,051,369
Inflammatory heart diseases		211,485
Other cardiovascular diseases		1,400,124

World Heath Risk, Part 3 of 4

Death classification	Number of deaths	
Respiratory diseases	1,880,654	
Chronic obstructive pulmonary disease		1,404,832
Asthma		136,119
Other respiratory diseases		339,703
Digestive diseases	878,632	
Peptic ulcer disease		106,804
Cirrhosis of the liver		261,547
Appendicitis		9,075
Other digestive diseases		501,206
Genitourinary diseases	434,599	
Nephritis and nephrosis		354,026
Other genitourinary system diseases		80,573
Skin diseases	40,614	
Musculoskeletal diseases	80,673	
Rheumatoid arthritis		18,387
Osteoarthritis		4,307
Gout		616
Low back pain		1,715
Other musculoskeletal disorders		56,264
Congenital anomalies	212,249	
Abdominal wall defect		2,066
Anencephaly		8,375
Anorectal atresia		816
Cleft lip		57
Cleft palate		704
Esophageal atresia		684
Renal agenesis		905
Down syndrome		14,108
Congenital heart anomalies		109,112
Spina bifida		11,646
Other congenital anomalies		66,942
Oral conditions	1,471	
Dental caries		30
Periodontal disease		180
Other oral diseases		1,471

World Heath Risk, Part 4 of 4

Death Classification	Number of Deaths	
Injuries	**1,847,296**	
Unintentional injuries	1,385,839	
Road traffic accidents		330,790
Poisonings		123,655
Falls		163,949
Fires		189,975
Drownings		125,060
Other unintentional injuries		452,410
Intentional injuries	461,457	
Self-inflicted injuries		315,616
Violence		114,573
War and civil conflict		28,780
Other intentional injuries		2,488

Index